Microsensors, MEMS, and Smart Devices

Microsensors, MEMS, and Smart Devices

Julian W. Gardner
University of Warwick, UK

Vijay K. Varadan
Osama O. Awadelkarim
Pennsylvania State University, USA

JOHN WILEY & SONS, LTD
Chichester • New York • Weinheim • Brisbane • Singapore • Toronto

Other Wiley Editorial Offices

John Wiley & Sons, Inc., 605 Third Avenue,
New York, NY 10158-0012, USA

Wiley-VCH Verlag GmbH
Pappelallee 3, D-69469 Weinheim, Germany

John Wiley, Australia, Ltd, 33 Park Road, Milton,
Queensland 4064, Australia

John Wiley & Sons (Canada) Ltd, 22 Worcester Road
Rexdale, Ontario M9W 1L1, Canada

John Wiley & Sons (Asia) Pte Ltd, 2 Clementi Loop #02-01,
Jin Xing Distripark, Singapore 129809

Library of Congress Cataloguing-in-Publication Data

Gardner, J. W. (Julian W.), 1958-
    Microsensors, MEMS, and smart devices / Julian W. Gardner, Vijay K. Varadan.
        p. cm.
    Includes bibliographical references and index.
    ISBN 0-471-86109-X
        1.Microelectromechanical systems. 2. Detectors. 3.Intelligent control systems. I.
    Varadan, V. K., 1943-II. Title.

    TK7875 G37 2001
    621.381-dc21

                                                                            2001024353

British Library Cataloguing in Publication Data

A catalogue record for this book is available from the British Library

ISBN 0-471- 86109X

Typeset in 10/12pt Times by Laser Words Private Limited, Chennai, India
Printed and bound in Great Britain by Antony Rowe, Ltd, Chippenham, Wiltshire
This book is printed on acid-free paper responsibly manufactured from sustainable forestry,
in which at least two trees are planted for each one used for paper production.

Contents

Appendices

Preface

The miniaturisation of sensors has been made possible by advances in the technologies originating in the semiconductor industry, and the emergent field of microsensors has grown rapidly during the past 10 years. The term *microsensor* is now commonly used to describe a *miniature* device that converts a nonelectrical quantity, such as pressure, temperature, or gas concentration, into an electrical signal. This book basically reports on the recent developments in, firstly, the miniaturisation of a sensor to produce a microsensor; secondly, the integration of a microsensor and its microelectronic circuitry to produce a so-called *smart sensor*; and thirdly, the integration of a microsensor, a microactuator, and their microelectronic circuitry to produce a *microsystem*.

Many of the microsystems being fabricated today employ silicon microtechnology and are called *microelectricalmechanical systems* or *MEMS* in short. Consequently, the first part of this book concentrates on the materials and processes required to make different kinds of microsensors and MEMS devices. The book aims to make the reader familiar with these processes and technologies. Of course, most of these technologies have been derived from those currently employed in the semiconductor industry and so we also review the standard microelectronics technologies used today to produce silicon wafers, process them into discrete devices or very large-scale integrated circuits, and package them. These *must* be used when the microelectronics is being integrated to form either a *hybrid* device, such as a *multichip module* (MCM), or a fully integrated device, such as a smart sensor. We then describe the new techniques that have been developed to make microsensors and microactuators, such as bulk and surface silicon micromachining, followed by the emerging technology of microstereolithography that can be used to form true three-dimensional micromechanical structures.

The reader is now fully prepared for our description of the different types of microsensors made today and the way in which they can be integrated with the microelectronics to make a smart device (e.g. an electronic eye, electronic nose, or microtweezers) or integrated with a microactuator to make a microsystem. Several of these chapters have been dedicated to the important topic of IDT microsensors, that is, surface acoustic wave devices that possess an interdigital transducer and so can be used to sense a wide variety of signals from mechanical to chemical. This type of microsensor is attractive, not only because it offers both high sensitivity and compatibility with the microelectronics industry but also because it can be operated and even powered by a wireless radio frequency link. The latter overcomes the initial constraints of communicating with small, low energy budget, and even mobile MEMS – now referred to as micromachines!

Our aim has been to write a book that serves as a text suitable both for an advanced undergraduate course and for a master's programme. Some of the material may well be familiar to students of electrical engineering or electronics. However, our comprehensive treatment will make it equally familiar to mechanical engineers, physicists, and materials scientists.

We have provided more than 10 appendices to aid the reader and serve as a source of reference material. These appendices explain the key abbreviations and terms used in the book, provide suggestions for further reading, give tables of the properties of materials important in microsensors and MEMS, and finally provide a list of the web sites of major journals and active institutions in this field. In addition, this book is aimed to be a valuable reference text for anyone interested in the field of microsensors and MEMS (whether they are an engineer, a scientist, or a technologist) and the technical references at the end of each chapter will enable such readers to trace back the original material.

Finally, much of the material for this book has been taken from short courses prepared by the authors and presented to students and industrialists in Europe, North America, and the Far East. Their many valuable comments have helped us to craft this book into its final form and so we owe them our thanks. The authors are also grateful to many of their students and colleagues, in particular Professor Vasundara V. Varadan, Dr. K. A. Jose, Dr. P. Xavier, Mr. S. Gangadharan, Mr. William Suh, and Mr. H. Subramanian for their valuable contributions.

Julian W. Gardner
Vijay K. Varadan
Osama O. Awadelkarim
September 2001

About the Authors

Julian W. Gardner is the Professor of Electronic Engineering at Warwick University, Coventry, UK. He has a B.Sc. in Physics (1979) from Birmingham University, a Ph.D. in Physical Electronics (1983) from Cambridge University, and a D.Sc. in Electronic Engineering (1997) from Warwick University. He has more than 15 years of experience in sensor engineering, first in industry and then in academia, in which he specialises in the development of microsensors and, in collaboration with the Southampton University, electronic nose instrumentation. Professor Gardner is currently a Fellow of the Institution of Electrical Engineers (UK) and member of its professional network on sensors. He has authored more than 250 technical papers and 5 books; the textbook *Microsensors: Principles and Applications* was first published by Wiley in 1994 and has enjoyed some measure of success, now being in its fourth reprint.

Vijay K. Varadan is Alumni Distinguished Professor of Engineering at the Pennsylvania State University, USA. He received his Ph.D. degree in Engineering Science from the Northwestern University in 1974. He has a B.E. in Mechanical Engineering (1964) from the University of Madras, India and an M.S. in Engineering Mechanics (1969) from the Pennsylvania State University. After serving on the faculty of Cornell University and Ohio State University, he joined the Pennsylvania State University in 1983, where he is currently Alumni Distinguished Professor of Engineering science, Mechanics, and Electrical Engineering. He is involved in all aspects of wave-material interaction, optoelectronics, microelectronics, photonics, microelectromechanical systems (MEMS): nanoscience and technology, carbon nanotubes, microstereolithography smart materials and structures; sonar, radar, microwave, and optically absorbing composite media; EMI, RFI, EMP, and EMF shielding materials; piezoelectric, chiral, ferrite, and polymer composites and conducting polymers; and UV conformal coatings, tunable ceramics materials and substrates, and electronically steerable antennas. He is the Editor of the *Journal of Wave-Material Interaction* and the Editor-in-Chief of the *Journal of Smart Materials and Structures* published by the Institute of Physics, UK. He has authored more than 400 technical papers and six books. He has eight patents pertinent to conducting polymers, smart structures and smart antennas, and phase shifters.

Osama O. Awadelkarim is a Professor of Engineering Science and Mechanics at the Pennsylvania State University. Dr. Awadelkarim received a B.Sc. Degree in Physics from the University of Khartoum in Sudan in 1977 and a Ph.D. degree from Reading University in the United Kingdom in 1982. He taught courses in soild-state device physics, microelectronics, material science, MEMS/Smart structures, and mechanics. Prior to joining the Pennsylvania State University in 1992, Dr. Awadelkarim worked as a senior scientist at Linkoping University (Sweden) and the Swedish Defence Research Establishment. He was also a visiting researcher at the University of Oslo (Norway), Kammerlingh Onnes Laboratories (Netherlands), and the International Centre for Theoretical Physics (Italy). Dr. Awadelkarim's research interests include nanoelectronics, power semiconductor devices, and micro-electromechanical systems. Dr. Awadelkarim has authored/co-authored over 100 articles in journals and conference proceedings.

Acknowledgments

The authors wish to thank the following people for helping in the technical preparation of this book: Dr. Marina Cole, Dr. Duncan Billson, and especially Dr. William Edward Gardner. We also wish to thank Mrs. Marie Bradley for her secretarial assistance in typing many of the chapters and John Wiley & Sons, Ltd for producing many of the line drawings. We also thank various researchers who have kindly supplied us with the original or electronic copies of photographs of their work.

1
Introduction

1.1 HISTORICAL DEVELOPMENT OF MICROELECTRONICS

The field of microelectronics began in 1948 when the first transistor was invented. This first transistor was a point-contact transistor, which became obsolete in the 1950s following the development of the bipolar junction transistor (BJT). The first modern-day junction field-effect transistor (JFET) was proposed by Shockley (1952). These two types of electronic devices are at the heart of all microelectronic components, but it was the development of integrated circuits (ICs) in 1958 that spawned today's computer industry.

IC technology has developed rapidly during the past 40 years; an overview of the current bipolar and field-effect processes can be found in Chapter 4. The continual improvement in silicon processing has resulted in a decreasing device size; currently, the minimum feature size is about 200 nm. The resultant increase in the number of transistors contained within a single IC follows what is commonly referred to as *Moore's law*. Figure 1.1 shows that in just 30 years the number of transistors in an IC has risen from about 100 in 1970 to 100 million in 2000. This is equivalent to a doubling of the number per chip every 18 months. Figure 1.1 plots a number of different common microprocessor chips on the graph and shows the clock speed rising from 100 kHz to 1000 MHz as the chip size falls. These microprocessors are of the type used in common personal computers costing about €1000 in today's prices[1].

Memory chips consist of transistors and capacitors; therefore, the size of dynamic random access memories (DRAM) has also followed Moore's law as a function of time. Figure 1.2 shows the increase of a standard memory chip from 1 kB in 1970 to 512 MB in 2000. If this current rate of progress is maintained, it would be possible to buy for €1000 a memory chip that has the same capacity as the human brain by 2030 and a memory chip that has the same brain capacity as everyone in the whole world combined by 2075! This phenomenal rise in the processing speed and power of chips has resulted first in a computer revolution and currently in an information revolution. Consequently, the world market value of ICs is currently worth some 250 billion euros, that is, about 250 times their processing speed in hertz.

---

[1] 1 euro (€) is currently worth about 1 US dollar.

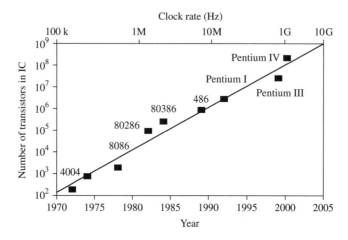

Figure 1.1 Moore's law for integrated circuits: exponential growth in the number of transistors in an IC during the past 30 years

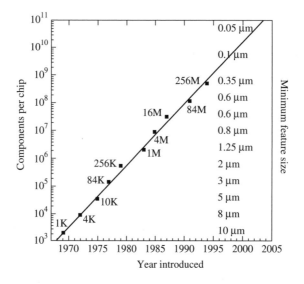

Figure 1.2 Size of memory chips (DRAM) and minimum feature as a function of time. From Campbell (1996)

1.2 EVOLUTION OF MICROSENSORS

The microelectronics revolution has led to increasingly complex signal–data processing chips; this, remarkably, has been associated with falling costs. Furthermore, these processing chips are now combined with sensors and actuators[2] to make an information-processing triptych (see Figure 1.3). These developments follow the recognition in the

---

[2] A sensor is a device that normally converts a nonelectrical quantity into an electrical quantity; an actuator is the converse. See Appendix C for the definition of some common terms.

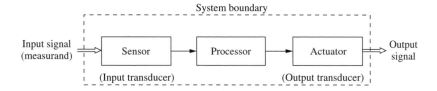

Figure 1.3   The information-processing triptych. From Gardner (1994)

1980s that the price-to-performance ratio of both sensors and actuators had fallen woefully behind processors. Consequently, measurement systems tended to be large and, more importantly, expensive. Work therefore started to link the microelectronic technologies and use these to make silicon sensors, the so-called microsensors.

Working definition of the term sensor:

'A microsensor is a sensor that has at least one physical dimension at the submillimeter level.'

This work was inspired by the vision of microsensors being manufactured in volumes at low cost and with, if necessary, integrated microelectronic circuitry. Chapters 5 and 6 describe in some detail the silicon micromachining technologies used today to make microsensors and microactuators. An overview of the field of microsensors is given in Chapter 8.

Figure 1.4 shows the relative market for ICs and microsensors in the past 10 years. It is evident that the market for microsensors lags well behind the market for ICs; nevertheless, it is worth 15 to 20 billion euros. The main cause has been the relatively stable price–performance (p/p) ratio of sensors and actuators since 1960, as illustrated in Figure 1.5. This contrasts markedly with the p/p ratio of ICs, which has fallen enormously between 1960 and 2000 and is now significantly below that for sensors and actuators. As a consequence of these changes, the cost of a measurement system is, in general, dominated first by the cost of the microactuator and second by the cost of the microsensor.

However, despite the cost advantages, there are several major technical advantages of making microsensors with microsystems technology (MST); the main ones are as follows:

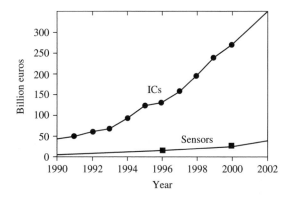

Figure 1.4   World market for ICs and microsensors from 1990 to 2000. From various sources

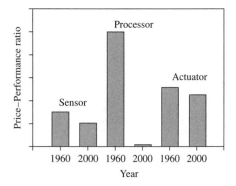

Figure 1.5   Price–performance indicators for ICs, sensors, and actuators

- The employment of well-established microtechnology
- The production of miniature sensors
- The production of less bulky and much lighter sensors
- The batch production of wafers for high volume
- The integration of processors

The UK marketplace for microsensors is diverse, as shown in Figure 1.6, and includes processing plants – environment and medical. However, the largest sector of the world (rather than UK) sensor market[3] is currently automotive; in 1997, the sales of pressure

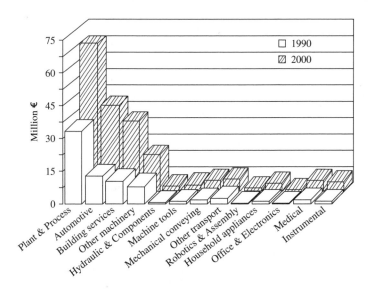

Figure 1.6   Sensor market by application for the United Kingdom. From Gardner (1994)

[3] These figures relate to the sensor market and hence exclude the larger markets for disk and ink-jet printer heads.

sensors was about 700 million euros and that for accelerometers was about 200 million euros (see Tables 8.10 and 8.11).

As the market for automotive sensors has matured, the price has fallen from €100 to €10 for a pressure sensor. In addition, the sophistication of the chips has increased and so has the level of integration. How this has led to the development of 'smart' sensors is discussed in Chapter 15.

Working definition of the term smart sensor:

'A smart sensor is a sensor that has part or its entire processing element integrated in a single chip.'

1.3 EVOLUTION OF MEMS

The next ambitious goal is to fabricate monolithic or integrated chips that can not only sense (with microsensors) but also actuate (with microactuators), that is, to create a microsystem that encompasses the information-processing triptych. The technology employed to make such a microsystem is commonly referred to as MST. Figure 1.7 provides an overview of MST together with some of the application areas. Work to achieve this goal started in the late 1980s, and there has been enormous effort to fabricate microelectromechanical systems (MEMS) using MST.

Working definition of the term MEMS:

'A MEMS is a device made from extremely small parts (i.e. microparts).'

Early efforts focused upon silicon technology and resulted in a number of successful micromechanical devices, such as pressure sensors and ink-jet printer nozzles. Yet, these are, perhaps, more accurately described as devices rather than as MEMS. The reason

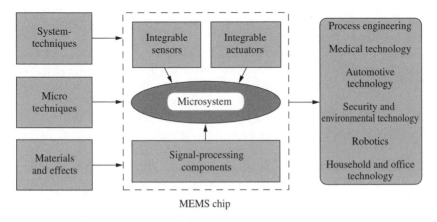

Figure 1.7 Overview of microsystems technology and the elements of a MEMS chip. From Fatikow and Rembold (1997)

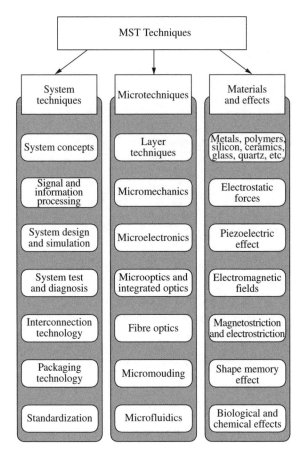

Figure 1.8 Some of the many fundamental techniques required to make MEMS devices. From Fatikow and Rembold (1997)

for the relatively slow emergence of a complete MEMS has been the complexity of the manufacturing process. Figure 1.8 details some new materials for MEMS and the various microtechnologies that need to be developed.

In Chapter 3, some of the new materials for MEMS have been introduced and their fundamental properties have been described. One attractive solution to the development of MEMS is to make all the techniques compatible with silicon processing. In other words, conventional complementary metal oxide semiconductor (CMOS) processing is combined with a pre-CMOS or post-CMOS MST. Because of the major significance of this approach, Chapters 12 to 14 have been dedicated to the topic of interdigitated transducers (IDTs) and their use in microsensors and MEMS devices.

The present MEMS market is relatively staid and mainly consists of some simple optical switches for the communications industry, pressure sensors, and inertial sensors for the automotive industry, as shown in Figure 1.9. This current staidness contrasts with the potential for MEMS, which is enormous. Table 1.1 is taken from a recent report on the world market for MEMS devices. The major growth areas were identified as microfluidics and photonics and communications. However, there have been some exciting

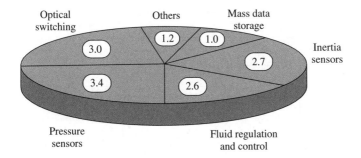

Optical switching   Others   Mass data storage   Inertia sensors   Pressure sensors   Fluid regulation and control

Figure 1.9 Pie chart showing the relative size of the current world MEMS market. The units shown are billions of euros

Table 1.1 Sales in millions of euros of MEMS devices according to the System Planning Corporation Market Survey (1999)

Devices and applications	1996	2003
Ink-jet printers, mass-flow sensors, biolab chips: microfluidics	400–500	3000–4450
Pressure sensors: automotive, medical, and industrial	390–760	1100–2150
Accelerometers and gyroscopes: automotive and aerospace	350–540	700–1400
Optical switches and displays: photonics and communications	25–40	440–950
Other devices such as microrelays, sensors, disk heads	510–1050	1230–2470
TOTAL IN MILLION €	1675–2890	6470–11 420

developments in methods to fabricate true three-dimensional structures on the micron scale. Chapter 7 describes the technique of microstereolithography and how it can be used to make a variety of three-dimensional microparts, such as microsprings, microgears, microturbines, and so on.

There are two major challenges facing us today: first, to develop methods that will manufacture microparts in high volume at low cost and, second, to develop microassembly techniques. To meet these challenges, certain industries have moved away from the use of silicon to the use of glasses and plastics, and we are now seeing the emergence of chips in biotechnology that include microfluidic systems (Chapter 15), which can truly be regarded as MEMS devices.

1.4 EMERGENCE OF MICROMACHINES

Natural evolution will then lead to MEMS devices that move around by themselves. Such chips are commonly referred to as micromachines and the concepts of microplanes, microrobots, microcars, and microsubmarines have been described by Fujimasa (1996). Figure 1.10 shows the scales involved and compares them with the size of a human flea!

Micromachines, if developed, will need sophisticated microsensors so that they can determine their location and orientation in space and proximity to other objects. They should also be able to communicate with a remote operator and hence will require a wireless communication link – especially if they are asked to enter the human body. Wireless communication has already been realised in certain acoustic microsensors, and

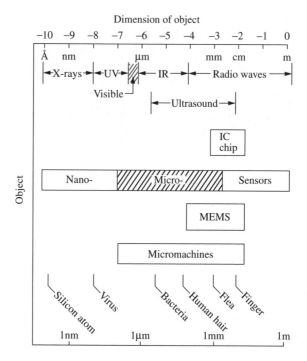

Figure 1.10   Dimensions of microsensors, MEMS, and micromachines; they are compared with some everyday objects. The horizontal axis has a logarithmic scale. Modified from Gardner (1994)

MEMS devices are described in Chapters 13 and 14. Associated with this development, there is a further major problem to solve, namely, miniaturisation of a suitable power source. Moving a micromachine through space requires significant energy. If it is to then do something useful, such as removing a blood clot in an artery, even more power will be required. Consequently, the future of MEMS devices may ultimately be limited by the communication link and the size of its 'battery pack!'

The road to practicable micromachines appears to be long and hard but the first steps toward microsensors and MEMS devices have been taken, and this book provides an overview of these initial steps.

REFERENCES

Campbell, S. A. (1996). *Science and Engineering of Microelectronic Fabrication*, Oxford University Press, Oxford, p. 536.

Fatikow, S. and Rembold, U. (1997). *Microsystem Technology and Microrobotics*, Springer, Berlin, p. 408.

Fujimasa, I. (1996). *Micromachines: A New Era in Mechanical Engineering*, Oxford University Press, Oxford, p. 156.

Gardner, J. W. (1994). *Microsensors*, Wiley, Chichester, p. 331.

Shockley, W. (1952). "A unipolar field-effect transistor," *Proc. IRE* **40**, 1365.

2

Electronic Materials and Processing

2.1 INTRODUCTION

Integrated circuit (IC) processing is a mature technology for the fabrication of electronic devices and systems. Steady advances in IC fabrication technology have been the basis for the microelectronic revolution, starting with the first discrete germanium transistors in the 1950s to the 64-MB dynamic random access memories (DRAMs) of today. The main purpose of this chapter is to introduce the reader to the basic terminology and to provide a basic overview of the processing steps required to process silicon wafers. Therefore, this chapter begins with the introduction of a special group of materials that may be referred to as *electronic materials*. These materials are commonly used in conventional IC technologies and some of them are used as microelectromechanical system (MEMS) materials (see following chapter). Electronic materials have no common physical or chemical properties: for instance, their electrical properties span the range from near-ideal insulators to excellent conductors, and their chemical composition may consist of one atom in simple materials to several atoms in compound electronic materials. Therefore, the term *electronic materials* has no physical or chemical meaning; it solely describes materials used in IC fabrication.

A more detailed discussion of conventional silicon IC processing is presented in Chapter 4, which describes the additional steps required to package electronic chips. There are also a number of excellent textbooks that describe in full the processing of conventional electronic IC chips, such as microprocessors and DRAMs (for further information see Sze (1985, 1988); Fung *et al.* (1985)).

2.2 ELECTRONIC MATERIALS AND THEIR DEPOSITION

Many different kinds of bulk materials and thin films are used in the fabrication of ICs. The bulk materials are predominantly semiconducting. The most important semiconductors in IC fabrication are silicon and gallium arsenide. There are four important thin-film materials (or class of materials) in IC fabrication:

1. Thermal silicon oxide

2. Dielectric layers

3. Polycrystalline silicon (poly-Si)

4. Metal films (predominantly aluminum)

The dielectric layers include deposited silicon dioxide (SiO_2) (sometimes referred to as oxide) and silicon nitride (Si_3N_4). These dielectrics are used for insulation between conducting layers, for diffusion and ion-implantation masks, and for passivation to protect devices from impurities, moisture, and scratches. Poly-Si is used as a gate electrode in metal oxide semiconductor (MOS) devices, as a conductive material for multilevel metallisation, and as a contact material for devices with shallow junctions. Metal films are used to form low-resistance ohmic connections, both to heavily doped n^+/p^+ regions and to poly-Si layers, and rectifying (nonohmic) contacts in metal semiconductor barriers.

The thermal oxide is usually a better-quality oxide (compared with deposited oxide) and is used for the gate oxide layers in field-effect transistors (FETs). A detailed description of FET devices and their electrical characteristics is given in Chapter 4.

As shall become apparent in the following chapters, electronic materials are of major importance in MEMS devices. Therefore, the methods used for growing thermal SiO_2 and for depositing dielectric poly-Si and metallic layers are reviewed in the following sections.

2.2.1 Oxide Film Formation by Thermal Oxidation

Thermal oxidation is the method by which a thin film of SiO_2 is grown on top of a silicon wafer. It is the key method of producing thin SiO_2 layers in modern IC technology. The basic thermal oxidation apparatus is shown in Figure 2.1. The apparatus comprises a resistance-heated furnace, a cylindrical fused quartz tube that contains the silicon wafers held vertically in slotted quartz boat, and a source of either pure dry oxygen or pure water vapour. The loading end of the furnace tube protrudes into a vertical flow hood, wherein a filtered flow of air is maintained. The hood reduces dust in the air that surrounds the wafers and minimises contamination during wafer loading.

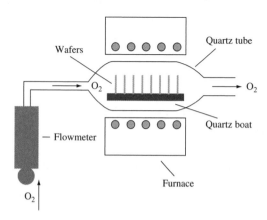

Figure 2.1 Basic furnace arrangement for the thermal oxidation of silicon wafers

Thermal oxidation of silicon in oxygen or water vapour can be described by the following two chemical reactions:

$$Si \text{ (solid)} + O_2 \text{ (gas)} \xrightarrow{900-1200\,°C} SiO_2 \text{ (solid)} \tag{2.1}$$

and

$$Si \text{ (solid)} + 2H_2O \text{ (gas)} \xrightarrow{900-1200\,°C} SiO_2 \text{ (solid)} + 2H_2 \text{ (gas)} \tag{2.2}$$

The silicon–silicon dioxide interface transverses the silicon during the oxidation process. Using the densities and molecular weights of silicon and SiO_2, it can be shown that growing an oxide of thickness x consumes a layer of silicon that is $0.44x$ thick.

The basic structural unit of thermal SiO_2 is a silicon atom surrounded tetrahedrally by four oxygen atoms, as shown in Figure 2.2(a). The silicon–oxygen and oxygen–oxygen interatomic distances are 1.6 and 2.27 Å, respectively. SiO_2 or silica has either a crystalline structure (e.g. quartz in Figure 2.2(b)) or an amorphous structure (Figure 2.2(c)). Typically, amorphous SiO_2 has a density of ~ 2.2 gm/cm^3, whereas quartz has a density of ~ 2.7 gm/cm^3. Thermally grown oxides are usually amorphous in nature.

Oxidation of silicon in a high-pressure atmosphere of steam (or oxygen) can produce substantial acceleration in the growth rate and is often used to grow thick oxide layers. One advantage of high-pressure oxide growth is that oxides can be grown at significantly lower temperatures and at acceptable growth rates.

2.2.2 Deposition of Silicon Dioxide and Silicon Nitride

There are three deposition methods that are commonly used to form a thin film on a substrate. These methods are all based on chemical vapour deposition (CVD) and are as follows:

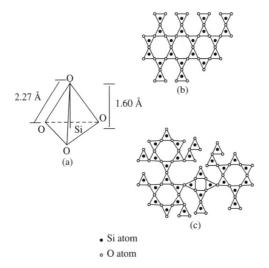

• Si atom

∘ O atom

Figure 2.2 Atomic structure of (a) single unit of thermal oxide; (b) regular array of quartz; and (c) disordered array of amorphous SiO_2

1. Atmospheric pressure chemical vapour deposition (APCVD)

2. Low-pressure chemical vapour deposition (LPCVD)

3. Plasma-enhanced chemical vapour deposition (PECVD)

The latter method is an energy-enhanced CVD method. The appropriate method from among these three deposition methods is determined by the substrate temperature, the deposition rate and film uniformity, the morphology, the electrical and mechanical properties, and the chemical composition of the dielectric films.

A schematic diagram of a typical CVD system is shown in Figure 2.3; the only exception is that different gases are used at the gas inlet. Figures 2.3(a) and (b) show a LPCVD reactor and PECVD reactor, respectively. In Figure 2.3(a), the quartz tube is heated by a three-zone furnace and gas is introduced (gas inlet) at one end of the reactor and is pumped out at the opposite end (pump). The substrate wafers are held vertically in a slotted quartz boat. The type of LPCVD reactor shown in Figure 2.3(a) is a hot-wall LPCVD reactor, in which the quartz tube wall is hot because it is adjacent to the furnace; this is in contrast to a cold-wall LPCVD reactor, such as the horizontal epitaxial reactor that uses radio frequency (RF) heating. Usually, the parameters for the LPCVD process in the reaction chamber are in the following ranges:

1. Pressure between 0.2 and 2.0 torr

2. Gas flow between 1 to 10 cm^3/s

3. Temperatures between 300 and 900 °C

Figure 2.3(b) shows a parallel-plate, radial-flow PECVD reactor that comprises a vacuum-sealed cylindrical glass chamber. Two parallel aluminum plates are mounted in the chamber with an RF voltage applied to the upper plate while the lower plate is grounded. The RF voltage causes a plasma discharge between the plates (electrodes). Wafers are placed in the lower electrode, which is heated between 100 and 400 °C by resistance heaters. Process gas flows through the discharge from outlets that are located along the circumference of the lower electrode.

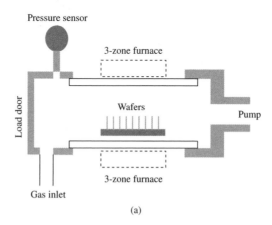

(a)

Figure 2.3   (a) Typical layout of an LPCVD reactor; (b) two PECVD reactors

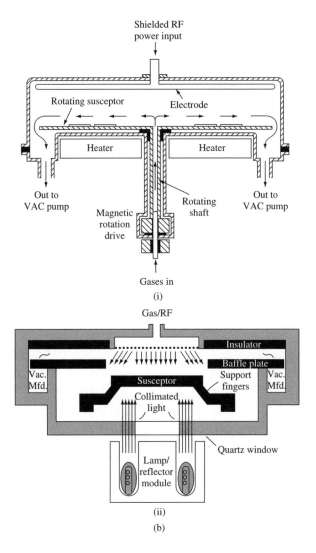

Figure 2.3   (*continued*)

CVD is used extensively in depositing SiO_2, Si_3N_4, and polysilicon. CVD SiO_2 does not replace thermally grown SiO_2 that has superior electrical and mechanical properties as compared with CVD oxide. However, CVD oxides are instead used to complement thermal oxides and, in many cases, to form oxide layers that become much thicker in relatively short times than do thermal oxides. SiO_2 can be CVD-deposited by several methods. It can be deposited by reacting silane and oxygen at 300 to 500 °C in an LPCVD reactor wherein

$$SiH_4 + O_2 \xrightarrow{\;500\,°C\;} SiO_2 + 2H_2 \tag{2.3}$$

It can also be LPCVD-deposited by decomposing tetraethylorthosilicate, $Si(OC_2H_5)_4$. The compound, abbreviated as TEOS, is vaporised from a liquid source. Alternatively, dichlorosilane can be used as follows:

$$SiCl_2H_2 + 2H_2O \xrightarrow{900\,°C} SiO_2 + 2H_2 + 2HCl \qquad (2.4)$$

A property that relates to CVD is known as *step coverage*. Step coverage relates the surface topography of the deposited film to the various steps on the semiconductor substrate. Figure 2.4(a) shows an ideal, or conformal, film deposition in which the film thickness is uniform along all surfaces of the step, whereas Figure 2.4(b) shows a nonconformal film (for a discussion of the physical causes of uniform or nonuniform thickness of deposited films, see Fung *et al.* (1985)).

Table 2.1 compares different SiO_2 films deposited by different methods and contrasts them with thermally grown oxides. Similarly, Si_3N_4 can be LPCVD-deposited by an intermediate-temperature process or a low-temperature PECVD process. In the LPCVD process, which is the more common process, dichlorosilane and ammonia react according to the reaction

$$3SiCl_2H_2 + 4NH_3 \xrightarrow{\sim 800\,°C} Si_3N_4 + 6HCl + 6H_2 \qquad (2.5)$$

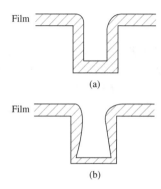

Figure 2.4   (a) Conformal (i.e. ideal); (b) nonconformal deposition of a film

Table 2.1   Properties of deposited and thermally grown oxide films (Sze 1985)

Property	Composition	Step coverage	Density ρ (g/cm^3)	Refractive index n_r	Dielectric strength (V/cm)
Thermally grown at 1000 °C	SiO_2	–	2.2	1.46	$>10^{-5}$
Deposited by $SiH_4 + O_2$ at 450 °C	$SiO_2(H)$	Nonconformal	2.1	1.44	8×10^{-6}
Deposited by TEOS at 700 °C	SiO_2	Conformal	2.2	1.46	10^{-5}
Deposited by $SiCl_2H_2 + N_2O$ at 900 °C	SiO_2	Conformal	2.2	1.46	10^{-5}

Table 2.2 Properties of some selected electronic materials

Material property	Si	GaAs	SiO$_2$	Si$_3$N$_4$	Al	Au	Ti
Density (kg/m^3)	2330	5316	1544	3440	2699	19 320	4508
Melting point (°C)	1410	1510	1880	1900	660	1064	1660
Electrical conductivity^a (10^3 W^{-1} cm^{-1})	4×10^{-5}	10^{-11}	–	–	377	488	26
Thermal conductivity (W/m/K)	168	47	6.5–11	19	236	319	22
Dielectric constant	11.7	12	4.3–4.5	7.5	–	–	–
Young's modulus (GPa)	190	–	380	380	70	78	~40
Yield strength (GPa)	6.9	–	14	14	50	200	480

^{a}Measured at room temperature. Some other properties will vary with temperature

2.2.3 Polysilicon Film Deposition

Polysilicon is often used as a structural material in MEMS. Polysilicon is also used in MEMS for electrode formation and as a conductor or as a high-value resistor, depending on its doping level. A low-pressure reactor, such as the one shown in Figure 2.3(a), operating at temperatures between 600 and 650 °C is used to deposit polysilicon by pyrolysing silane according to the following reaction:

$$SiH_4 \xrightarrow{600\,°C} Si + 2H_2 \tag{2.6}$$

The most common low-pressure processes used for polysilicon deposition are the ones that operate at pressures between 0.2 and 1.0 torr using 100 percent silane. Another process for polysilicon deposition involves a diluted mixture of 20 to 30 percent silane in nitrogen.

The properties of electronic materials are summarised in Appendices F (metals) and G (semiconductors), and some of the properties of common electronic materials used in MEMS are summarised in Table 2.2.

2.3 PATTERN TRANSFER

2.3.1 The Lithographic Process

Lithography is the process of imprinting a geometric pattern from a mask onto a thin layer of material called a *resist*, which is a radiation-sensitive material. Figure 2.5 shows

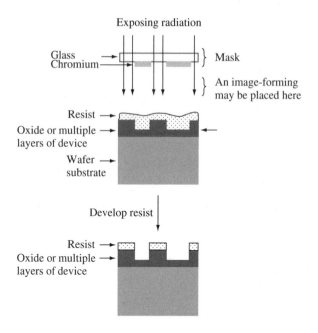

Figure 2.5   Basic steps in a lithographic process used to fabricate a device

schematically the lithographic process that is used to fabricate a circuit element. First, a resist is usually spin-coated or sprayed onto the wafers and then a mask is placed above it. Second, a selected radiation (see Figure 2.5) is transmitted through the 'clear' parts of the mask. The circuit patterns of opaque material[1] (mask material) block some of the radiation. The radiation is used to change the solubility of the resist in a known solvent.

The pattern-transfer process is accomplished by using a lithographic exposure tool that emits radiation. The performance of the tool is determined by three properties: resolution, registration, and throughput. *Resolution* is defined as the minimum feature size that can be transferred with high fidelity to a resist film on the surface of the wafer. *Registration* is a measure of how accurately patterns of successive masks can be aligned with respect to the previously defined patterns on a wafer. *Throughput* is the number of wafers that can be exposed per hour for a given mask level. Depending on the resolution, several types of radiation, including electromagnetic (e.g. ultraviolet (UV) and X rays) and particulate (e.g. electrons and ions), may be employed in lithography.

Optical lithography uses UV radiation ($\lambda \sim 0.2$–0.4 μm). Optical exposure tools are capable of approximately 1 μm resolution, 0.5 μm registration, and a throughput of 50 to 100 wafers per hour. Because of backscattering, electron-beam lithography is limited to a 0.5 μm resolution with 0.2 μm registration. Similarly, X-ray lithography typically has 0.5 μm resolution with 0.2 μm registration. However, both electron-beam and X-ray

---

[1] The circuit pattern may be defined alternatively by the transparent part, depending on the choice of resist polarity and film process (see later).

lithographies require complicated masks. The vast majority of lithographic equipment used for IC fabrication is optical equipment. Optical lithography uses two methods for imprinting the desired pattern on the photoresist. These two methods are shadow printing and projection printing.

In shadow printing, the mask and wafer are in direct contact during the optical exposure (contact printing is shown in Figure 2.6(a)) or are separated by a very small gap g that is on the order of 10 to 50 µm (proximity printing is shown in Figure 2.6(b)).

The minimum line width ($L_{\min}$) that can be achieved by using shadow printing is given by

$$L_{\min} \approx \sqrt{\lambda g} \qquad (2.7)$$

The intimate contact between the wafer and mask in contact printing offers the possibility of very high resolution, usually better than 1 µm. However, contact printing often results in mask damage caused by particles from the wafer surface that become attached to the mask. These particles may end up as opaque spots in regions of the mask that are supposed to be transparent.

Projection printing is an alternative exposure method in which the mask damage problem associated with shadow printing is minimised. Projection printing exposure tools are used to project images of the mask patterns onto a resist-coated wafer several centimeters away from the mask (Figure 2.7). To increase resolution in projection printing, only a small portion of the mask is exposed at a time. A narrow arc-shaped image field, about 1 mm in width, serially transfers the slit image of the mask onto the wafer. Typical resolutions achieved with projection printing are on the order of 1 µm.

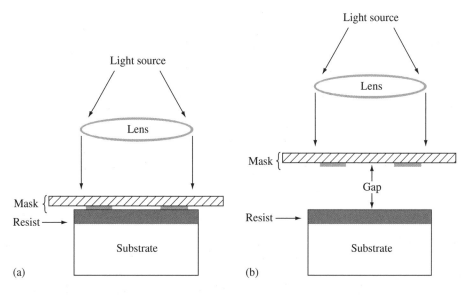

Figure 2.6 Basic lithographic mask arrangements: (a) shadow printing and (b) proximity printing (not to scale as chrome layer on glass mask is exaggerated)

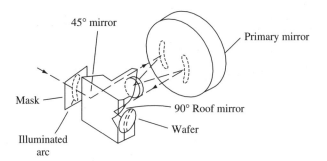

Figure 2.7 Basic lithographic arrangement for mask projection (Sze 1985)

2.3.2 Mask Formation

For discrete devices, or small-scale-to-medium-scale ICs (typically up to 1000 components per chip), a large composite layout of the mask set is first drawn. This layout is a hundred to a few thousand times the final size. The composite layout is then broken into mask levels that correspond to the IC process sequence such as isolation region on one level, the metallisation region on another, and so on. Artwork is drawn for each masking level. The artwork is reduced to 10× (ten times) glass reticule by using a reduction camera. The final mask is made from the 10× reticule using a projection printing system.

The schematic layout of a typical mask-making machine is shown in Figure 2.8. It consists of the UV light source, a motorised x-y stage sitting on a vibration-isolated table, and optical accessories. The operation of the machine is computer-controlled. The information that contains the geometric features corresponding to a particular mask is electrically entered with the aid of a layout editor system. The geometric layout is then broken down into rectangular regions of fixed dimensions. The fractured mask data is stored on a tape, which is transferred to the mask-making machine. A reticule mask plate, which consists of one glass plate coated with a light-blocking material (e.g. chromium) and a photoresist coating, is placed on the positioning stage. The tape data are then read by the equipment and, accordingly, the position of the stage and the aperture of the shutter blades are specified.

The choice of the mask material, just like radiation, depends on the desired resolution. For feature sizes of 5 μm or larger, masks are made from glass plates covered with a soft surface material such as emulsion. For smaller feature sizes, masks are made from low-expansion glass covered with a hard surface material such as chromium or iron oxide.

2.3.3 Resist

The method used for resist-layer formation is called *spin casting*. Spin casting is a process by which one can deposit uniform films of various liquids by spinning them onto a wafer. A typical setup used for spin casting is shown in Figure 2.9. The liquid is injected onto

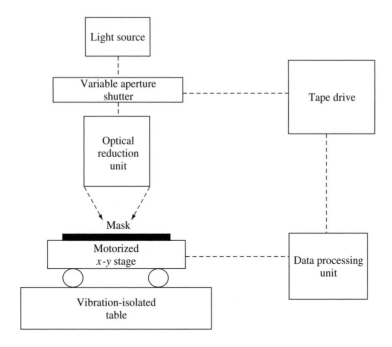

Figure 2.8   Typical arrangement of a mask-making machine

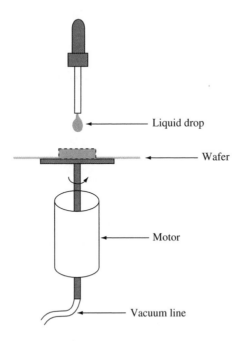

Figure 2.9   Basic setup for the spin casting of a photoresist layer onto a silicon wafer

the surface of a wafer, which is pressure-attached to a wafer holder through holes in the holder that are connected to a vacuum line, and continuously pumped during the process. The wafer holder itself is attached to and spun by a motor. The thickness x of the spin-on material is related to the viscosity η of the liquid and the solid content f in the solution as well as the spin speed ω:

$$x \propto \frac{\eta f}{\sqrt{\omega}} \tag{2.8}$$

Typical spin speeds are in the range 1000–10 000 rpm to give material thickness in the range of 0.5 to 1 μm. After the wafer is spin-coated with the resist solution, it is dried and baked at temperatures in the range of 90 to 450 °C, depending on the type of the resist. Baking is necessary for further drying of the resist and for strengthening the resist adhesion to the wafer (Table 2.3).

A resist is a radiation-sensitive material that can be classified as positive or negative, depending on how it responds to radiation. The positive resist is rendered soluble in a developer when it is exposed to radiation. Therefore, after exposure to radiation, a positive resist can be easily removed in the development process (dissolution of the resist in an appropriate solvent, which is sometimes called the *developer*). The net effect is that the patterns formed (also called *images*) in the positive resist are the same as those formed on the mask (Figure 2.10). A negative resist, on the other hand, is rendered less soluble in a developer when it is exposed to radiation. The patterns formed in a negative resist are thus the reverse of those formed on the mask patterns (Figure 2.10). Table 2.4 lists a few of the commercially available resists, the lithographic process, and their polarity (see Table 4.3).

Table 2.3   Some properties of the common spin-on materials

Material	Thickness (μm)	Bake temperature (°C)	Solvent
Photoresist	0.1–10	90–150	Weak base
Polyimide	0.3–100	350–450	Weak base
Silicon dioxide	0.1–0.5	500–900	HF
Lead titanate	0.1–0.3	650	HNO_3

Table 2.4   Commercially available resists

Resist	Lithography	Type
Kodak 747	Optical	Negative
AZ-1350J	Optical	Positive
PR102	Optical	Positive
Poly(methyl methacrylate) (PMMA)	E-beam and X ray	Positive
Poly[(glycidyl methacrylate)-co-ethylacrylate] (COP)	E-beam and X ray	Negative
Dichloropropyl acrylate and glycidyl methacrylate-co-ethyl acrylate (DCOPA)	X ray	Negative

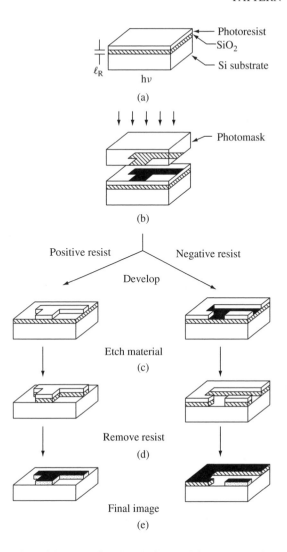

Figure 2.10   Formation of images after developing positive and negative resists (Sze 1985)

2.3.4 Lift-off Technique

The pattern-transfer technique, referred to as *lift-off*, uses a positive resist to form the resist pattern on a substrate. The steps of the technique are shown in Figure 2.11. The resist is first exposed to radiation via the pattern-carrying mask (Figure 2.11(a)) and the exposed areas of the resist are developed as shown in Figure 2.11(b). A film is then deposited over the resist and substrate, as shown in Figure 2.11(c). The film thickness must be smaller than that of the resist. Using an appropriate solvent, the remaining parts of the resist and the deposited film atop these parts of the resist are lifted off, as shown in Figure 2.11(d). The lift-off technique is capable of high resolution and is often used for the fabrication of discrete devices.

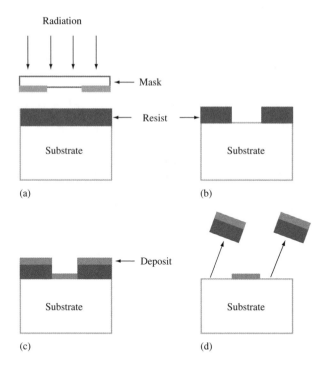

Figure 2.11   Four basic steps involved in a "lift-off" process to pattern a film

2.4 ETCHING ELECTRONIC MATERIALS

Etching is used extensively in material processing for delineating patterns, removing surface damage and contamination, and fabricating three-dimensional structures. Etching is a chemical process wherein material is removed by a chemical reaction between the etchants and the material to be etched. The etchant may be a chemical solution or a plasma. If the etchant is a chemical solution, the etching process is called *wet chemical etching*. Plasma-assisted etching is generally referred to as *dry etching*, and the term *dry etching* is now used to denote several etching techniques that use plasma in the form of low-pressure discharges.

2.4.1 Wet Chemical Etching

Wet chemical etching involves three principal steps:

1. The reactants are transported by diffusion[2] to the surface to be etched.

2. Chemical reactions take place at the surface.

3. Reaction products are again transported away from the surface by diffusion.

---

[2] Under some circumstances, reactions can be reaction-rate-limited rather than diffusion-rate-(mass-transport) limited.

Let us consider, as an example, etching of silicon. For silicon, the most commonly used etchants are mixtures of nitric acid (HNO_3) and hydrofluoric acid (HF) in water or acetic acid (CH_3COOH). Wet chemical etching usually proceeds by oxidation. Initially, silicon is oxidised in the presence of holes as follows:

$$Si + 2H^+ \longrightarrow Si^{2+} + H_2 \tag{2.9}$$

Water dissociates according to the reaction

$$H_2O \longrightarrow (OH)^- + H^+ \tag{2.10}$$

The hydroxyl ions $(OH)^-$ recombine with positively charged silicon ions to form SiO_2 in two steps:

$$Si^{2+} + 2(OH)^- \longrightarrow Si(OH)_2 \tag{2.11}$$

and

$$Si(OH)_2 \longrightarrow SiO_2 + H_2 \tag{2.12}$$

SiO_2 dissolves in HF acid according to the reaction

$$SiO_2 + 6HF \longrightarrow H_2SiF_6 + 2H_2O \tag{2.13}$$

where H_2SiF_6 is soluble in water. The reactions of (2.9) to (2.13) may be represented with HNO_3 by the following overall reaction:

$$Si + HNO_3 + 6HF \longrightarrow H_2SiF_6 + HNO_2 + H_2O + H_2 \tag{2.14}$$

The chemical solution used for gallium arsenide (GaAs) etching is a combination of hydrogen peroxide (H_2O_2) and sulfuric acid (H_2SO_4) dissolved in water. Dielectrics and metals are etched using the same chemicals that dissolve these materials in bulk form and involve their conversion into soluble salts or complexes. Generally, film materials will etch more rapidly than their bulk counterparts.

Etching processes are characterised by three parameters:

1. Etch rate

2. Etch selectivity

3. Etch uniformity

The etch rate is defined as the material thickness etched per unit time. Etch selectivity is a measure of how effective the etch process is in removing the material to be etched without affecting other materials or films present in the wafer. Quantitatively, etch selectivity can be expressed as the ratio between the etch rate of the material to be etched and etch-mask materials on the wafer. Table 2.5 lists the properties of different wet etchants for different materials.

2.4.2 Dry Etching

The basic concept of dry or plasma etching is very simple. A glow discharge is used to generate chemically reactive species (atoms, radicals, and ions) from a relatively inert

Table 2.5   Wet etchants used in etching some selected electronic materials

Material	Etchant composition	Etch rate (Å/min)
Si	3 ml HF + 5 ml HNO_3	3.5×10^5
GaAs	8 ml H_2SO_4 + 1 ml H_2O_2 + 1 ml H_2O	0.8×10^5
SiO_2	28 ml HF + 170 ml H_2O + 113 g NH_4F	1000
	or	
	15 ml HF + 10 ml HNO_3 + 300 ml H_2O	120
Si_3N_4	Buffered HF or H_3PO_4	5 or 100
Al	1 ml HNO_3 + 4 ml CH_3COOH + 4 ml H_3PO_4 + 1 ml H_2O	350
Au	4 g KI + 1 g I_2 + 40 ml H_2O	1.0×10^5

molecular gas. The etching gas is chosen so as to produce species that react chemically with the material to be etched to form a reaction product that is volatile. The etch product then desorbs from the etched material into the gas phase and is removed by the vacuum pumping system. The most common example of the application of plasma etching is in the etching of carbonaceous materials, for example, resist polymers, in oxygen plasma – a process referred to as *plasma ashing* or *plasma stripping*. In this case, the etch species are oxygen atoms and the volatile etch products are CO, CO_2, and H_2O gases.

In etching silicon and silicon compounds, glow discharges of fluorine-containing gases, such as CF_4, are used. In this case, the volatile etch product is SiF_4 and the etching species are mainly fluorine atoms. In principle, any material that reacts with fluorine atoms to form a volatile product can be etched in this way (e.g. W, Ta, C, Ge, Ti, Mo, B, U, etc.). Chlorine-containing gases have also been used to etch some of the same materials, but the most important uses of chlorine-based gases have been in the etching of aluminum and poly-Si. Both aluminum and silicon form volatile chlorides. Aluminum is not etched in fluorine-containing plasmas because its fluoride is nonvolatile.

The characteristic of etching processes, which is becoming more and more important as the lateral dimensions of the lithography become smaller, is the so-called directionality (anisotropy) of the etch process. This characteristic is illustrated in Figure 2.12 in which the lithographic pattern is in the x-y plane and the z-direction is normal to this plane. If the etch rates in the x and y directions are equal to the etch rate in the z-direction, the etching process is said to be isotropic (or nondirectional) and the shape of the sidewall of the etched feature will be as shown in Figure 2.12(a). Etch processes that are anisotropic or directional have etch rates in the z-direction and are larger than the lateral (x or y) etch rates. The extreme case of directional etching in which the lateral etch rate is zero (to be referred to here as *vertical etch process*) is shown in Figure 2.12(b).

Plasma etching, as described in the preceding discussion, is predominantly an isotropic process. However, anisotropy in dry etching can be achieved by means of the chemical reaction preferentially enhanced in a given direction to the surface of the wafer by some mechanism. The mechanism used in dry etching to achieve etch anisotropy is ion bombardment. Under the influence of an RF field, the highly energised ions impinge on the surface either to stimulate reaction in a direction perpendicular to the wafer

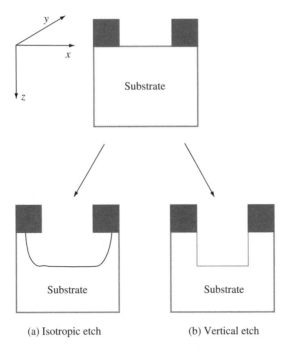

(a) Isotropic etch   (b) Vertical etch

Figure 2.12   Characteristic profile of an (a) isotropic and (b) vertical etching process

surface or to prevent inhibitor species from coating the surface and hence reenhance etching in the direction perpendicular to the wafer surface. Therefore, the vertical sidewalls, being parallel to the direction of ion bombardment, are little affected by the plasma.

Figure 2.13 is a schematic diagram of a planar etching system, which comprises a vacuum chamber, two RF-powered electrodes, an etching gas inlet, and a pumping mechanism. The planar systems are also called *parallel-plate systems* or *surface loaded systems*. These systems have been used in two distinct ways: *(1)* the wafers are mounted on a grounded surface opposite to the RF-powered electrode (cathode) or *(2)* the wafers are mounted on the RF-powered electrode (cathode) directly. This latter approach has been called *reactive ion etching* (RIE). In this approach, ions are accelerated toward the wafer surface by a self-bias that develops between the wafer surface and the plasma. This bias is such that positively charged ions are attracted to the wafer surface, resulting in surface bombardment. It has been demonstrated that a planar etching system, when operated in the RIE mode, is capable of highly directional and high-resolution etching.

To illustrate the mechanisms involved in reactive ion etching, consider the example of poly-Si etched in chlorine plasma:

1. Ions, radicals, and electron generations:

$$n\text{Cl (or Cl}_2) \longrightarrow n\text{Cl}^+(\text{or Cl}_2{}^+) + ne \qquad (2.15)$$

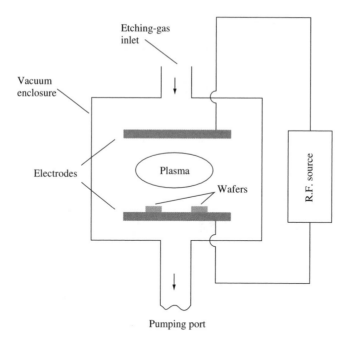

Figure 2.13   Schematic cross section of a plasma-etching system

2. Etchant formation:

$$e + Cl_2 \xrightarrow{\text{Energy supplied by electron}} 2Cl + e \qquad (2.16)$$

3. Adsorption of etchant on poly-Si:

$$nCl \text{ (or } Cl_2) \longrightarrow Si \text{ (surface)} + nCl \qquad (2.17)$$

4. Ion-bombardment-assisted reaction to form product:

$$Si \text{ (surface)} + nCl \xrightarrow{\text{Ion-bombardment}} SiCl_n \text{(adsorbed)} \qquad (2.18)$$

Table 2.6   Etch gases used for various electronic materials

Material	Gases
Crystalline Si and poly-Si	CF_4, CF_4/O_2, CF_3, Cl, SF_6/Cl, Cl_2/H_2, C_2ClF_5/O_2, SF_6/O_2, SiF_4/O_2, NF_3, $C_2Cl_3F_5$, CCl_4/He, Cl_2/He, $HBr/Cl_2/He$
SiO_2	CF_4/H_2, C_2F_6, C_3F_8, CHF_3
Si_3N_4	CF_4/O_2, CF_4/H_2, C_2F_6, C_3F_8, SF_6/He
Organic solids	O_2, O_2/CF_4, O_2/CF_6
Al	BCl_3, CCl_4, $SiCl_4$, BCl_3/Cl_2, CCl_4/Cl_2, $SiCl_4/Cl_2$
Au	$C_2Cl_2F_4$, Cl_2

5. Product desorption:

$$\text{SiCl}_n \text{ (adsorbed)} \longrightarrow \text{SiCl}_n \text{ (gas)} \qquad (2.19)$$

The final gas product is pumped out of the etching chamber. Table 2.6 provides a list of etch gases used for dry-etching various electronic materials.

2.5 DOPING SEMICONDUCTORS

When impurities are intentionally added to a semiconductor, the semiconductor is said to be 'doped.' Figure 2.14 shows a hypothetical two-dimensional silicon crystal in which one silicon atom is replaced (or substituted) by an atom – in this example, a Group V element in the periodic table, namely, phosphorus. Phosphorus has five valence electrons, whereas silicon has only four. The phosphorus atom shares four of its electrons with four neighboring silicon atoms in covalent bonds. The remaining fifth valence electron in phosphorus is loosely bound to the phosphorus nucleus.

The ionisation energy of an impurity atom of mass m in a semiconductor crystal can be estimated from a one-electron model. If this ionisation energy is denoted by the symbol

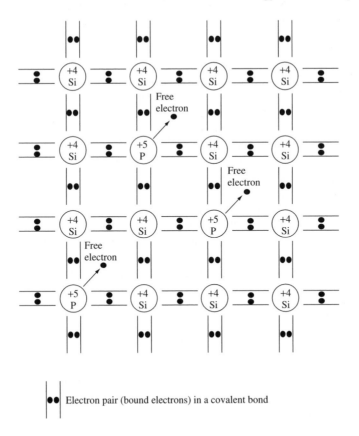

•• Electron pair (bound electrons) in a covalent bond

Figure 2.14   Hypothetical 2D silicon crystal doped with phosphorus (*n*-type semiconductor)

E_d, then

$$E_d = \left(\frac{\varepsilon_0}{\varepsilon_r}\right)^2 \left(\frac{m^*}{m}\right) E_n \qquad (2.20)$$

where ε_0 is the permittivity of free space, ε_r is that of the semiconductor, and m^* is the effective electron mass in the semiconductor crystal. The energy E_n is discussed in Section 3.1 of the following Chapter. When the phosphorus atom in silicon is ionised, the released electron becomes a free electron that is available for conduction. The phosphorus atom is, hence, called a *donor atom* because it donates a free electron to the crystal. All atoms with five valence electrons, that is, Group V elements, can behave in a similar manner to phosphorus in silicon, that is, donate a free electron to the semiconductor crystal. However, the amount of energy required, E_d, for this process to occur may differ from one type of donor atom to another. All Group V atoms will donate electrons if they substitute for host atoms in crystals of Group IV elemental semiconductors. Consequently, Group V elements, such as phosphorus or arsenic, are called *donor atoms* or simply *donors*, and the doped semiconductor is now referred to as an *extrinsic semiconductor*. This may be contrasted to an intrinsic (undoped) semiconducting material.

Now consider the introduction of a large concentration of phosphorus atoms in an otherwise pure silicon crystal, for example, a phosphorus atom concentration of $\sim 10^{15}$ cm^{-3}. With a minimal energy supply, each of these phosphorus atoms will donate an electron to the crystal, amounting to a concentration of electrons in the conduction band on the order of 10^{15} cm^{-3} at room temperature. This concentration of electrons is to be contrasted with the concentration of conduction electrons in intrinsic silicon at room temperature, which is on the order of 10^{10} cm^{-3}. Thus, with this doping level, a five-order-magnitude increase in the free-electron concentration has been achieved. Note that there are about 10^{22} to 10^{23} atoms/cm^3 in a solid and that a doping level of 10^{15} cm^{-3} is equivalent to merely replacing one silicon atom in every 10^7 to 10^8 atoms/cm^3 by a phosphorus atom. Obviously, this level of doping introduces a very insignificant change in the overall crystal structure but its effect on the free-electron concentration is clearly very significant. Note that conduction in this phosphorus-doped silicon will therefore be dominated by electrons. This type of extrinsic (Group IV) semiconductor, or more specifically, silicon, is called an *n-type semiconductor* or *n-type Si*. The term *n-type* indicates that the charge carriers are the negatively charged electrons. The example discussed in the preceding text was specific to silicon doped with phosphorus; however, the conclusion arrived at will apply generally to all elemental semiconductors doped with a higher group element. The values of the ionisation energies E_d for several Group V donors in silicon are given in Table 2.7 together with those for some acceptors.

Table 2.7 Common donor and acceptor atoms in silicon

Atom	Atomic number	Type	Ionisation energy in Si (eV)
Boron	5	Acceptor	0.045
Aluminum	13	Acceptor	0.057
Phosphorus	15	Donor	0.044
Gallium	31	Acceptor	0.065
Arsenic	33	Donor	0.049
Indium	49	Acceptor	0.16
Antimony	51	Donor	0.039

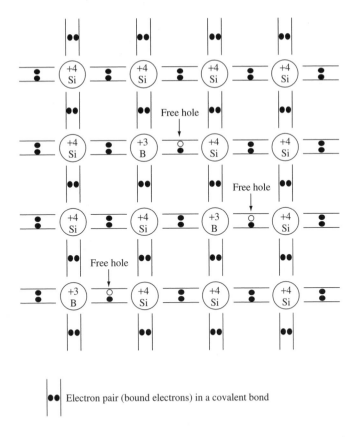

Figure 2.15 Hypothetical 2D silicon crystal doped with boron (p-type semiconductor)

Now consider the situation in which a Group IV semiconductor is doped with atoms from an element in Group III of the periodic table, that is, atoms that have only three valence electrons. To be more specific, let us take silicon doped with boron as an example, as is shown in the hypothetical two-dimensional silicon lattice in Figure 2.15. As can be seen from Figure 2.15, the net effect of having a boron atom that substitutes for silicon is the creation of a free hole (an electron deficiency in a covalent bond). This hole is generated as follows: because boron has three valence electrons, three neighbouring silicon atoms will be bonded covalently with boron. However, the fourth nearest-neighbour silicon atom has one of its four valence electrons sitting in a dangling bond; that is, the whole system of the boron atom and the four neighbouring silicon atoms has one electron missing. An electron from a neighbouring Si–Si covalent bond may replace the missing electron, thereby creating an electron deficiency (a hole) at the neighbouring bond. The net effect is, hence, the generation of a free hole in the silicon crystal. Therefore, this type of extrinsic semiconductor, silicon in this particular example, is called a *p-type semiconductor* or *p-type Si*. It is p-type because electrical conduction is carried out by positively charged free holes. Common acceptor atoms to silicon are given in Table 2.7.

Diffusion and ion implantation are the two key processes used to introduce controlled amounts of dopants into semiconductors. These two processes are used to dope selectively the semiconductor substrate to produce either an n-type or a p-type region.

2.5.1 Diffusion

In a diffusion process, the dopant atoms are placed on the surface of the semiconductor by deposition from the gas phase of the dopant or by using doped oxide sources. Diffusion of dopants is typically done by placing the semiconductor wafers in a furnace and passing an inert gas that contains the desired dopant through it. Doping temperatures range from 800 to 1200 °C for silicon. The diffusion process is ideally described in terms of Fick's diffusion equation

$$\frac{\partial C}{\partial t} = D \frac{\partial^2 C}{\partial x^2} \tag{2.21}$$

where C is the dopant concentration, D is the diffusion coefficient, t is time, and x is measured from the wafer surface in a direction perpendicular to the surface (Figure 2.16).

The initial conditions of the concentration $C(x, 0) = 0$ at time $t = 0$ and the boundary conditions are that surface concentration $C(0, t) = C_s$ at surface and that a semi-infinite medium has $C(\infty, t) = 0$. The solution of Equation (2.21) that satisfies the initial and boundary conditions is given by

$$C(x, t) = C_s \operatorname{erfc}\left(x/2\sqrt{Dt}\right) \tag{2.22}$$

where erfc is the complementary error function and the diffusion coefficient D is a function of temperature T expressed as

$$D = D_0 \exp\left(-E_a/kT\right) \tag{2.23}$$

where E_a is the activation energy of the thermally driven diffusion process, k is Boltzmann's constant, and D_0 is a diffusion constant. The diffusion coefficient is

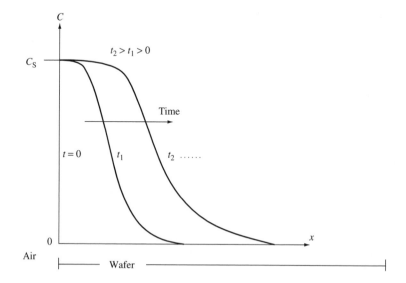

Figure 2.16   Theoretical diffusion profile of dopant atoms within a silicon wafer

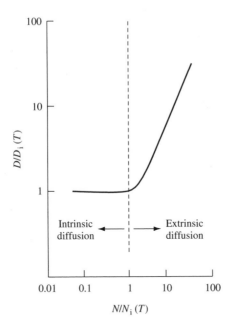

Figure 2.17 Relationship between the diffusion coefficient and doping concentration of a charge carrier in a semiconducting material

independent of dopant concentration when the doping concentration is low. However, when the doping concentration exceeds some temperature-dependent characteristic value, called the *intrinsic carrier concentration* $[N_i(T)]$, the diffusion coefficient becomes dependent on dopant concentration. When D is independent of dopant concentration, the diffusion process is called *intrinsic diffusion*, whereas when D is dependent on the doping concentration, the diffusion process is called *extrinsic diffusion* (Figure 2.17). In intrinsic diffusion the dopant diffusion profiles are complementary error functions as given by Equation (2.22); however, extrinsic diffusion profiles are somewhat complex and deviate from the basic linear theory. Instead, more complex models or empirical lookup tables are used to predict the diffusion depth.

The diffusion coefficients of commonly used dopants are considerably smaller in silicon dioxide than in silicon. Hence, while doping silicon, silicon dioxide can be used as an effective diffusion barrier or mask. Typical diffusion coefficients in the oxide at ~900 °C are 3×10^{-19} cm²/s for boron and 10^{-18} cm²/s for phosphorus. This is to be contrasted with diffusion coefficients at ~900 °C for the same dopants in silicon that are on the order of 10^{-14} cm²/s.

2.5.2 Ion Implantation

Ion implantation is induced by the impact of high-energy ions on a semiconductor substrate. Typical ion energies used in ion implantations are in the range of 20 to 200 keV

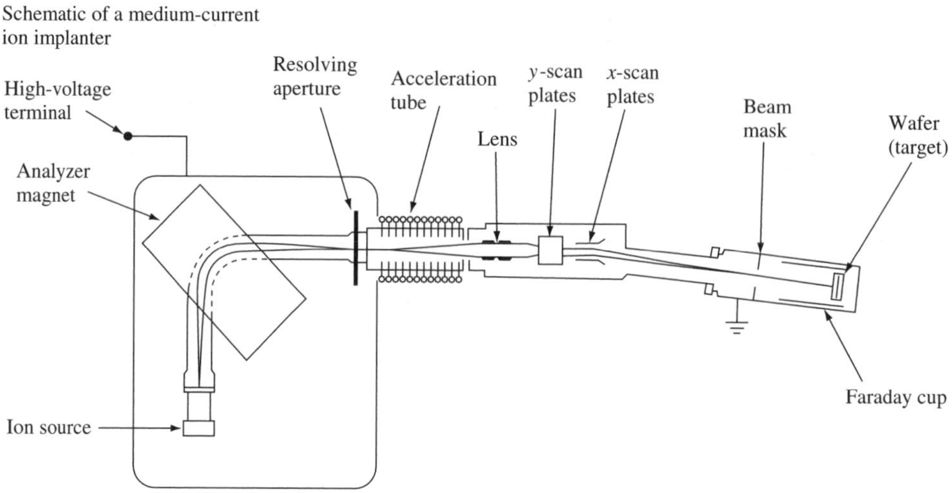

Schematic of a medium-current
ion implanter

Figure 2.18 Schematic arrangement of an ion implanter for precise implantation of a dopant into a silicon wafer (Sze 1985)

and ion densities could be between 10^{11} and 10^{16} ions/cm^2 incident on the wafer surface. Figure 2.18 shows the schematics of a medium-current ion implanter. It consists of an ion source, a magnet analyser, resolving aperture and lenses, acceleration tube, x- and y-scan plates, beam mask, and Faraday cup. After ions are generated in the ion source, the magnetic field in the analyser magnet is set to the appropriate value, depending on charge-to-mass ratio of the ion, so that desired ions are deflected toward the resolving aperture where the ion beam is collimated. These ions are then accelerated to the required energy by an electric field in the acceleration tube. The beam is then scanned in the x-y plane using the x- and y-deflection plates before hitting the wafer that is placed in the Faraday cup.

Commonly implanted elements are boron, phosphorus, and arsenic for doping elemental semiconductors, n- or p-type. After implantations, wafers are given a rapid thermal anneal to activate electrically the dopants. Oxygen is also implanted in silicon wafers to form buried oxide layers. The implanted ion distribution is normally Gaussian in shape and the average projected range of ions is related to the implantation energy (Figure 2.19).

2.6 CONCLUDING REMARKS

This chapter has introduced the topic of electronic materials and has described some basic processing steps. The next chapter describes another important class of materials, namely, those relating to the field of MEMS. These two chapters should help acquaint the reader with the materials commonly used in both microsensors and MEMS devices. Later chapters deal with more specialised processing and fabrication techniques, such as bulk and surface micromachining and stereomicrolithography.

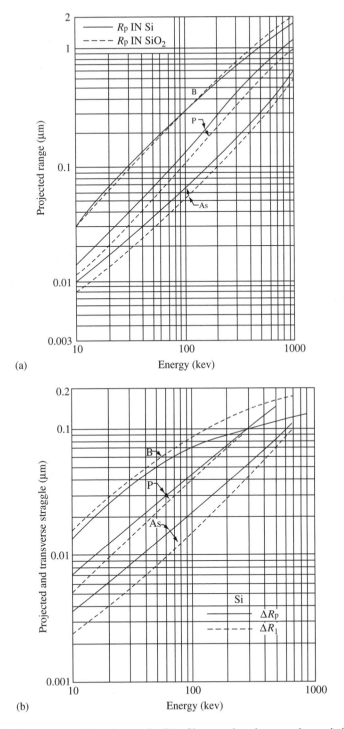

Figure 2.19 Projected range (a) and straggle; (b) of boron, phosphorus and arsenic ions implanted in silicon at different ion energies (Sze 1985)

REFERENCES

Fung C. D., Cheung P. W., Ko W. H. and Fleming D. C., eds. (1985). *Micromachining and Micropackaging of Transducers*, Elsevier, Amsterdam, The Netherlands.

Sze S. M. (1985). *Semiconductor Devices: Physics and Technology*, Wiley & Sons, New York.

Sze S. M. (1988). *VLSI Technology*, McGraw-Hill, New York.

3
MEMS Materials and their Preparation

3.1 OVERVIEW

Microelectromechanical system (MEMS) materials may be classified into five main types: metals, semiconductors, ceramics, polymers, and composites. This chapter first introduces the basic nature of each type of material and then discusses the different ways in which they can be prepared. It also provides the reader with a brief overview of the underlying physical structure of these materials and how they relate to their material properties[1]. More advanced readers may wish to omit parts of this chapter as the details given here on the preparation of electronic-grade silicon (EGS) are also presented in Chapter 4, *Standard Microelectronic Technologies*, that discusses the central role of silicon processing within the fields of microsensors and smart devices.

3.1.1 Atomic Structure and the Periodic Table

To understand the classification of materials, it is necessary to understand the quantum mechanical structure of an atom and relate it to the nature of the macroscopic bulk material. Every atom in an element is made up of a small, positively charged nucleus, which is balanced in charge by the negatively charged electrons surrounding the nucleus. The positive charge in the nucleus arises from nuclear particles called *protons*. Each proton possesses a positive charge that is equal in magnitude to the negative charge of an electron e, and in a neutral atom the number of protons in the nucleus is equal to the number of electrons outside the nucleus. The number of electrons around the nucleus, or the number of protons in the nucleus, of a neutral atom is equal to its atomic number Z.

If we consider a one-electron atomic model, such as that for hydrogen, then the energy E_n of its single electron can only have values given by the following equation (Tuck and Christopoulos (1986)):

$$E_n = -\frac{Z^2 m_e e^4}{32\pi^2 \varepsilon_0 \hbar^2} \frac{1}{n^2} \tag{3.1}$$

---

[1] See Moseley and Crocker (1996) for a review of sensor materials.

where n is a positive nonzero integer, m_e is the free-electron mass, e is its electronic charge, $\hbar$ is the Planck's constant divided by 2π, and ε_0 is the permittivity of free space[2]. These are the only allowed energy values for the electron. These energy levels are characterised by the integer n, which is called the *principal quantum number*. These levels, or the orbits corresponding to them, are called *shells*; shells corresponding to $n = 1, 2, 3, \ldots$, are labeled the K, L, M, $\ldots$, shells, respectively, where the K shell is the one nearest to the nucleus. Some shells have subshells determined by the angular momentum of the electron in its orbit. We observe that these energy values are negative, meaning that the electron is restricted to move close to the nucleus. Such an electron is generally referred to as a bound electron, that is, the electron is bound to the nucleus. If this electron acquires energy from an external source such that its energy E_n becomes equal to or greater than zero, then the electron frees itself from the nucleus and is then called a *free electron*. As we shall see later, these free electrons are moving charges and are therefore important in defining the properties of electronically conducting materials.

The possible values of the orbital angular momentum for the electron in an atom are given by $[l(l + 1)]^{1/2}\hbar$, where the quantum number l takes the values $l = 0, 1, 2, 3, \ldots$, $(n - 1)$; that is, $l \le (n - 1)$. Here again, just as the energy of the electron in a one-electron atom is quantised, so is its orbital angular momentum. The orbital quantum number l governs the angular momentum of the electron. In an external magnetic field, we also find that the possible (allowed) values of the z-component of the orbital angular momentum are denoted by m, where the quantum number m is called the *magnetic quantum number*. It is called as such because in the absence of a magnetic field, all the states defined by $m\hbar$ have the same energy values, that is, the states are degenerate. However, the application of a magnetic field lifts the degeneracy and each of these states would have a different energy. The value of m can only be an integer between $-l$ and $+l$, that is, m is $-l, (-l + 1), (-l + 2), \ldots, 0, 1, 2, \ldots, (l - 2), (l - 1), l$ or $l \le m \le +l$.

To sum up, while in a specific shell n, the electron can have any one of a number of allowed orbital angular momenta characterised by the quantum number $l = 0, 1, 2, \ldots$, $(n - 1)$. These different l values correspond to different subshells, or orbitals, each of angular momentum value of $[l(l + 1)]^{1/2}\hbar$. Hence, although an electron occupies a certain shell n of certain energy E_n, its orbital angular momentum can assume different values. Within the same orbital, the projection of the angular momentum along the z-axis can take certain values as described earlier. The quantum number for the z-component of the angular momentum is m, where $-l \le m \le +l$. Therefore, at this point it can be said that the state of the electron in a hydrogen atom is characterised by n, l, and m.

So far, we have implied that the electronic distribution in an isolated atom can be characterised by only three quantum numbers n, l, and m. However, in addition to orbital angular momentum, it was found that each individual electron also has an intrinsic spin angular momentum. The value of this spin angular momentum is characterised by the quantum number S that assumes only one value $-\frac{1}{2}$. The magnitude of this spin angular momentum is $[S(S + 1)]^{1/2}\hbar = \sqrt{3}\hbar/2$ and is called the *total spin*. Actually, the spin of an electron in a magnetic quantum number has two states identified by the spin quantum

---

[2] The values of the fundamental constants are given in Appendix D.

Table 3.1 Quantum numbers, their allowed values, and the parameters they quantise

Quantum number	Quantised parameter	Allowed values
n	Total energy	1, 2, 3, 4, etc.
l	Orbital angular momentum	$0, 1, 2, 3, \ldots, (n-1)$
m	Orbital angular momentum component along the axis of quantisation	$-l, -l+1, -l+2, \ldots, -1,$ $0, 1, \ldots, l-2, l-1, l$
m_s	Spin angular momentum component along the axis of quantisation	$+\frac{1}{2}$ or $-\frac{1}{2}$

number m_s, which takes two values $+\frac{1}{2}$ and $-\frac{1}{2}$. Therefore, four quantum numbers n, l, m, and m_s are required for a complete description of the electronic states in an isolated one-electron atom. The four quantum numbers are listed in Table 3.1 along with the parameters that they represent and the limitations on the values of each of these quantum numbers.

When an atom containing more than one electron is treated quantum mechanically, a useful first-approximation is that the electrons do not exert forces on one another. Thus, in this approximation, the states occupied by these electrons are still characterised by the quantum numbers in Table 3.1. However, the arrangement of these electrons in the atom satisfies Pauli's exclusion principle. This principle states that in a multielectron system no two electrons can have identical sets of quantum numbers. Stated differently, the Pauli's exclusion principle requires that no two electrons may have the same spatial distribution and spin orientation and that no more than one electron may have the same wave function when spin is included. Therefore, four quantum numbers n, l, m, and m_s are required to describe the exact state of an electron in an atom.

The lowest-energy, or ground-state, electron configuration for any atom can be explained using the results summarised in Table 3.1 and the Pauli's exclusion principle. The number of combinations of m and m_s for a given subshell or orbital (n, l) gives the maximum number of electrons in that subshell. For each value of l, there are $(2l + 1)$ values of m, and for each value of l and m, there are two values of m_s ($\pm\frac{1}{2}$). Therefore, the maximum number of electrons that can be placed in a given subshell, in accordance with Pauli's exclusion principle, is $2(2l + 1)$. As stated earlier, the shells associated with $n = 1, 2, 3, 4, 5$, and 6 are labeled as the K, L, M, N, O, and P shells, respectively. The orbitals associated with $l = 0, 1, 2$, and 3 are labeled as the s, p, d, and f orbitals, respectively. The quantum number l specifies the shape of the envelope in which the electron is likely to be found. Figure 3.1 shows the calculated (from quantum theory) envelopes for s, p, and d electrons. In the case of s electrons, the envelope is spherical; for p electrons, it is dumbbell-shaped; and for d electrons, it is clover-shaped in four cases and dumbbell-shaped in one.

Using the results on the atomic structure outlined earlier, we can now proceed to set up the periodic table of elements. The first few elements of the periodic table are shown in Table 3.2. The first element is hydrogen with only one electron in the lowest energy state defined by $n = 1$, $l = 0$, and $m = 0$. This configuration of the hydrogen atom is designated as $1s^1$. The number 1 to the left stands for the shell $n = 1$, s specifies the orbital $l = 0$,

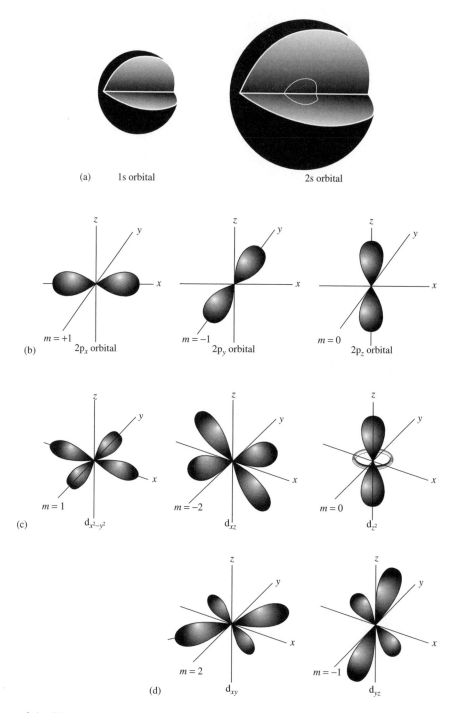

Figure 3.1   Theoretical envelopes for the (a) s; (b) p; and (c) d electrons calculated from quantum theory

Table 3.2 The first 10 elements in the periodic table

Z	Element	Notation	n	l	m	m_s
1	Hydrogen	$1s^1$	1	0	0	$+\frac{1}{2}$ or $-\frac{1}{2}$
2	Helium	$1s^2$	1	0	0	$+\frac{1}{2}$ or $-\frac{1}{2}$
3	Lithium	$1s^2 2s^1$	2	0	0	$+\frac{1}{2}$ or $-\frac{1}{2}$
4	Beryllium	$1s^2 2s^2$	2	0	0	$+\frac{1}{2}$ or $-\frac{1}{2}$
5	Boron	$1s^2 2s^2 2p^1$	2	1	-1, or 0, or 1	$+\frac{1}{2}$ or $-\frac{1}{2}$
6	Carbon	$1s^2 2s^2 2p^2$	2	1	-1, or 0, or 1	$+\frac{1}{2}$ or $-\frac{1}{2}$
7	Nitrogen	$1s^2 2s^2 2p^3$	2	1	-1, or 0, or 1	$+\frac{1}{2}$ or $-\frac{1}{2}$
8	Oxygen	$1s^2 2s^2 2p^4$	2	1	-1, or 0, or 1	$+\frac{1}{2}$ or $-\frac{1}{2}$
9	Fluorine	$1s^2 2s^2 2p^5$	2	1	-1, or 0, or 1	$+\frac{1}{2}$ or $-\frac{1}{2}$
10	Neon	$1s^2 2s^2 2p^6$	2	1	-1, or 0, or 1	$+\frac{1}{2}$ or $-\frac{1}{2}$

and the superscript 1 stands for the number of electrons occupying this state. Note that the electron can take on spin quantum number of either $+\frac{1}{2}$ or $-\frac{1}{2}$, as the two spin states have the same energy in the absence of magnetic field and are equally accessible to the electron. The second element helium has two electrons that occupy the two lowest energy states. These states are (n, l, m, m_s) equal to $(1, 0, 0, +\frac{1}{2})$ and $(1, 0, 0, -\frac{1}{2})$, that is, the ground state of the helium atom can be labeled as $1s^2$. Note that for helium the lowest shell ($n = 1$) is full. The chemical activity of an element is determined primarily by the valence electrons that are electrons in the outermost unfilled shell. Because the valence energy shell of helium is full, helium does not react with other elements and is an inert element.

The third element lithium has three electrons, two of which will occupy the states (n, l, m, m_s) equal to $(1, 0, 0, +\frac{1}{2})$, $(1, 0, 0, -\frac{1}{2})$, and the third will occupy one of the other eight states in the next shell with $n = 2$. From among these eight states, the third electron in the lithium atom occupies one of the two states $(2, 0, 0, +\frac{1}{2})$ or $(2, 0, 0, -\frac{1}{2})$. Therefore, the ground state configuration of lithium is written as $1s^2 2s^1$.

In theory, we may continue to build up the periodic table following the method illustrated in the preceding discussion, as shown in Table 3.2. However, in practice, electrons will start to interact with each other as the atomic number of the element increases. This electron–electron interaction is not taken into account in the model that was presented earlier for the atomic structure, and the buildup of the periodic table will, therefore, somewhat deviate from the predicted one. This deviation is shown in Table 3.3 and Figure 3.2, which gives the energy scheme for the first six different shells and orbitals.

Proceeding in the same manner as described earlier, we can arrange the 105 or so elements that are known to us presently to complete the periodic table, as shown in Figure 3.3. In the periodic table, elements are arranged in horizontal rows in order of atomic number. A new row is begun after each noble gas (e.g. He, Ne, Ar, Kr, Xe, and Rn) is encountered. Notice that the elements in each vertical column have similar properties. Also notice that the elements on the left-hand side of the diagonal dividing band are metals, whereas those on the right-hand side are nonmetals. The elements within the bands are semimetals, which are more commonly known as *semiconductors*.

Table 3.3 The first six shells of the periodic table in which electron–electron interaction is significant

Shell	Subshell or orbital	Electrons in subshell	Total number of electrons
K	1s	2	2
L	2s	2	
	2p	6	10
M	3s	2	
	3p	6	18
N	4s	2	
	3d	10	
	4p	6	36
O	5s	2	
	4d	10	
	5p	6	54
P	6s	2	
	4f	14	
	5d	10	
	6p	6	86

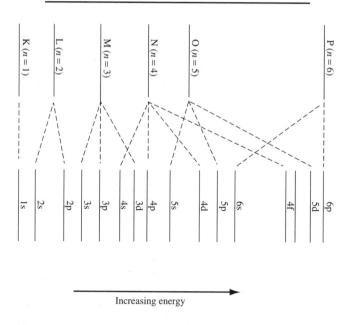

Increasing energy

Figure 3.2 Energy scheme for different atomic shells and orbitals

3.1.2 Atomic Bonding

Materials are composed of elements that are connected to each other at the atomic level. The connections between elements in a material are known as *bonds*. The properties and applications of materials depend upon the nature and strength of the bond. There are several different types of bond and these are described briefly in the following text.

1a	2a	3b	4b	5b	6b	7b	8	8	8	1b	2b	3a	4a	5a	6a	7a	0
1 **H** 2.1																	**2** **He**
3 **Li** 1.0	**4** **Be** 1.5											**5** **B** 2.0	**6** **C** 2.5	**7** **N** 3.0	**8** **O** 3.5	**9** **F** 4.0	**10** **Ne**
11 **Na** 0.9	**12** **Mg** 1.2											**13** **Al** 1.5	**14** **Si** 1.8	**15** **P** 2.1	**16** **S** 2.5	**17** **Cl** 3.0	**18** **Ar** -
19 **K** 0.8	**20** **Ca** 1.0	**21** **Sc** 1.3	**22** **Ti** 1.5	**23** **V** 1.6	**24** **Cr** 1.6	**25** **Mn** 1.5	**26** **Fe** 1.8	**27** **Co** 1.8	**28** **Ni** 1.8	**29** **Cu** 1.9	**30** **Zn** 1.6	**31** **Ga** 1.6	**32** **Ge** 1.8	**33** **As** 2.0	**34** **Se** 2.4	**35** **Br** 2.8	**36** **Kr** -
37 **Rb** 0.8	**38** **Sr** 1.0	**39** **Y** 1.2	**40** **Zr** 1.4	**41** **Nb** 1.6	**42** **Mo** 1.8	**43** **Tc** 1.9	**44** **Ru** 2.2	**45** **Rh** 2.2	**46** **Pd** 2.2	**47** **Ag** 1.9	**48** **Cd** 1.7	**49** **In** 1.7	**50** **Sn** 1.8	**51** **Sb** 1.9	**52** **Te** 2.1	**53** **I** 2.5	**54** **Xe** -
55 **Cs** 0.7	**57** **Ba** 0.9	**57-71** **La-Lu** 1.1 1.2	**72** **Hf** 1.3	**73** **Ta** 1.5	**74** **W** 1.7	**75** **Re** 1.9	**76** **Os** 2.2	**77** **Ir** 2.2	**78** **Pt** 2.2	**79** **Au** 2.4	**80** **Hg** 1.9	**81** **Tl** 1.8	**82** **Pb** 1.8	**83** **Bi** 1.9	**84** **Po** 2.0	**85** **At** 2.2	**86** **Rn** -
87 **Fr** 0.7	**88** **Ra** 0.9	**89-103** **Ac-Lr** 1.1 1.7	**104** **(Rf)**	**105** **(Ha)**													

Figure 3.3 Periodic table of elements showing the elements in order of their atomic number Z

3.1.2.1 Ionic bonding

In ionic bonding, one element gives up its outer-shell electron(s) to uncover a stable inner shell of eight electrons (resembling the nearest noble element). The electrons are attracted to a second element in which they can serve to complete its outer shell of eight (again resembling the nearest noble element). An example of ionic bonding is the sodium chloride (NaCl) molecule in which the sodium atom donates its $3s^1$ electron, leaving an L shell of eight (as in Ne), whereas the chlorine atom with an outer shell of seven ($3s^2 3p^5$), attracts that electron to form an outer shell of eight (as in Ar). As a result of this electron-transfer process, two ions, Na$^+$ and Cl$^-$, are formed (see Figure 3.4).

The ions formed in an ionic bond are attracted by a coulombic force that is proportional to the product of the charge on the ions and inversely proportional to the square of their separation when the ions can be regarded as distinct points in space. However, when the ions are close to one another, the force field changes and a repulsive force develops between the electron fields. On the basis of these observations, a graph of potential energy versus interatomic distance looks like the one shown in Figure 3.5. The minimum in the curve corresponds to the equilibrium distance between the two ions.

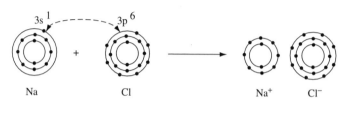

Figure 3.4   Ionic bonding of sodium (Na) and chlorine (Cl) atoms to form sodium chloride (NaCl)

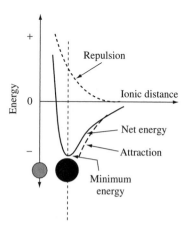

Figure 3.5   Variation of potential energy with interatomic distance for an ionic bond

3.1.2.2 Covalent bonding

In covalent bonding, atoms share their outermost shell electrons to attain a stable group of eight electrons. In the case of chlorine gas molecule (Cl_2), one electron from each chlorine atom is used to form a common covalent bond, whereby each atom is surrounded by a stable group of eight electrons (Figure 3.6). This is called a *molecular bond* because the stable molecule Cl_2 is formed. The Cl_2 molecular bond is pictorially depicted in Figure 3.7, in which the merging of two of the 3p electron envelopes is shown. Instead of drawing the entire ring, it suffices to only show the pair of shared electrons as two dots or as a single line.

There are situations in which the covalent bonding becomes more complex than in a simple molecular bond. The most important of these complex covalent bonds are those associated with a tetrahedral structure, such as the methane (CH_4) molecule. The atomic structure of carbon is $1s^2 2s^2 2p^2$. As CH_4 is about to be formed, a two-process step occurs within the carbon atom. First, one of the 2s electrons is promoted to a 2p state. The energy required to achieve this first step is provided during the formation of the C–H bonds. Next, the 2s electron and three 2p electrons hybridise to form a hybridised group of four electrons with orbits along four evenly spaced tetrahedral axes (Figure 3.8). Four equal C–H bonds are then formed to produce the tetrahedral structure of the CH_4 molecule (Figure 3.8).

3.1.2.3 Metallic bonding

In metallic bonding, encountered in pure metals and metallic alloys, the atoms give up their outer-shell electrons to a distributed electron cloud for the whole block of metal (see

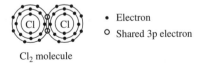

- • Electron
- ○ Shared 3p electron

Cl_2 molecule

Figure 3.6   Covalent bonding between chlorine atoms to form a chlorine molecule (Cl_2)

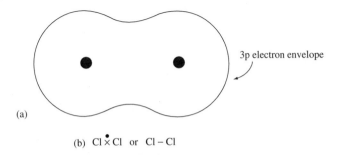

3p electron envelope

(a)

(b) Cl ẋ Cl   or   Cl – Cl

Figure 3.7   (a) Pictorial; (b) written expression of a chlorine molecule

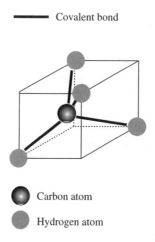

Figure 3.8   Tetrahedral structure of a methane (CH_4) molecule

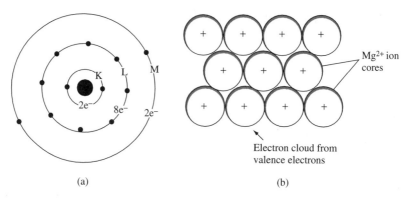

Figure 3.9   Metallic bonding in magnesium. The outer two electrons in the M shell (a) become mobile and free to move within a distributed electron cloud (b)

Figure 3.9). This model shows that good electrical conductivity would be characteristic of metallic bonding, owing to the ability of a high concentration of electrons to move freely, meaning that they possess a high mobility.

3.1.3 Crystallinity

Materials occur in either a crystalline or an amorphous state. The crystalline state refers to the organisation of $\sim 10^{22}$ atoms/cm^3 arranged in a regular manner in a three-dimensional structure. This regular array of atoms may be obtained by repeating in three dimensions an elementary arrangement of building blocks called *unit cells*, which contain atoms placed at fixed positions. If the periodic arrangement occurs throughout the volume of a sample material, this constitutes a single crystal. However, if the regular structure occurs only in portions of a material and the different portions are aligned arbitrarily with respect to each other, the material is said to be *polycrystalline*. The individual regular portions are referred to as *crystallites* or *grains* and are separated from each other by grain boundaries. If the individual crystallites are reduced in size to the point where they approach the size of a unit cell, periodicity is lost and the material is called *amorphous* or *glassy*.

The geometric shape of a unit cell is a three-dimensional parallel-piped structure and contains one or a few of the same[3] atom in simple crystals such as copper, sodium, or silver but may contain thousands of atoms in complex organic crystals. The length of an edge of the unit cell is called the *lattice constant*. The variety of crystal structures can be defined by arranging atoms systematically about a regular or periodic arrangement of points in space called a *space lattice*. A lattice is defined by three fundamental translational vectors **a**, **b**, and **c**, so that the arrangement of atoms in a crystal of infinite extent looks identical when observed from any point that is displaced a distance **r** from an origin, as viewed from the point **R**, where

$$\mathbf{R} = \mathbf{r} + n_1\mathbf{a} + n_2\mathbf{b} + n_3\mathbf{c} \tag{3.2}$$

n_1, n_2, and n_3 are integers. This is schematically shown in Figure 3.10. The points described by the position vector **R** constitute the space lattice.

Parallel planes of atoms in a crystal are identified by a set of numbers called *Miller indices*. These numbers can be obtained in the following way: choose the origin of the coordinate system x, y, z to coincide with a lattice point in one of these parallel planes. Find the intercepts of the next parallel plane on the x-, y-, and z-axes as x_1, y_1, and z_1, respectively. Now take the reciprocals of these numbers and multiply by a common factor so as to obtain the three lowest integers h, k, and l; these integers are called *Miller indices* and are normally written within parentheses (h, k, l). An example of this method for plane identification is shown in Figure 3.11. If one of the intercepts happens to be a negative number, an overbar is added to the corresponding Miller index to indicate that the plane intercepts with the negative axis.

A crystallographic direction in a crystal is denoted by a square-bracket notation $[h, k, l]$. The numbers h, k, and l correspond, respectively, to the x, y, and z components of a vector that defines a particular direction. Again these numbers h, k, and l are the smallest integers, the ratios for which are the same as those of the vector length ratios. An overbar on any of the integers denotes a negative vector component on the associated axis. For example, the direction of the negative x-axis is $[\bar{1}, 0, 0]$, whereas the positive x-axis is denoted by the direction of $[1,0,0]$.

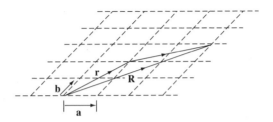

Figure 3.10   Representation of a two-dimensional crystal lattice in terms of fundamental translation vectors. A third unit cell vector **c** provides the third dimension (not shown)

---

[3] The number depends on whether the unit cell is SC, BCC, or FCC in a cubic crystal lattice (see later).

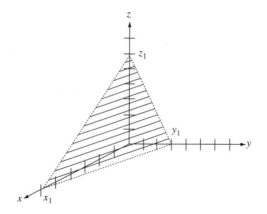

Figure 3.11   Plane identification in terms of the intercepts on the x, y, and z axes

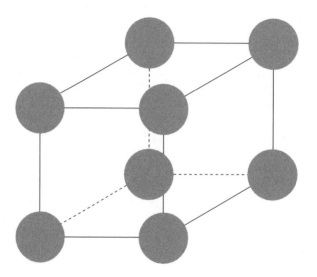

Figure 3.12   Unit cell of a simple cubic (SC) crystal lattice structure

The following subsections describe some of the common crystal structures encountered and those that are used to describe MEMS materials.

3.1.3.1 Cubic structures

The geometric shape of the unit cell in cubic structures is a cube. If a single atom or a group of atoms are placed at the corners of the cube, this constitutes a simple cubic (SC) crystal (Figure 3.12). Each one of the cube corners constitutes a site, or a lattice point, which can be occupied by one building block (an atom or a group of atoms). Because each corner of a given cube is shared by other seven cubes, it turns out that each unit cell contains, in essence, one site or one lattice point in SC structure. In addition, in an

SC structure, each atom or a group of atoms at some lattice point is surrounded by six nearest neighbours.

A second cubic crystal structure is the body-centred cubic (BCC). The BCC crystal lattice not only contains a single site at each corner of the cube but also one at the centre of the cubic cell, as shown in Figure 3.13. The number of nearest neighbours to any particular lattice point in this case is eight. The third cubic structure is face-centred cubic (FCC) lattice (see Figure 3.14), which contains one site in the centre of each of the six cube faces in addition to the eight positions at the corners of the cube. The number of nearest-neighbour sites to any given lattice point in this case is 12. Using arguments similar to those used in SC structures, it is found that the number of lattice points in a BCC and an FCC unit cell are two and four, respectively.

The conventional unit cell of diamond is basically cubic, as shown in Figure 3.15. This structure is more loosely packed than the previously discussed cubic structures.

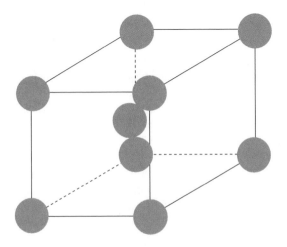

Figure 3.13 Unit cell of a body-centred cubic (BBC) crystal lattice structure

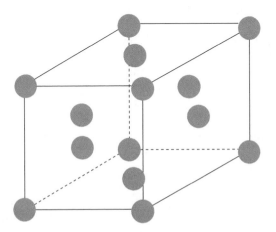

Figure 3.14 Unit cell of a face-centred cubic (FCC) crystal lattice structure

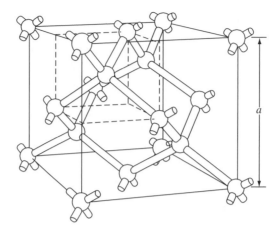

Figure 3.15   Tetrahedral structure of carbon in its diamond state

Each carbon atom has four nearest neighbours forming a tetrahedral bond. This diamond structure could be visualised as placing an atom at the centre of the cube, two atoms at the opposite corners of the top face of this cube, and two atoms placed at opposite corners at the bottom face of the cube, but twisted 90° with respect to the top-face atoms. This configuration is shown in Figure 3.15, but it is not the unit cell. The tetrahedral bonds of the four corner atoms to the central atom are very strong and highly directional, occurring at angles of ~109.5°. In essence, the diamond structure can be viewed as two interpenetrating FCC lattices – one displaced from the other by one-fourth the length and along a cube diagonal. The diamond structure, which is a special type of cubic structure, is of particular interest because some of the electronic materials (semiconductors) have diamond-like crystal structures. Moreover, diamond itself has been used as a functional material in microdevices.

3.1.3.2 Hexagonal close-packed structure

The hexagonal close-packed (HCP) structure ranks in importance with the BCC and FCC lattices; more than 30 elements crystallise in the HCP form. Underlying the HCP structure is hexagonal lattice geometry (see Figure 3.16). To describe hexagonal structures, a few simple modifications of the Miller indices of directions and planes are required. Instead of three axes, x, y, and z, four axes are used – three in the horizontal (x, y) plane at 120° to each other, called a_1, a_2, a_3, and the fourth, c, in the z-direction. The use of the extra axis makes it easier to distinguish between similar planes in the hexagonal structure. Figure 3.16 shows some planes located using this four-axes reference frame. Using either three axes $(a_1, a_2, \text{ and } c)$ or four axes develops the notation for a direction. It is noted that a_1 and a_2 are at 120° even in this instance. Figure 3.16 shows directions specified using the three-coordinate system.

  We have now reviewed all the necessary basic background information that will enable us to describe different classes of materials. We broadly classify MEMS materials into five categories: metals, semiconductors, ceramics, polymers, and composites. In the course of

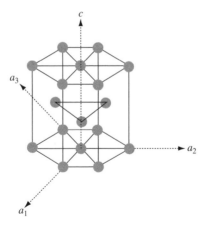

Figure 3.16 Hexagonal close-packed (HCP) crystal structure

discussing these different MEMS materials, several material-preparation techniques are described. These techniques are described within different material sections according to the frequency of their use in preparing the particular material under consideration. However, it should be pointed out that several of the preparation techniques described in the following sections are used to prepare more than one type of material. For instance, the sputtering technique is described in the metals section; however, it is also used to deposit semiconductor and ceramic films.

3.2 METALS

3.2.1 Physical and Chemical Properties

Metals are inorganic substances that are composed of one or more metallic elements. Examples of metallic materials with one element are iron, aluminum, copper, and cobalt. When a metallic material is composed of two or more metallic elements, it is called an *alloy*. Some metallic materials may contain nonmetallic elements that are added intentionally to improve the material's engineering qualities. An example of such a metallic material is steel, in which the nonmetallic element carbon is added to iron. Metals and alloys are commonly divided into two types: ferrous metals and alloys that contain high concentrations of iron and nonferrous metals and alloys that contain no or very low concentrations of iron.

Single-crystal metals are mostly found in the three simple types of cells: BCC, FCC, and HCP. Under different conditions of temperature and pressure, different crystal structures (that is, different unit cells) or phases for the same metal are formed. For example, a bar of iron at room temperature has a BCC structure. However, if the bar is heated above 900 °C, the structure changes to FCC[4]. The BCC iron and the FCC iron are called the *α-phase* and *γ-phase*, respectively.

---

[4] The phase change of a material is sometimes used as the sensing or actuating principle of a microdevice. One example is a shape-memory alloy.

Table 3.4   The atomic properties and crystal structures of selected metals

Atomic number (Z)	Symbol	Atomic radius (Å)	Lattice structure	Interatomic distance (Å)
13	Al	1.43	FCC	2.86
22	Ti	1.47	HCP	2.90
24	Cr	1.25	BCC (α)	2.49
		1.36	HCP (β)	2.71
26	Fe	1.24	BCC (α)	2.48
		1.26	FCC (γ)	2.52
27	Co	1.25	HCP (α)	2.49
		1.26	FCC (β)	2.51
28	Ni	1.25	HCP (α)	2.49
		1.25	FCC (β)	2.49
29	Cu	1.28	FCC	2.55
30	Zn	1.33	HCP	2.66
47	Ag	1.44	FCC	2.97
78	Pt	1.38	FCC	2.77
79	Au	1.44	FCC	2.88
82	Pb	1.75	FCC	3.49

Metals are, in general, good thermal and electrical conductors. They are somewhat strong and ductile at room temperature and maintain good strength both at room and elevated temperatures. Table F.1 in Appendix F gives some important physical properties of metals that are commonly used in microelectronics and MEMS.

Table 3.4 provides atomic and crystal structure information on 12 selected metals, and these illustrate the three principal lattice structures described earlier.

3.2.2 Metallisation

Metallisation is a process in which metal films are formed on the surface of a substrate. These metallic films are used for interconnections, ohmic contacts, and so on[5]. Metal films can be formed using various methods, the most important being physical vapour deposition (PVD). PVD is performed under vacuum using either the evaporation or the sputtering technique.

3.2.2.1 Evaporation

Thin metallic films can be evaporated from a hot source onto a substrate, as shown in Figure 3.17. An evaporation system consists of a vacuum chamber, pump, wafer holder, crucible, and a shutter. A sample of the metal to be deposited is placed in an inert crucible, and the chamber is evacuated to a pressure of 10^{-6} to 10^{-7} torr. The crucible is then heated using a tungsten filament or an electron beam to flash-evaporate the metal from the crucible and condense it onto the cold sample. The film thickness is determined

---

[5] Copper-based printed circuit board and other interconnect technologies are discussed in Section 4.5.

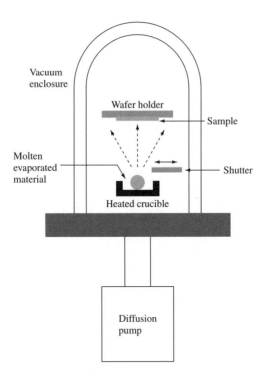

Figure 3.17   Schematic view of a thermal evaporation unit for depositing materials

by the length of time that the shutter is opened and can be measured using a quartz microbalance (QMB)–based film thickness monitor. The evaporation rate is a function of the vapour pressure of the metal. Therefore, metals that have a low melting point T_{mp} (e.g. 660 °C for aluminum) are easily evaporated, whereas refractory metals require much higher temperatures (e.g. 3422 °C for tungsten) and can cause damage to polymeric or plastic samples. In general, evaporated films are highly disordered and have large residual stresses; thus, only thin layers of the metal can be evaporated. In addition, the deposition process is relatively slow at a few nanometres per second.

3.2.2.2 Sputtering

Sputtering is a physical phenomenon, which involves the acceleration of ions through a potential gradient and the bombardment of a 'target' or cathode. Through momentum transfer, atoms near the surface of the target metal become volatile and are transported as a vapour to a substrate. A film grows at the surface of the substrate through deposition.

Figure 3.18 shows a typical sputtering system that comprises a vacuum chamber, a sputtering target of the desired film, a sample holder, and a high-voltage direct current (DC) or radio frequency (RF) power supply. After evacuating the chamber down to a pressure of 10^{-6} to 10^{-8} torr, an inert gas such as helium is introduced into the chamber at a few millitorr of pressure. A plasma of the inert gas is then ignited. The energetic ions of the plasma bombard the surface of the target. The energy of the bombarding ions ($\sim$keV) is sufficient to make some of the target atoms escape from the surface. Some

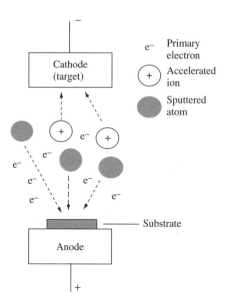

Figure 3.18 Basic components in a physical sputtering unit for depositing materials

of these atoms land on the sample surface and form a thin film. Sputtered films tend to have better uniformity than evaporated ones, and the high-energy plasma overcomes the temperature limitations of evaporation. Most elements from the periodic table, including both inorganic and organic compounds, can be sputtered. Refractory materials can be sputtered with ease, whereas the evaporation of materials with very high boiling points is problematic. In addition, materials from more than one target can be sputtered at the same time. This process is referred to as *cosputtering*.

The structure of sputtered films is mainly amorphous, and its stress and mechanical properties are sensitive to specific sputtering conditions. Some atoms of the inert gas can be trapped in the film, causing anomalies in its mechanical and structural characteristics. Therefore, the exact properties of a thin film vary according to the precise conditions under which it was made. Consequently, values given for the bulk material, such as those given in Appendix F, serve only as an approximate guide to the film values.

3.3 SEMICONDUCTORS

3.3.1 Semiconductors: Electrical and Chemical Properties

Semiconductors are commonly inorganic materials made from elements in the fourth column (Group IV) of the periodic table. The most important among these elements is silicon that can be modified in several ways to change its electrical, mechanical, and optical properties. The use of silicon in solid state and microelectronics has shown a spectacular growth since the early 1970s, and this growth pattern is still continuing[6]. Other

---

[6] Chapter 1 describes the recent emergence of microtechnologies.

semiconductor materials, from Group IV elements in the periodic table, are germanium and carbon (diamond). Semiconductor materials can also be made from a combination of elements either from Group III and Group V or from Group II and Group VI. Examples of these are gallium arsenide and zinc telluride materials. The name *semiconductor* is given to these materials because at certain regimes of temperatures they are able to exhibit good electrical conduction properties, and outside these temperature regimes they behave as insulators[7].

Semiconductor crystals can be made from both single elements and compounds. Semiconductors that are made from single elements are called *elemental semiconductors*. Elemental semiconductors are found in Group IV of the periodic table, for example, silicon (Si), and germanium (Ge). Compound semiconductors are made up of special combinations either of Group III and Group V elements or of Group II and Group VI elements, as stated earlier. Table 3.5 lists a few of the elemental and compound semiconductors. Properties of some common elemental and compound semiconductors are given in Appendix G.

Among the elemental semiconductors, silicon is by far the most commonly used material. Silicon is the most important material for microelectronics and integrated circuit technology. In addition, silicon-based compounds and technologies are becoming the major cornerstones for the rapidly developing fields of MEMS and nanofabrication. For this reason, we will be emphasising silicon and using it to demonstrate the general properties of semiconductor materials. Table 3.6 lists a few of the mechanical, electrical, and thermal properties of single crystalline silicon. Gallium arsenide (GaAs) is the most commonly used among the compound semiconductors, especially in fabricating optical and high-speed devices.

The crystal structure of many semiconductors, including silicon and gallium arsenide, is based on the cubic crystalline system[8] (see Figure 3.15). Diamond itself could be

Table 3.5 Structure and lattice properties for some common elemental and compound semiconductors. The lattice constants and band gaps are given at a temperature of 27 °C

Material	Lattice structure[a]	Lattice constant (Å)	Energy gap (eV)
Ge	Diamond structure	5.66	0.66
Si	Diamond structure	5.43	1.12
GaAs	Zinc-blende structure	5.64	1.44
GaSb	Zinc-blende structure	6.12	0.78
InSb	Zinc-blende structure	6.46	0.18
InAs	Zinc-blende structure	6.04	0.33
InP	Zinc-blende structure	5.86	1.25
PbSe	Zinc-blende structure	6.14	0.27
PbTe	Zinc-blende structure	6.34	0.30

[a]For more precise classification of structures, use alphanumeric system, that is, A3 is diamond.

[7] See Pierret (1988) for a review of fundamentals of semiconductors.
[8] All crystal lattice structures can be classified according to an alphanumeric system to avoid confusion. In order to aid clarity, it has not been used here.

Table 3.6 Electrical, mechanical, and thermal properties of crystalline silicon

Electrical		Mechanical		Thermal	
Resistivity (P-doped)	1–50 Ωcm	Yield strength	7×10^9 N/m^2	Thermal conductivity	1.57 W/cm °C
Resistivity (Sb-doped)	0.005–10 Ωcm	Young's modulus	1.9×10^{11} N/m^2	Thermal expansion	2.33×10^{-6}/°C
Resistivity (B-doped)	0.005–50 Ωcm	Density	2.3 g/cm^3	–	–
Minority-carrier lifetime	30–300 µs	Dislocations	<500/cm^2	–	–

thought of as a semiconductor with a wide band gap of $\sim$6 eV, and its structure is that of two interleaved FCC arrays, in which one array is about a fraction of the interatomic distance from the other. In the gallium arsenide–type of compound, one of the two arrays is composed entirely of gallium atoms, whereas the other array is composed of arsenic atoms. This particular class of the diamond structure is called the *zinc-blende structure*. In the diamond lattice, each atom has four nearest neighbours. In both elemental and compound semiconductors, there is an average of four valence electrons per atom. Each atom is thus held in the crystal by four covalent bonds, wherein two electrons participate in each bond. In a perfect semiconductor crystal and at absolute zero temperature, the number of electrons available would exactly fill the inner atomic shells and the covalent bonds. At temperatures above absolute zero, some of these electrons gain enough thermal energy to break loose from these covalent bonds and become free electrons. Free electrons are responsible for electrical conduction across the semiconductor crystal. Some of the physical properties of selected semiconductor crystals are given in Table 3.6.

3.3.2 Semiconductors: Growth and Deposition

To demonstrate the methods of growing semiconductors, let us consider crystal growth of silicon in detail. Silicon is used as an example because it is the most utilised semiconductor in microelectronics and MEMS. In fact, the next three chapters are devoted to conventional silicon microtechnology (Chapter 4), bulk micromachining (Chapter 5), and surface (Chapter 6) micromachining techniques.

Section 3.3.2.1 briefly outlines silicon crystal growth from the melt – a technique that is widely used in growing bulk silicon wafers. This is followed by the epitaxial growth of thin crystalline silicon layers in Section 3.3.2.2. A variation of the method for silicon growth from the melt is the Bridgman technique that is used for growing gallium arsenide wafers. The Bridgman technique is not discussed in this chapter (for a description of the Bridgman technique see Tuck and Christopoulous (1986)). A more detailed description of the way in which silicon wafers are made is given in Section 4.2. However, a brief overview is presented in the following subsections.

3.3.2.1 Silicon crystal growth from the melt

Basically, the technique used for silicon crystal growth from the melt is the Czochralski technique. The technique starts when a pure form of sand (SiO$_2$) called *quartzite* is placed in a furnace with different carbon-releasing materials such as coal and coke. Several reactions take place inside the furnace and the net reaction that results in silicon is

$$SiC + SiO_2 \longrightarrow Si + SiO \text{ (gas)} + CO \text{ (gas)} \tag{3.3}$$

The silicon so produced is called *metallurgical-grade silicon* (MGS), which contains up to 2 percent impurities. Subsequently, the silicon is treated with hydrogen chloride (HCl) to form trichlorosilane ($SiHCl_3$):

$$Si + 3HCl \longrightarrow SiHCl_3 \text{ (gas)} + H_2 \text{ (gas)} \tag{3.4}$$

$SiHCl_3$ is liquid at room temperature. Fractional distillation of the $SiHCl_3$ liquid removes impurities, and the purified liquid is reduced in a hydrogen atmosphere to yield electronic-grade silicon (EGS) through the reaction

$$SiHCl_3 + H_2 \longrightarrow Si + 3HCl \tag{3.5}$$

EGS is a polycrystalline material of remarkably high purity and is used as the raw material for preparing high-quality silicon wafers.

The Czochralski technique uses the apparatus shown in Figure 3.19 called the *puller*. The puller comprises three main parts:

1. A furnace that consists of a fused-silica (SiO_2) crucible, a graphite susceptor, a rotation mechanism, a heating element, and a power supply.

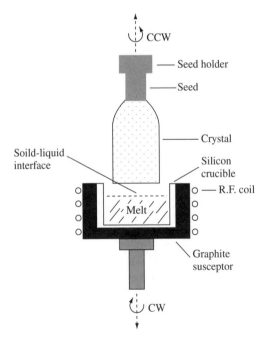

Figure 3.19 A crystal puller (Czochralski) for growing silicon boules, which are then sawn up to make crystal wafers

2. A crystal pulling mechanism, which is composed of a seed holder and a rotation mechanism.

3. An atmosphere control, which includes a gas source (usually an inert gas), a flow control, and an exhaust system.

In crystal growing, the EGS is placed in the crucible and the furnace is heated above the melting temperature of silicon. An appropriately oriented seed crystal (e.g. [100]) is suspended over the crucible in a seed holder. The seed is lowered into the melt. Part of it melts but the tip of the remaining seed crystal still touches the liquid surface. The seed is then gently withdrawn. Progressive freezing at the solid–liquid interface yields a large single crystal. A typical pull rate is a few millimeters per minute.

After a crystal is grown, the seed and the other end of the ingot, which is last to solidify, are removed. Next, the surface is ground so that the diameter of the material is defined. After that, one or more flat regions are ground along the length of the ingot. These flat regions mark the specific crystal orientation of the ingot and the conductivity type of the material (Figure 3.20). Finally, the ingot is sliced by a diamond saw into wafers. Slicing determines four wafer parameters: surface orientation, thickness, taper (which is the variation in the wafer thickness from one end to another), and bow (i.e. surface curvature of the wafer, measured from the centre of the wafer to its edge). Typical specifications for silicon wafers are given in Table 3.7.

3.3.2.2 Epitaxial growth

The method for growing a silicon layer on a substrate wafer is known as an *epitaxial* process in which the substrate wafer acts as a seed crystal. Epitaxial processes are different

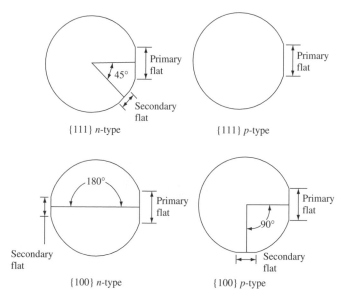

Figure 3.20 Flat regions are machined into the silicon wafers and used for the subsequent identification of crystal orientation and dopant type

Table 3.7 A list of specifications for silicon wafers

Diameter Parameter	100 mm	125 mm	150 mm
Thickness (mm)	0.50–0.55	0.60–0.65	0.65–0.70
Primary flat[a] length (mm)	30–35	40–45	55–60
Secondary flat length (mm)	16–20	25–30	35–40
Bow (mm)	60	70	60
Total thickness variation (μm)	50	65	50
Surface orientation	(100) or (111)	(100) or (111)	(100) or (111)

[a]Wafer flats are defined in Section 4.2.

from crystal growth from the melt in that the epitaxial layer can be grown at a temperature very much below the melting point. Among various epitaxial processes, vapour-phase epitaxy (VPE) is the usual process for silicon layer growth.

A schematic of the VPE apparatus is shown in Figure 3.21. The figure shows a horizontal susceptor made from graphite blocks. The susceptor mechanically supports the wafer, and, being an induction-heated reactor, it also serves as the source of thermal energy for the reaction.

Several silicon sources are usually used: silicon tetrachloride ($SiCl_4$), dichlorosilane (SiH_2Cl_2), trichlorosilane ($SiHCl_3$), and silane (SiH_4). Typical reaction temperature for $SiCl_4$ is $\sim$1200 °C. The overall reaction in the case of $SiCl_4$ is reduction by hydrogen,

$$SiCl_4 \text{ (gas)} + 2H_2 \text{ (gas)} \longrightarrow Si \text{ (solid)} + 4HCl \text{ (gas)} \tag{3.6}$$

A competing reaction that would occur simultaneously is

$$SiCl_4 \text{ (gas)} + Si \text{ (solid)} \longrightarrow 2SiCl_2 \text{ (gas)} \tag{3.7}$$

In reaction (3.6), silicon is deposited on the wafer, whereas in reaction (3.7), silicon is removed (etched). Therefore, if the concentration of $SiCl_4$ is excessive, etching rather than growth of silicon will take place.

An alternative epitaxial process for silicon layer growth is molecular beam epitaxy (MBE), which is an epitaxial process that involves the reaction of a thermal beam of silicon

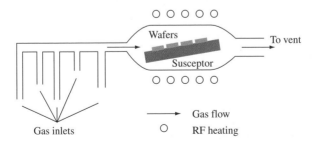

Figure 3.21 Apparatus used to grow a silicon layer by vapour-phase epitaxy (VPE)

atoms with a silicon wafer surface under ultrahigh vacuum conditions ($\sim 10^{-10}$ torr). MBE can achieve precise control in both chemical composition and impurity profiles (if introduced intentionally). Single-crystal multilayer structures with dimensions on the order of atomic layers can be made using MBE.

3.4 CERAMIC, POLYMERIC, AND COMPOSITE MATERIALS

Ceramics are inorganic materials that consist of metallic and nonmetallic elements that are chemically bonded together. Examples of these are alumina (Al_2O_3), salts, such as sodium chloride (NaCl) and calcium fluoride (CaF_2), and ceramic superconductors such as $YBa_2Cu_3O_{6.5}$.

Most ceramic materials, whether single-crystalline or noncrystalline (glass), have high hardness and high-temperature strength but tend to be mechanically brittle. Several new ceramic materials have been developed for engine applications; they are lightweight and have high strength of hardness, good resistance to heat and wear, reduced friction, and insulating properties[9]. These properties are highly desirable for engine applications. An important application for ceramics is their use in tiles for the NASA space shuttle. These ceramic tiles thermally protect the aluminum internal structure of the space shuttle during ascent out of and reentry into the earth's atmosphere. Table 3.8 gives the structure and uses of some selected engineering ceramics.

Polymers are materials that are organic (containing carbon) compounds, which have long molecular chains or networks. Examples of polymeric materials include epoxies, polyesters, nylons, and silicones. The strength and ductility of polymers vary greatly, and because of their atomic structure, most polymers are poor electrical conductors. In fact, some polymers are excellent insulators and are used in electrical insulation applications. In general, polymers have low densities and relatively low softening or decomposition temperatures. The physical properties of some common polymers are given in Table H.2 of Appendix H.

Because polymers are either long snakelike structures or three-dimensional random networks, polymeric materials are usually noncrystalline. However, one may still find some single-crystal polymers. Single-crystal properties are encountered only in linear

Table 3.8 Ceramics and some of their application areas

Ceramic	Structure	Applications			
Material		Structural	Electrical	Optical	Magnetic
Silicates of Li, Al, etc.	Crystalline	X	X		
Alumina	Hexagonal	X	X	X	
Zirconia	Cubic	X			
Nonmetallic nitrides, e.g. Si	Glass	X	X		
Metallic silicides	Tetrahedral		X		
Spinel	Cubic				X
Garnet	Cubic				X

[9] The physical properties of some common ceramic materials are listed in Appendix H.

polymers and the single-crystallinity in these polymers is only partial. The development of single-crystallinity is important because of its effects on mechanical, thermal, and optical properties. The single-crystallinity in polymers is achieved by forming long molecules such as chains of many thousands of carbon atoms by chemical reaction. On cooling from liquid, these molecules bend back and forth to form individual crystals. The application of these materials is discussed in various chapters in this book (for a review on the use of polymers for electronics and optoelectronics, see Chilton and Goosey (1995)).

The last type of MEMS material is a composite material, which is a combination of two or more materials from the other four categories. Most composites consist of a selected filler or reinforcing material and a compatible resin binder to obtain the desired material characteristics. Usually, the component materials of a composite do not dissolve in one another and are physically identifiable by an interface between the components.

Ceramic, polymer, and composite materials are important because they are often used as passive materials in microsensors and MEMS devices. In other words, they are used either to provide inert substrates or to form inert structures (e.g. shuttles) in micromechanical devices. These materials are also used as active materials in microsensors and MEMS devices; sometimes they are referred to as *smart materials* (Culshaw (1996)). An example is the use of an electroactive polymer, such as polypyrrole, to make chemoresistive sensors in an electronic nose (Gardner and Bartlett (1999)). Further details on such smart devices are given in Chapter 15.

REFERENCES

Chilton, J. A. and Goosey, M. T., eds. (1995). *Special Polymers for Electronics and Optoelectronics*, Chapman and Hall, London, p. 351.

Culshaw, B. (1996). *Smart Structures and Materials*, Artech House, Boston, p. 209.

Gardner, J. W. and Bartlett, P. N. (1999). *Electronic Noses: Principles and Applications*, Oxford University Press, Oxford, p. 245.

Moseley, P. T. and Crocker, A. J. (1996). *Sensor Materials*, Institute of Physics Publishing, Bristol, p. 227.

Pierret, R. F. (1988). *Semiconductor Fundamentals*, Addison-Wesley, Reading, Massachusetts, p. 146.

Tuck, B. and Christopoulos, C. (1986). *Physical Electronics*, Edward Arnold, London, p. 114.

4

Standard Microelectronic Technologies

4.1 INTRODUCTION

A large number of different microelectronic technologies exist today, and they are used for making conventional microelectronic components, such as operational amplifiers, logic gates, and microprocessors. Many technologies, and in some cases, all of a standard process are used to fabricate the type of microdevices that are of interest to us here. For example, a thermal microsensor can be made by using a standard bipolar process in which, for instance, the forward voltage of a p-n diode (under constant current) or the base-emitter voltage of an n-p-n transistor is proportional to the absolute temperature (details may be found in Section 8.2.1). In this case, there is no significant difference between the processing of a thermodiode and a conventional diode, although there will be a difference in the device design and, perhaps, package. In most practical situations, it is highly desirable to employ, wherever possible, the standard technologies and then 'bolt-on' one or more nonstandard pre- or postprocessing steps. The part or full integration of standard microelectronics into a microsensor or microactuator is also often required to enhance its functionality and, in doing so, make a so-called smart microtransducer[1]. The degree of integration is a critical design issue; hence, designers should first consider the hybrid solutions that, once again, draw upon standard microelectronic process and packaging technologies. Because of the complexity of the end product, the integration of microelectronics into a microelectromechanical system (MEMS) device is an even more difficult problem. Consequently, it is covered in some detail in other chapters, wherein we consider the different possible approaches, such as microelectronics first or MEMS first. For all these reasons, this chapter seeks to provide a brief account of the standard microelectronic technologies and thus provide a sound foundation on which to add other nonstandard materials and processes – such as those described in later chapters on new silicon-processing technologies (Chapters 5 and 6) and interdigitated transducer surface acoustic wave (IDT SAW) microsensors (Chapter 12). We shall not attempt to describe all the standard technologies, or indeed, even the subtle details of employing some of them. Instead, our aim is to make the reader aware of the common choices currently available and provide an appreciation of the nature of the technologies involved.

---

[1] We will use the term *microtransducer* to describe a microsensing (or microactuating) device.

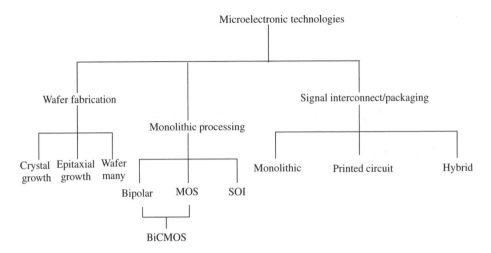

Figure 4.1 Overview of the microtechnologies used to make standard microelectronic devices today

Figure 4.1 shows the basic technologies that are used to make a standard microelectronic device and essentially provides an overview of the contents of this chapter. The growth and production of the single-crystal wafers are presented first, and this is followed by their processing to make monolithic bipolar or metal oxide semiconductor (MOS) devices, such as standard components (e.g. transistor–transistor logic (TTL), programmable gate arrays (PGAs), and microprocessors) or application-specific integrated circuits (ASICs). Then, some of the common signal interconnect and packaging technologies are discussed. These have been divided into monolithic processes that include bonding and making the electrical connections of the die onto a substrate and wiring and packaging processes that include the different printed circuit board (PCB) and hybrid solutions. Of particular importance to microtransducers and MEMS is the encapsulation process. Standard components are usually plastic-encapsulated – a process that costs little and provides protection from both mechanical stress and chemical attack from the environment. However, some components are hermetically sealed inside a metal can in an inert atmosphere that permits a higher power output through a higher device-operating temperature. In some types of microsensors, great care must often be taken in the selection of both the mounting technologies and encapsulation process. Consider, for example, a chemical microsensor, such as an ion-sensitive field-effect transistor (ISFET), which needs to be operated in an ionic solution (or perhaps even in a river or reservoir) for long periods of time without fouling, or a mechanical microsensor, such as a barometric pressure sensor, which needs to operate in ambient conditions again for long periods of time. It is our intention to introduce the reader to some of the different packaging technologies and to emphasise their importance while designing a microtransducer or MEMS device.

In this chapter, we concentrate upon the monolithic processing of *silicon* because it is the most commonly used semiconductor material today. Silicon is also the most important material for making microtransducers and for making the integrated circuits (ICs) that form the processing unit in a smart transducer or MEMS device.

4.2 WAFER PREPARATION

4.2.1 Crystal Growth

As stated in the preceding text, silicon is the most commonly used semiconductor material for making microelectronic devices. Other semiconductor materials are employed in certain important niche areas, for example, gallium arsenide (GaAs) for optoelectronic devices and silicon carbide and gallium nitride (GaN) for high-power or high-temperature devices, but silicon is by far the dominant material not only for standard microelectronic components but also for microtransducers and MEMS devices.

Silicon is one of the earth's most abundant elements forming about 25 percent of its surface crust. However, silicon reacts readily with oxygen and other materials in the earth's atmosphere and hence is generally found in the form of sand (i.e. silicon dioxide (SiO_2)) or silicates. Sand can be found with an impurity level of less than 1 percent, and this composition is usually the starting point for making single-crystal silicon wafers. The process has three main steps: first, refining the sand into polycrystalline silicon rods; second, growing the single-crystal silicon rods, known as *boules*; and third, producing the wafers suitable for monolithic processing (see Section 4.3).

The first step is achieved by placing this relatively pure form of sand, quartzite, into a furnace with a common source, such as coal, coke, or wood chips, of carbon. The silicon dioxide (SiO_2) is reduced by carbon and the condensed silicon vapour to form metallurgical-grade silicon (98 to 99 percent pure):

$$SiO_2(s) + 2C(s) \xrightarrow{\text{heat}} Si(s) + 2CO(g) \tag{4.1}$$

Next, the metallurgical-grade silicon is further purified by heating it up in an atmosphere of hydrogen chloride (HCl) to form the compound trichlorosilane ($SiHCl_3$):

$$Si(s) + 3HCl(g) \xrightarrow{\text{heat}} SiHCl_3(g) + H_2(g) \tag{4.2}$$

The trichlorosilane is cooled to form a liquid at room temperature (boiling point is $32\,^\circ C$) and is easily purified to semiconductor standards by a fractional distillation procedure. The distillation procedure removes the unwanted chlorides of metallic and dopant impurities, such as copper, iron, phosphorous, and boron. Finally, the thermal reduction of trichlorosilane gas in a chemical vapour deposition (CVD) reactor produces electronic-grade polycrystalline silicon, which is deposited onto a slim pure silicon rod (see Figure 4.2).

$$SiHCl_3(g) + H_2(g) \xrightarrow{\text{heat}} Si(s) + 3HCl(g) \tag{4.3}$$

The electronic-grade silicon rods are typically 20 cm in diameter and several metres in length, and these are used to grow single-crystalline silicon by either the Czochralski or the zone-melt process.

The present-day Czochralski process is based on the process invented by Czochralski in 1917 for the growth of single crystals of metals. It consists of dipping a single-crystal seed

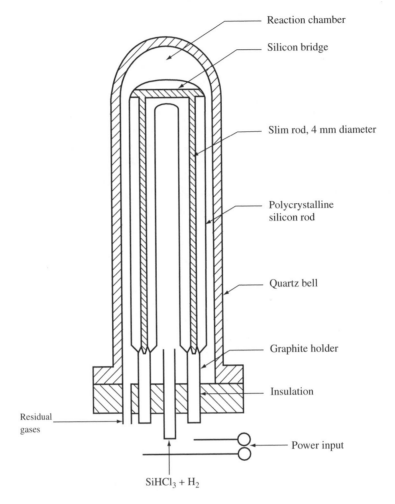

Reaction chamber

Silicon bridge

Slim rod, 4 mm diameter

Polycrystalline
silicon rod

Quartz bell

Graphite holder

Insulation

Residual
gases

Power input

$SiHCl_3 + H_2$

Figure 4.2   Schematic diagram of a CVD reactor used to make electronic-grade polycrystalline silicon

of silicon into a graphite crucible that contains molten silicon under an argon atmosphere. The seed is then slowly withdrawn at a typical rate of millimeters per minute while simultaneously rotating the seed. Figure 4.3 shows a schematic diagram of a Czochralski puller for crystal growth. Either a radio frequency (RF) or an electrical source supplies the heat, and a known amount of the dopant is added to the melt to obtain the desired n- or p-type doping in the grown crystal. The doping level is reasonably constant radially across the boule but varies axially in a predefined manner; this does not usually pose a problem for standard microelectronic devices. Czochralski crystals are grown along either the (100) or the (111) crystal axis. Silicon crystals with a (111) axis grow more slowly because of the smaller separation of these crystal planes of 4.135 Å, but they are easier to grow and are thus less expensive. This orientation is widely used in many microelectronic devices. In some cases, such as in the fabrication of power devices, it is desirable to have a high carrier mobility, and this requires a higher purity of silicon than that produced by

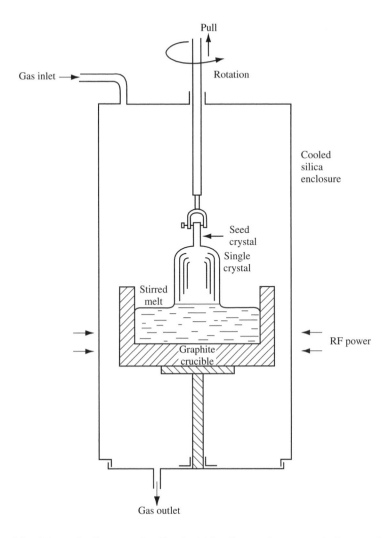

Figure 4.3   Schematic diagram of a Czochralski puller used to grow single-crystal silicon

the Czochralski process. An alternative method is the float-zone technique, which is more expensive but results in a higher level of crystal purity.

Figure 4.4 shows a schematic diagram of the apparatus used to grow a crystal by this technique. Once again, the crystals are grown in either the (100) or the (111) orientation, but in this case an RF field produces a melt zone at the bottom of a polycrystalline rod. The seed is brought into contact with the molten zone and slowly rotated and pulled as the molten zone is moved up the length of the polycrystalline rod until the process is complete. Crystals with bulk resistivities up to 30 kΩ·cm have been obtained using this technique, and they show a higher purity than the Czochralski crystals (typically less than 25 kΩ·cm). The main drawbacks of this technique are that the crystals have smaller diameter (less than 75 mm) and show a greater radial variation in crystal resistivity, leading to a higher manufacturing cost.

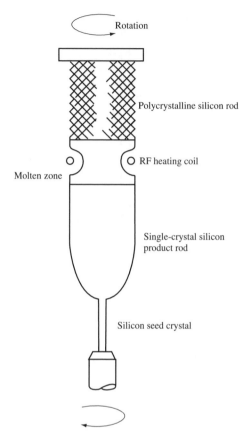

Figure 4.4   Schematic diagram of the apparatus used in a float-zone process to grow single-crystal silicon

4.2.2 Wafer Manufacture

The conventional base, or substrate, *within* which microelectronic devices and ICs are fabricated, is the silicon wafer. Other substrates (e.g. glass, dielectric) *upon* which hybrid circuits are built are covered later in this chapter. There are a number of production steps that are required to turn a single-crystal silicon boule produced by Czochralski or float-zone crystal growth into a silicon wafer suitable for monolithic processing, namely,

1. Diameter sizing

2. Orientation

3. Slicing

4. Etching

5. Polishing

6. Cleaning

First, the boule is ground down to a standard diameter for use in the automated processing of wafers. This is done by a variety of wheel or belt grinders with a bias of about +0.4 mm, but it leaves a damaged surface on the crystal. The outer damaged material is then removed later by a less harsh treatment to give the desired wafer diameter (see following text). The standard wafer sizes vary according to the microelectronic component that is manufactured and the desired yield. Low volume production (e.g. specialist microsensors/MEMS fabrication) may still use a 3″ 4″ or 6″ wafer, whereas high volume production of memory or microprocessor chips would use the highest available diameter, that is, 8″ or higher.

Crystal orientation and identification flats are ground into the silicon boule, and these are used as an alignment reference in subsequent processing. Figure 4.5 shows both the standard (larger) orientation flat ($\pm 0.5°$) and the doping-type identification flat for silicon wafers. Now, the damaged material is removed by an isotropic silicon-polishing etch of hydrofluoric, nitric, and acetic acids to leave the boule diameter to the nominal standard (± 1 mm).

The boule is then sliced into wafers using a saw blade with a diamond or nickel cutting edge mounted on the inside of the diameter. The blade is made as thin as possible but is still about 100 μm in thickness and therefore results in the loss of about 125 μm of silicon per slice that represents a quarter of the silicon! Once again, the sawing processing damages the surface of the silicon wafer, and so a further 50 to 80 μm of silicon is removed by an isotropic etch using the same acid mixture as before.

The final steps for producing the wafer are chemo-mechanical polishing using a colloidal suspension of fine SiO_2 particles in an aqueous, alkaline solution, followed by cleaning in various detergents and water. The desired flatness of the polished wafer depends on the subsequent process but would typically be around ± 5 μm. Wafers used for making microtransducers and MEMS devices often require an etching from the back of the wafer and therefore, in this case, both sides of the wafer need to be polished to a fine finish.

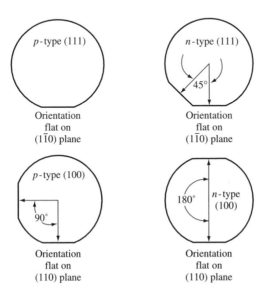

Figure 4.5   Standard orientations and identification flats found on silicon wafers

4.2.3 Epitaxial Deposition

Epitaxial deposition is a process in which a thin, single-crystal layer can be deposited onto the surface of a substrate wafer. The substrate wafer acts as a seeding layer and usually has the same crystal structure as that of the epitaxial layer. Epitaxial growth is now increasingly used in the production of both discrete devices and ICs. It is most commonly used to isolate junctions from the substrate, that is, in junction isolation (JI) technology and to allow formation of buried layers. JI technology permits the fabrication of circuits with better characteristics, such as lower parasitic capacitances and faster switching speeds. The most common types of epitaxial deposition are vapour-phase epitaxy (VPE) and molecular beam epitaxy (MBE).

There are many different designs of epitaxial reactors, and Figure 4.6 shows three common types: a horizontal epitaxial reactor, a vertical epitaxial reactor, and a barrel epitaxial reactor. In all cases, there is a reaction chamber in which the chemicals react to form the epitaxial layer, a heat source (e.g. RF coils) to stimulate the chemical reaction, and a gas supply and venting system.

In the horizontal reactor, the wafers are placed on the susceptor at a slight angle in order to compensate for the depletion of the reactive gases as they pass along the surface. In contrast, the susceptor is rotated in the vertical reactor and the gases can circulate above the wafers to keep the gas partial pressures constant. Finally, in the barrel reactor, the wafers are mounted vertically and the susceptor is again rotated to ensure an even supply of gases to the wafers.

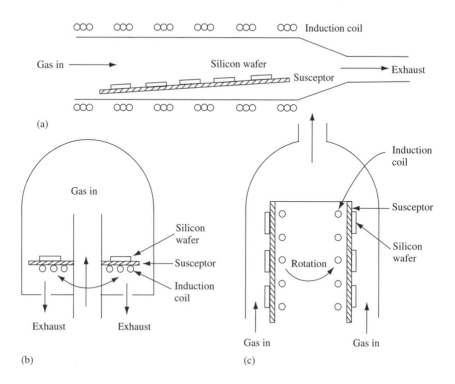

Figure 4.6   Some common types of epitaxial reactor: (a) horizontal; (b) vertical; and (c) barrel

Table 4.1 Silicon sources and dopant gases used in vapour epitaxial deposition

Source/gas	Formula	Deposition temperature $T(°C)$	Deposition rate $\dot{L}(\mu m/min)$
Silicon sources:			
Silicon tetrachloride	$SiCl_4$	1150–1225	0.2–1.0
Dichlorosilane	SiH_2Cl_2	1025–1100	0.1–1.0
Trichlorosilane	$SiHCl_3$	1100–1175	0.2–2.0
Silane	SiH_4	950–1050	0.1–0.25
Dopant gases:		*Dopant type*	
Arsine	AsH_3	n	N/A
Phosphine	PH_3	n	N/A
Diborane	B_2H_6	p	N/A

A number of different sources of both silicon and the dopant gas are available for epitaxial deposition, as shown in Table 4.1.

A common silicon source is silicon tetrachloride ($SiCl_4$), and a typical reaction temperature is 1200 °C. The gas is reduced by hydrogen in the following reaction:

$$SiCl_4(g) + 2H_2(g) \xrightarrow{\text{heat}} Si(s) + 4HCl(g) \qquad (4.4)$$

There is competition with a second reaction that removes the silicon layer; hence, the concentration of silicon tetrachloride must be kept low ($\sim$0.02 mole fraction) to grow rather than etch the silicon layer through

$$Si(s) + SiCl_4(g) \xrightarrow{\text{heat}} 2SiCl_2(g) \qquad (4.5)$$

The growth of an epitaxial layer of silicon onto a single-crystal silicon wafer is relatively straightforward as the lattice spacing is matched, leaving little stress between the epi-layer and support. In this case, the deposited material is the same material as the substrate and is called *homoepitaxy*. When the deposited material is different from the substrate (but close in lattice spacing and thermal expansivity), it is called *heteroepitaxy*. Heteroepitaxy is an important technique used in microelectronics to produce a number of specialist devices; for example, silicon-on-insulator (SOI), silicon on glass (SiO_2) for thin-film transistors, silicon on sapphire (Al_2O_3), and gallium nitride (GaN) for power devices.

Epitaxial layers can be grown using CVD or physical vapour deposition (PVD); the latter is known as *MBE*. In MBE, (see Figure 4.7) the single-crystal substrate is held at a temperature of only 400 to 800 °C and in an ultrahigh vacuum of 10^{-11} torr. The deposition process is much slower than CVD epitaxy at about 0.2 nm/s but provides precise control of layer thickness and doping profile. MBE can be used to make specialised structures, such as heterojunction transistors, quantum devices, and so on. However, the ultrahigh vacuum needed and the slow deposition rate make this a very expensive technique to employ when compared with CVD epitaxy.

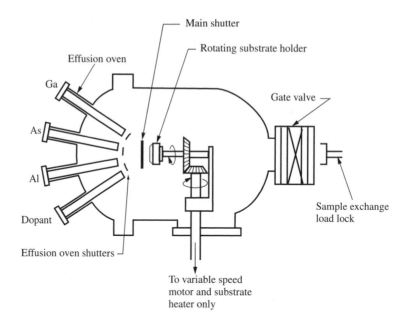

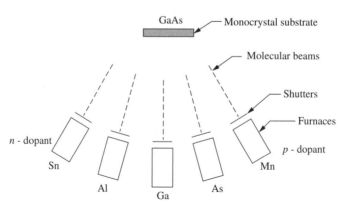

Figure 4.7   Schematic diagram of a molecular beam epitaxy reactor

4.3 MONOLITHIC PROCESSING

A large number of technologies are used to fabricate discrete devices and ICs; therefore, it is only possible here to cover some of the most commonly used ones and to simply give the reader an appreciation of the typical steps involved. There are numerous books that cover, in detail, semiconductor fabrication technologies (Interested readers should see, for example, Colclaser (1980), Gise and Blanchard (1986), and Sze (1985)).

Figure 4.8 illustrates the common technologies used today to fabricate standard microelectronic devices. The two main classes of technology are bipolar and MOS. These two technologies can be combined in a bipolar complementary metal oxide semiconductor (BiCMOS) process to produce, for example, a mixture of high-speed, high-drive

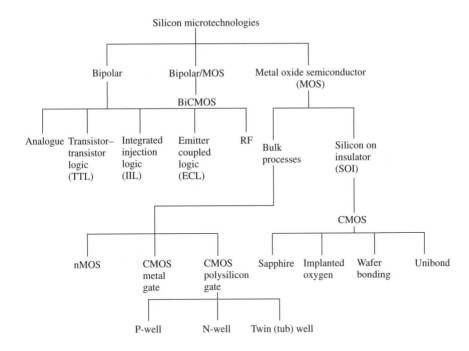

Figure 4.8 Common silicon processes used to make standard semiconductor devices and ICs

analogue circuitry with low-power switching digital circuitry. Clearly, this dual process is more difficult and expensive than a single process.

The bipolar process is used to make many standard analogue components (see Section 4.3.1), for example, *p-n* diodes and bipolar junction transistors (BJTs), and to make some popular logic families, such as TTL, integrated injection logic (IIL), and emitter coupled logic (ECL) for fast switching. The MOS process can be divided into those creating an IC within the silicon wafer (i.e. a bulk process) and those creating an IC on top of an insulator (i.e. an SOI process). The *n*-type metal oxide semiconductor (nMOS) process only makes *n*-channel metal oxide semiconductor field-effect transistor (MOSFETs) and is simpler than the complementary MOS (CMOS) process that makes both nMOS and *p*-type metal oxide semiconductor (pMOS) devices. The lower cost, higher packing density, and low power dissipation per gate of the CMOS process makes it one of the most commonly used technologies today.

Figure 4.9 shows the difference between the main technologies in gate-switching speed and power dissipation per gate. GaAs has traditionally been a niche technology producing very high-speed devices and switching circuits but at a much higher cost from a lower yield process. However, the power dissipation of GaAs does compare favorably with bipolar TTL and ECL, and, in some cases, with CMOS when the cost difference is diminishing; therefore, GaAs is becoming more competitive.

A summary of the relative performances of basic bipolar and MOS processes is given in Table 4.2. Most of the devices available for circuit design are described in the following two sections.

There is the option to use a polysilicon gate instead of a metal gate in a CMOS process with an N- or P-well (or both) implanted into a *p*-type or *n*-type silicon wafer

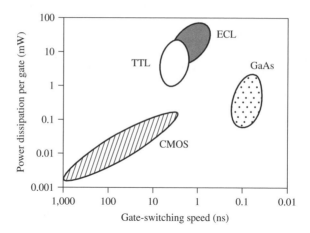

Figure 4.9   Relative power dissipation (per gate) and packing densities of standard microelectronic processes

Table 4.2   Relative performances of standard IC processing technologies

Source/gas	Bipolar	nMOS or pMOS	CMOS
Features of circuitry:			
Switching speed	High	Low	Low
Power dissipation	High	Intermediate	Low
Device density	Low	High	High
Current	High	Low	Low
Voltage	High	Low	Low
Complexity of process:			
Epitaxial depositions	1	0	0 or 1
Diffusion/oxidation cycles	5	2	3
Masks used in process	7	5	7
Possible devices:			
Transistors	*n-p-n, p-n-p* (lateral/vertical)	nMOS or pMOS	nMOS and pMOS (metal/poly-Si gate)
Diodes	5 or more	1	3
Resistors	5 or more	1	2 or 3
Capacitors	Dielectric/junction	Dielectric/junction	

to get both pMOS and nMOS devices. Polysilicon-gate CMOS has more process steps involved than the simpler metal-gate CMOS process, but the CMOS devices do possess lower power dissipation per gate and higher gate-switching speed and are hence attractive for many applications. The SOI CMOS process is being increasingly used to make very large-scale integrated devices because of the higher packing densities possible. These thin-film transistors have reduced latch-up and parasitic capacitance problems by using better device isolation. SOI technology is also of interest in microsensor technology because of the ability either to back-etch the SOI wafer to form a micromechanical silicon membrane or to use non-silicon, dielectric substrate, such as sapphire, which has an excellent thermal conductivity. In addition, SOI technology offers extremely low unwanted parasitic effects and excellent isolation between devices (see Section 4.3.5).

4.3.1 Bipolar Processing

The bipolar process has evolved over many years, as has its so-called standard process. Clearly, this is an important issue and the integration of a microsensor, or microactuator, will depend on the exact details of the process that is employed. As stated earlier, the possible approaches to microsensors and MEMS integration and the problems associated with compliance to a standard process are both discussed in some detail in later chapters.

This section presents what may be regarded as the standard bipolar process, which employs an epi-layer to make the two most important types of bipolar components; that is, *vertical* and *lateral* transistors. Bipolar *n-p-n* transistors are the most commonly used components in circuit design as both amplifiers and switches because of their superior characteristics compared with *p-n-p* transistors. Let us now consider in detail the process steps required to make a vertical *n-p-n* and lateral *p-n-p* transistors. A similar process can be defined to make vertical *p-n-p* transistors or the simpler substrate *p-n-p* transistors with slightly different device characteristics.

Worked Example E4.1: **Vertical and Lateral Bipolar Transistors**

The standard bipolar process begins by taking a *p*-type substrate (i.e. single-crystal silicon wafer) with the topside polished[2]. A buried *n*-layer is formed within the *p*-type substrate by first growing an oxide layer. The oxide is usually grown in an oxidation furnace using either oxygen gas (dry oxidation) or water vapour (wet oxidation) at a temperature in the range of 900 to 1300 °C. The chemical reactions for these oxidation processes are as follows:

$$Si(s) + O_2(g) \xrightarrow{\text{Heat}} SiO_2(s)$$

$$Si(s) + 2H_2O(g) \xrightarrow{\text{Heat}} SiO_2(s) + 2H_2(g) \qquad (4.6)$$

Other ways of forming an oxide layer, such as CVD, are discussed in Chapter 5.

The thermal oxide layer is then patterned using a process called *lithography*. A basic description of these processes is given in this chapter and a description about more advanced lithographic techniques is given in Chapter 5. Lithography is the name used to describe the process of imprinting a geometric pattern from a mask onto a thin layer of material, a resist, which is a radiation-sensitive polymer. The resist is usually laid down onto the substrate using a spin-casting technique (see Figure 4.10).

In spin-casting technique, a small volume of the resist is dropped onto the centre of the flat substrate, which is accelerated and spun at a constant low spin speed of about 2000 rpm to spread the resist uniformly. The spin speed is then rapidly increased to its final spin speed of about 5000 rpm, and this stage determines its final thickness of 1 to 2 μm. The thickness of the spun-on resist, d_R, is determined by the viscosity η of the

---

[2] Double-sided wafers are used if a back-etch is required to define a microstructure.

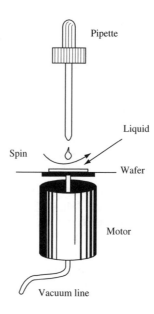

Figure 4.10   Apparatus used to cast a resist onto a substrate in preparation for optical lithography

resist and final spin speed v approximately according to

$$d_R = \frac{\eta}{\sqrt{v}} \times f_{sc} \tag{4.7}$$

where f_{sc} is the percentage solid content in solution. The resist-coated wafer is then cured by a soft-bake at a low temperature (80 to 100 °C for 10 to 20 minutes) and a process mask applied for shadow or projection printing[3]. The mask is generally a reticule mask plate and comprises a glass plate coated with a light-blocking material, such as a thin chromium film, that has itself been patterned using a wet-etching process and a second resist, but in this case the resist has been written on directly using a high-resolution electron-beam writer. Then the mask and substrate are exposed to a radiation source, usually ultraviolet (UV) light, and the radiation is transmitted through the clear parts of the mask but blocked by the chromium coating. The effect of the radiation depends on the type of resist – positive or negative. When a positive resist is exposed to the radiation, it becomes soluble in the resist developer and dissolves leaving a resist pattern of the same shape as that of the chromium film. Conversely, the negative resist becomes less soluble when exposed to the radiation and so leaves the negative of the chromium coating. Negative resists are used more commonly as they yield better results. Table 4.3 shows some commercially available resists for both optical and electron-beam lithography[4].

Figure 4.11 shows the lithographic patterning of a substrate with a negative resist and chrome mask plate (mask 1), and subsequent soft-baking and developing in chemicals, to leave the resist over predefined parts of the oxide layer. The resist is then hard-baked

---

[3] Optical and other lithographic techniques are described later in Section 5.3.
[4] These terms are defined later on under bipolar transistor characteristics.

Table 4.3   Some commercially available resists

Resist	Lithography	Type
Kodak 747	Optical	Negative
AZ-1350J	Optical	Negative
Shippley S-1813	Optical	Negative
PR102	Optical	Positive
COP	E-beam and X-ray	Negative
PMMA	E-beam and X-ray	Positive
PBS	E-beam and X-ray	Positive

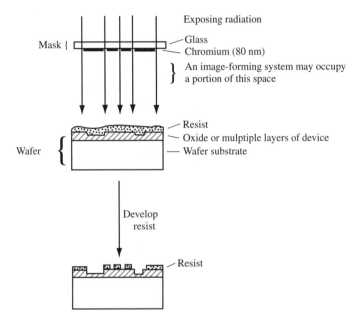

Figure 4.11   Use of radiation and a mask plate to create windows in a resist layer through which the oxide is etched and arsenic-doped to form buried *n*-regions in a *p*-type substrate. The buried regions are used to increase device performance

(110 to 130 °C for 10 to 20 minutes) and the oxide is selectively removed using either a wet-etching or a dry-etching process.

Table 4.4 shows the wet or liquid etchants commonly used to remove oxides and other materials during a standard process. Dry etching is becoming increasingly popular, and the details of different wet and dry etching processes, which are often referred to as *micromachining techniques*, are described in this Chapter.

Next, arsenic is introduced into the exposed *p*-type silicon regions. There are two different techniques used: thermal predeposition and ion implantation.

In thermal predeposition, a powder, liquid, or gas can be used as the source dopant material for the predeposition process. The solid solubility of the dopant in the material, predeposition time, and temperature determine how far the dopant diffuses into the wafer. For a constant source concentration C_s, the dopant concentration at distance x and at time

Table 4.4 Some wet etchants used in processing wafers for semiconductor devices

Material to etch	Composition of etchant	Etch rate (nm/min)	Temperature (°C)
Thermal SiO₂	Buffered oxide etch 4:1 to 7:1 NH₄F/HF (49%)	80–120	20–30
Deposited SiO₂	3:3:2 NH₄F/acetic acid/water	180–220	20–30
Polycrystalline silicon	1:50:20 HF/HNO₃/water or KOH	350–500	20–30
Aluminum	50:10:2:3 Phosphoric acid/acetic acid/nitric acid/water	200–600	20–40
Silicon nitride	Phosphoric acid	5–7.5	160–175

Table 4.5 Source materials for doping silicon substrates

Element	$\sqrt{D}$ at 1100°C (μm/$\sqrt{h}$)	Solid solubility at 1150°C	Compound name	State	Use
n-type:					
Antimony	0.110	7×10^{19}[a]	Antimony trioxide	Solid	Subcollector
Arsenic	0.090	1.8×10^{21}	Arsenic trioxide	Solid	Closed tube or source furnace; subcollector
			Arsine	Gas	Subcollector/emitter
Phosphorus	0.329	1.4×10^{21}	Phosphoric pentoxide	Solid	Emitters
			Phosphoric oxychloride	Liquid	Emitters
			Phosphine	Gas	Emitters
			Phosphoric oxychloride	Liquid	Emitters
			Silicon pyrophosphate	Solid	Wafer source
p-type:			Silicon pyrophosphate	Solid	Wafer source
Boron	0.329	5×10^{20}[a]	Boron trioxide	Solid	Base/isolation
			Boron tribromide	Liquid	Base/isolation
			Diborane	Gas	Base/isolation
			Boron nitride	Solid	Wafer source

[a] At 1250°C

t, $C(x, t)$, is determined by the following equation:

$$C(x, t) = C_s \, \mathrm{erfc} \left(x/2\sqrt{Dt} \right) \qquad (4.8)$$

where D is the diffusion coefficient.

Table 4.5 shows the different sources used to dope semiconductors. After predeposition, there is a drive-in step in which the existing dopant is driven into the silicon and a protective layer of oxide is regrown in an oxygen atmosphere. In practice, the diffusion coefficient D does vary itself with the level of doping, increasing somewhat linearly with arsenic and quadratically with phosphorus. As a result, calibrated charts, instead of a simple mathematical formula, are used to determine the doping profiles.

In the second doping technique, namely, ion implantation, the dopant element is ionised, accelerated to a kinetic energy of several hundred keV, and driven into the substrate. This alternative method is discussed in Section 2.5.2.

After predeposition and drive-in steps, the oxide protection layer is stripped off using a wet or dry etch to leave the n-regions defined in the p-type substrate. An n-type epi-layer of 4 to 6 μm in thickness is grown on top of the substrate (see Section 4.2.3) to create the buried n-type areas within the p-type substrate. The buried n-layer is used to minimise both the collector series resistance of the vertical n-p-n transistor that is formed later and the common-base (CB) current gain, α_F, of the parasitic p-n-p transistor formed by the collector and base of the lateral p-n-p transistor and the substrate. However, the buried n-regions can diffuse further into the epi-layer at elevated temperatures, and so caution is required in any subsequent processing.

The transistors need to be electrically separated from each other, and there are a number of techniques with which to do this. Common techniques used are oxide isolation, based on local oxide isolation of silicon (LOCOS), junction isolation (JI) based on a deep boron dope, or trench isolation, which is useful for minimising parasitic capacitance. At this stage, a second mask (mask 2) is used to define the regions into which boron is implanted and thus isolate one transistor from another (see Figure 4.12).

Then the deep n^+-type contacts of the vertical n-p-n collector and lateral p-n-p base are defined using a third mask in another patterning process, followed by an extrinsic p-type implant for the base of the vertical n-p-n transistor and for both the emitter and collector of the lateral p-n-p transistor (mask 3). The relatively thick, highly doped extrinsic layer is followed by a thinner, lighter doped intrinsic base implant (mask 4) below the emitter in the vertical n-p-n transistor. The lighter doping provides for a large common-emitter (CE) current gain β_F.

Next, the heavily doped n^+ contact to the emitter in the vertical n-p-n transistor is implanted (mask 5) to complete the transistor structures. Finally, the oxide layer is patterned (mask 6) to form contact holes through to the transistor contacts and substrate, and then the metal interconnect (normally 100 to 300 nm of aluminum) is deposited either by physical evaporation or by sputtering and is patterned (mask 7) to form the completed IC. Figure 4.13 shows the side view of two devices: the vertical n-p-n transistor and lateral p-n-p transistor.

In some cases, a passivation layer of SiO_2 or some other material is deposited and patterned (mask 8) to serve as a physical and chemical protective barrier over the circuit. This depends upon the proposed method of packaging of the die (see Section 4.4) and the subsequent use. The complete bipolar process described here is summarised in Figure 4.14.

It could be simplified by fabricating, for example, a pure n-p-n bipolar process, and, in fact, most of the transistors in monolithic ICs are n-p-n structures. However, although the characteristics of p-n-p transistors are generally inferior to an n-p-n transistor, as stated in the preceding text, they are used as active devices in operational amplifiers, and as the injector transistors in the IIL mentioned earlier. Similarly, a substrate p-n-p transistor

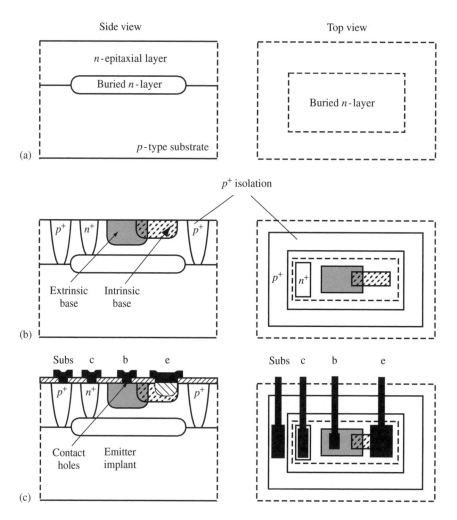

Figure 4.12 Formation of an isolation region (p^+) in the substrate to separate devices electrically in a bipolar process. The isolation regions are used to enhance packing densities

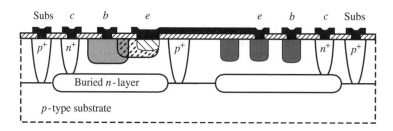

Figure 4.13 A vertical n-p-n transistor and lateral p-n-p transistor formed by a standard bipolar process

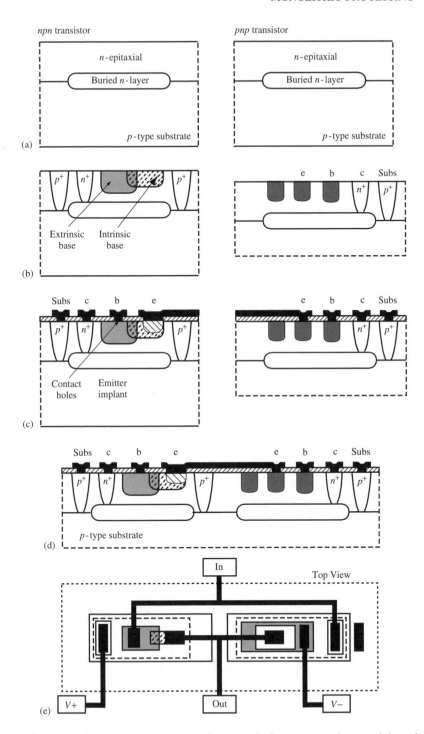

Figure 4.14   Standard bipolar process to make a vertical *n-p-n* transistor and lateral *p-n-p* transistor

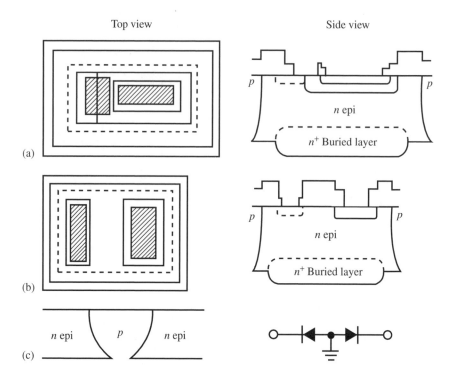

Figure 4.15 Various kinds of diodes available from a bipolar process: (a) emitter-base; (b) base-collector; and (c) epi-isolation

rather than a lateral p-n-p transistor could be fabricated by leaving out the buried n^+-layer. The problem that arises then is that it restricts the possible circuit configurations (because the collector is connected to the substrate that gives parasitic problems) and, therefore, the process can no longer be regarded as standard.

Various other components can also be formed using this bipolar process. For example, Figure 4.15 shows three different types of diode that can be formed, namely, the emitter-base diode that has a low reverse breakdown voltage of 6 to 60 V and can be used as a Zener diode; the base-collector diode that has a higher reverse breakdown voltage of 15 to 50 V; and the epi-isolation diode.

Figure 4.16 shows five different types of resistor that can be formed: the base resistor with a typical sheet resistance of 100 to 500 Ω/sq, the pinched-base resistor with a typical sheet resistance of 2000 to 10 000 Ω/sq, the emitter resistor with a typical sheet resistance of 4 to 20 Ω/sq, the epi-resistor with a sheet resistance that varies from 400 to 2000 Ω/sq, and a pinched epi-resistor that has a higher sheet resistance than an epi-resistor of 500 to 2000 Ω/sq and is often used in preference to the latter.

Finally, different capacitors can be formed; Figure 4.17 shows both the dielectric capacitor, in which the thermal oxide or thinner emitter oxide is used as the dielectric, and the junction capacitance, which is suitable when there is no requirement for low leakage

Top view   Side view

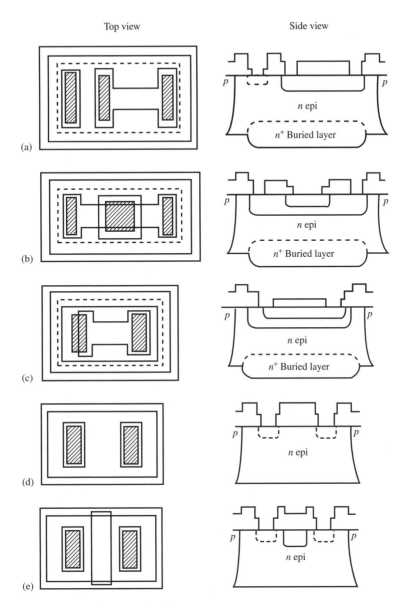

Figure 4.16   Various kinds of resistors available from a bipolar process: (a) base resistor; (b) pinched base resistor; (c) emitter resistor; (d) epi-resistor; and (e) pinched epi-resistor

currents or a constant capacitance. In general, these components tend to have inferior electrical properties compared with discrete devices that are fabricated by employing other technologies; therefore, extra care is required to design circuits using these components. It is common to use discrete components as external reference capacitors and resistors together with an IC to achieve the necessary performance.

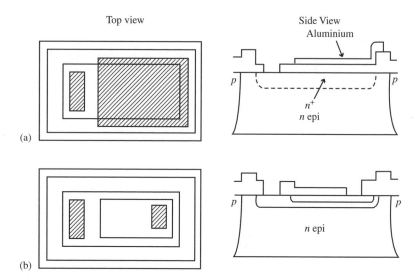

Figure 4.17   Two types of capacitors that are available from a bipolar process: (a) dielectric and (b) junction

4.3.2 Characteristics of BJTs

As noted earlier, there are a number of electronic devices available from a bipolar process, and these may be used either as discrete components or as part of an IC, such as an operational amplifier or logic switch. Here, a basic discussion of the characteristics of the bipolar transistor is presented. There are many textbooks that cover different aspects of the bipolar transistor from the basic (e.g. Sze 1985) through to an advanced treatment of the device physics, the construction of sophisticated models (e.g. Hart 1994), and bipolar circuitry. Most of this material is outside the scope of this book, because here we are mainly interested in the technologies that are relevant to the integration of standard ICs with microtransducers and MEMS devices. However, it is necessary to include some basic material on bipolar devices for three reasons: First, as a background material to the readers who are less familiar with electrical engineering topics and who want to know more about the basic electrical properties of a junction diode and transistor and the technical terms used, such as threshold voltage or current gain; second, to serve as a reminder to other readers of the typical characteristics of a bipolar device for use when designing an IC. Finally, and perhaps most important, to provide background information on microelectronic devices that can be exploited directly, or within an IC, as a microtransducer or MEMS device. For instance, a bipolar diode or bipolar transistor may be used to measure the ambient temperature (see Chapter 8). Therefore, for all these reasons, a brief discussion of the properties of the bipolar junction and FETs is given here.

The basic properties of a semiconducting material have already been discussed in Section 3.3 and they should be familiar to an electrical engineer or physicist. Therefore, we start our discussion with the properties of the junction diode before moving on to the BJT. As shown in the last section, a junction diode can be fabricated from a standard bipolar process by forming a contacting region between an n-type and p-type material.

Figure 4.18 shows (a) a schematic of the structure and (b) its symbol and implementation in a simple direct current (DC) circuit. The concentration of the donors and acceptors are shown in Figure 4.18(c).

In the absence of an external voltage V applied across the diode, there is a tendency for acceptors in the p-type material to diffuse across into the n-type region driven by the difference in electrochemical potential caused by the high concentration of acceptors p_p in p-type region compared with that in the n-type region n_p (i.e. $p_p \gg n_p$). Similarly, there is a tendency for the donors in the n-type material to move into the p-type material because of the mismatch in carrier concentration, namely $n_n \gg p_n$. Because no current can flow across the junction without an external voltage, a depletion region that is free of mobile charge carriers is formed. This region is called the *space charge region* and it is shown in Figure 4.18(c), where the region near the p-type material is left with a net negative charge and the region near the n-type material is left with a net positive charge. Together, these form a dipole layer and hence a potential barrier of height V_d (Figure 4.18(d)) that prevents the flow of majority carriers from one side to the other. The height of the potential barrier, which is also known as the *diffusion potential* or *contact potential*, is determined by the carrier concentrations

$$V_d = \frac{kT}{q} \ln \frac{n_n}{n_p} = \frac{kT}{q} \ln \frac{p_n}{p_p} \tag{4.9}$$

where T is the absolute temperature and q is the charge on an electron. The diffusion potential varies for different technologies and is about 0.5 to 0.8 V for silicon, 0.1 to 0.2 V for germanium, and about 1.5 V for GaAs.

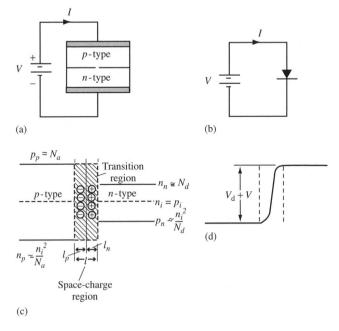

(a)   (b)

(c)   (d)

Figure 4.18 Schematic diagram of (a) a *p-n* junction diode; (b) equivalent symbol; (c) space-charge region; (d) contact potential

The depletion region has an associated depletion or junction capacitance that is given by

$$C_d = \frac{\varepsilon_r \varepsilon_0 A}{l} = \frac{A}{|l_p| + |l_n|} \tag{4.10}$$

where ε_0 is the dielectric permittivity of free space and ε_r is the relative dielectric constant that is about 12 for silicon. The space charge region l is related to the depletion voltage V_d and the external voltage V, and for an ideal junction it is given by

$$l = |l_p| + |l_n| = \sqrt{\frac{2\varepsilon}{q} \left(\frac{1}{N_d} + \frac{1}{N_a} \right)} \times \sqrt{[V_d - V]} \tag{4.11}$$

where N_d and N_a are the donor and acceptor concentrations in the semiconducting material. Thus, the capacitance of a junction diode is, in general, expressed by

$$C_d = K[V_d - V]^{-m} \tag{4.12}$$

where the constant K is a function of the impurity profile and area A, and m depends on the exact distribution of the impurities near the junction; in the ideal case, m takes a value of 0.5, that is, a plot of $1/C^2$ versus V would be a straight line. Typical characteristics of a silicon junction diode with the conductivity of the p-type and n-type regions being 100 S/cm and 1 S/cm, respectively, are a space charge layer l of 0.5 μm at $V = 0$ and 1.2 μm at $V = -5.0$ (reverse-biased) and a junction capacitance of 23 pF at 0 V and 8.3 pF at -5.0 V. The nonlinear $C-V$ characteristic of a junction diode, together with its significant leakage current, makes its use better suited to digital rather than analogue circuitry.

When the junction is forward-biased (i.e. $V > 0$), the height of the potential barrier is reduced, and more holes flow through diffusion from the p-type region to the n-type region and more electrons flow through diffusion from the n-type region to the p-type region. In a simple diffusion model, in which the transition width l is much smaller than the diffusion length, the recombination in the depletion region can be ignored and then the net current flow is determined from the continuity equation with an exponential probability of carriers crossing the barrier:

$$I = JA = qA \left(\frac{D_p p_n}{L_p} + \frac{D_n n_p}{L_n} \right) \left(\exp \left(\frac{qV}{kT} \right) - 1 \right) \tag{4.13}$$

where J is the current density, D denotes the diffusion coefficient of the carrier and L is its diffusion length. The $V-I$ characteristic of a p-n junction diode is usually rewritten in the slightly more general form of

$$I = I_s \left[\exp \left(\frac{\lambda q V}{kT} \right) - 1 \right] \tag{4.14}$$

where λ is an empirical scaling factor and would be 1.0 for an ideal diode. Figure 4.19 shows the typical $I - V$ characteristic of a discrete silicon p-n diode at room temperature. The scaling factor λ is about 0.58 here[5] and the saturation current I_s is about 1 nA. The simple theory (Equation (4.13)) gives a saturation current of about 1 fA, but recombination effects, thermal generation, and series resistance effects increase it by 6 orders of magnitude. Note that the saturation current is itself very temperature-dependent and increases by approximately 20 percent per °C. Therefore, in the *reverse*-bias regime, the shift in the $I - V$ characteristic of the diode could be used to create a nonlinear temperature sensor.

The basic theory ignores the reverse-bias breakdown of the diode, and this is shown in Figure 4.19 as occurring around −60 V because of avalanche breakdown[6]. In a zener diode, the breakdown voltage is reduced to below −10 V by higher doping levels and can be used as a reference voltage.

In the forward-bias region, the diode appears to switch on at a certain voltage and then becomes fully conducting. This voltage V_T will be referred to here as the *threshold voltage* but is also called the *cut-in* or *turn-on voltage*. The threshold voltage is determined by fitting a line to the high voltage values and extrapolating to the zero current axis, as

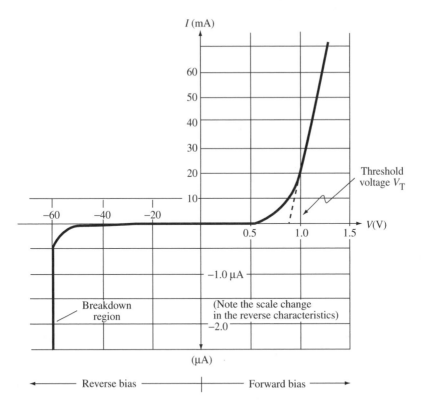

Figure 4.19 Typical $I - V$ characteristic of a silicon p-n junction diode showing the forward- and reverse-bias regions

---

[5] In normal operation λ is 1.0, but at low and very high levels of injection, it approaches 0.5.
[6] Diodes can be designed in silicon to have a breakdown voltage of up to 6.5 kV.

shown in Figure 4.19. The threshold voltage V_T of a typical silicon junction diode is 0.6 V and has a linear temperature coefficient of about -1.7 mV/°C at 20 °C. Therefore, operating a diode at constant current in the forward-bias regime produces a simple and linear temperature sensor.

The BJT consists of either a p-type region sandwiched between two n-type regions for an n-p-n transistor or an n-type region sandwiched between two p-type regions for a p-n-p transistor; hence, it could be regarded as an n-p and p-n diode back to back. In the last section, we saw how a bipolar process can be used to fabricate a vertical n-p-n transistor and lateral p-n-p transistor for an IC. All the transistor voltages and currents are defined in Figure 4.20 for both the n-p-n and p-n-p transistors. Bipolar transistors are basically current-controlled devices in contrast to MOS transistors that are voltage-controlled devices; therefore, we need to consider the currents in the transistor to characterise it.

A bipolar transistor can be configured in three different ways (see Figure 4.21 for n-p-n): (a) the CE, in which the emitter is the common terminal to both the base input voltage V_{BE} and the collector output voltage V_{CE}; (b) the CB, in which the base is the common terminal to both the emitter input voltage V_{EB} and the collector output voltage V_{CB}; and (c) the common-collector (CC) configuration in which the collector is the common terminal to both the base input voltage V_{BC} and the emitter output voltage V_{EC}. Bipolar transistors are normally operated in the CE configuration because it usually provides the largest power gain.

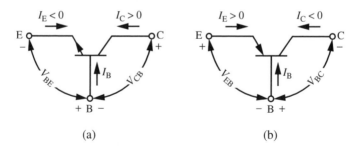

Figure 4.20   Definition of currents and voltages for (a) n-p-n and (b) p-n-p transistors

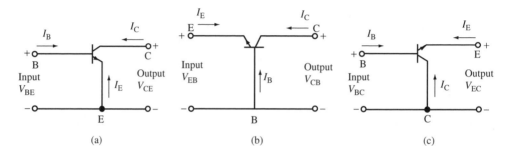

Figure 4.21   The three possible configurations of an n-p-n transistor: (a) common-emitter; (b) common-base; and (c) common-collector

In the normal mode of operation[7] of an n-p-n transistor in the CE configuration, the emitter-base junction is forward-biased (i.e. $V_{BE} > 0$) and the collector-base junction is reverse-biased (i.e. $V_{BC} < 0$). Therefore, the collector current I_C consists of two terms, the main term being a fraction α_F of the emitter current and the other being the reverse saturation current I_{CO} of the collector-base junction diode; hence

$$I_C = -\alpha_F I_E + I_{CO} \tag{4.15}$$

where the constant α_F depends on the doping and dimensions of the diode according to

$$\alpha_F \approx 1 - \frac{1}{2}\left(\frac{W}{L_n}\right)^2 \tag{4.16}$$

where L_n is the diffusion length of the injector carriers in the base of width W and is about 10 μm for silicon. As mentioned earlier, the reverse saturation current I_{CO} for a silicon diode is about 1 nA.

From Kirchoff's current law, the collector current can be related to the base current by

$$I_C = \frac{\alpha_F}{1 - \alpha_F} I_B + \frac{I_{CO}}{1 - \alpha_F} \quad \text{because } I_E + I_B + I_C = 0 \tag{4.17}$$

A second characteristic parameter is also defined and is called β_F. It is related to α_F by

$$\beta_F = \frac{\alpha_F}{1 - \alpha_F} \quad \text{or } \alpha_F = \frac{\beta_F}{1 + \beta_F} \tag{4.18}$$

These two parameters are used to describe the currents flowing through the CE-configured transistor in the forward-active region, and they can also be defined from Equations (4.15), (4.17), and (4.18) as

$$\alpha_F = \frac{\partial I_C}{\partial I_E} \quad \text{or } \beta_F = \frac{\partial I_C}{\partial I_B} \tag{4.19}$$

Ideally, α_F takes a value close to unity and β_F takes a value that is large.

The typical measured input characteristic $I_B - V_{BE}$ and output characteristic $I_C - V_{CE}$ of an n-p-n transistor are shown in Figure 4.22. The input characteristic is that of a diode with a threshold voltage that is about 0.7 V corresponding to a silicon transistor. The output characteristic shows all the possible regions of operation for the transistor.

As mentioned in the preceding text, the transistor is normally operated in the forward-active region ($V_{BE} > 0$; $V_{BC} < 0$) where the forward current parameters of α_F and β_F apply. In this region, the output characteristic may be described by a simple model (excluding the breakdown region) in which the collector current is given by

$$I_C = \beta_F I_B - \frac{I_{CO}}{1 - \alpha_F}\left[\exp\left(\frac{V_{BE} - V_{CE}}{kT/q}\right) - 1\right] \tag{4.20}$$

---

[7] All voltages and currents have the opposite sign in p-n-p transistors.

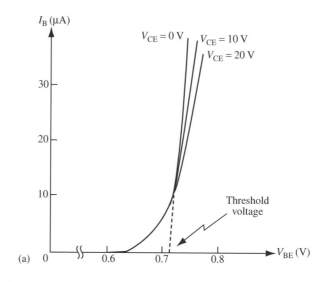

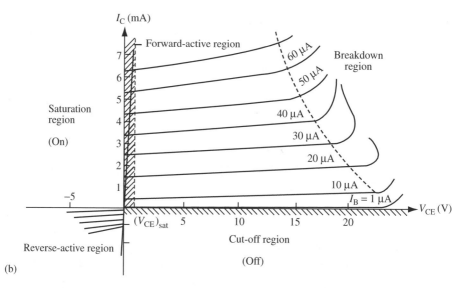

Figure 4.22 Input (a) and output (b) characteristic of a typical *n-p-n* transistor in the common-emitter configuration

When $(V_{BE} - V_{CE}) < -0.1$ V, the exponential term is small, and Equation (4.20) reduces to

$$I_C \approx \beta_F I_B + (\beta_F + 1)I_{CO} \tag{4.21}$$

In the saturated region ($V_{BE} > 0$, $V_{BC} > 0$), both junctions are forward-biased and the transistor is switched ON (i.e. closed), with the output voltage V_{BE} being close to zero. Conversely, in the cutoff region ($V_{BE} < 0$, $V_{BC} < 0$), both junctions are reverse-biased

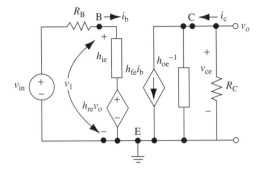

Figure 4.23 Hybrid-π model of an *n-p-n* transistor used to define the basic small-signal characteristics of the device

and the transistor is switched OFF (i.e. open), with the output current being close to zero. In these two regions, the transistor can be used as a binary switch for a digital mode of operation. The reverse-bias region ($V_{BE} < 0$, $V_{BC} > 0$) has the emitter-base junction reverse-biased and the collector-base junction forward-biased and is defined by the parameters α_R and β_R; however, this region is not used often.

Finally, it is worth noting that there are a number of low-frequency models of transistors used to characterise their behaviour. Treating the transistor as a nonlinear two-port device, the hybrid-π model is often used. Figure 4.23 gives the hybrid-π model for the CE configuration of the *n-p-n* transistor with an alternating current (AC) voltage v_{in} applied. In this case, the *h* parameters define the important characteristics of the transistor operating in the forward-active region (at the quiescent point), namely, the input impedance h_{ie}, the current gain h_{fe}, the output conductance h_{oe}, and voltage gain h_{ie}.

These parameters[8] are defined as follows:

$$h_{ie} = \frac{\partial v_{BE}}{\partial i_B}\bigg|_Q , h_{fe} = \frac{\partial i_C}{\partial i_B}\bigg|_Q , h_{oe} = \frac{\partial i_C}{\partial v_{CE}}\bigg|_Q , h_{re} = \frac{\partial v_{BE}}{\partial v_{CE}}\bigg|_Q \qquad (4.22)$$

The small-signal quantity h_{fe} is very similar to β_F, where

$$h_{fe} \cong \beta_{F0} + \frac{\Delta\beta_{F0}}{\Delta i_B} I_B \approx \beta_{F0} \qquad (4.23)$$

Typical values of the *h*-parameters for a transistor with the ideal values shown in brackets are as follows: $h_{ie} \sim 500\ \Omega$ (high), $h_{fe} \sim 100$ (high), $h_{oe} \sim 2\ \mu S$ (low), $h_{re} \sim 0$ (low). Higher input impedance and higher gain are realizable either in an MOS device or by the combination of several transistors in a circuit to give better characteristics and to be ideally linear – as in the operational amplifier. The important question to be asked here is how the integration of the sensor or actuator in a pre- or postprocess affects the characteristic parameters of the transistor and hence the circuit performance.

---

[8] Symbols in lower case denote AC and subscripts relate to input, forward-active region, output, and reverse.

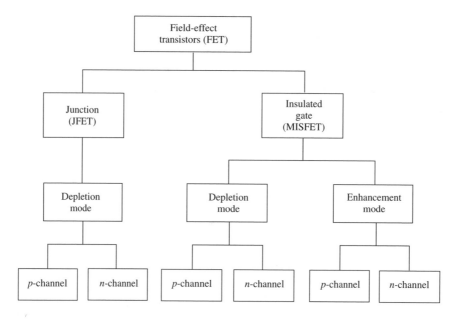

Figure 4.24   Family of field-effect transistors available from MOS technology

4.3.3 MOS Processing

As mentioned earlier, the two main semiconducting processes are bipolar and MOS. The most important microelectronic device available from an MOS process is the field-effect or unipolar transistor. There are two main kinds of FETs (see Figure 4.24): the junction field-effect transistor (JFET) and the metal insulator gate field-effect transistor (MISFET). Both types of transistors, namely, n-channel and p-channel, are available in advanced CMOS or BiCMOS technologies; however, it should be emphasised that the MISFET is an insulated gate control device, whereas JFET is a p-n junction control device. In silicon technology, an insulator is readily available, that is, an oxide for the MISFET, and this transistor is called a *MOSFET*. Here, we describe the process for making a MOSFET rather than the simpler JFET because of the greater use of MOSFETs in ICs and micro-transducers. Since the 1980s, MOSFETs have tended to be made with polysilicon rather than metal gates because of the better device characteristics (lower parasitic capacitance) and hence we show this process, although it is a slightly more complicated one.

As in the bipolar process, a transistor can be made in a number of different ways, depending on whether high frequency or high power is required. The small-signal planar MOSFET is fabricated using a simple substrate process, and Figure 4.25 summarises this process for an n-type enhancement-mode MOSFET.

Worked Example E4.2: n-type Enhancement-Mode MOSFET

The process starts with taking a p-type single-crystalline silicon wafer as the substrate and growing a thermal oxide layer, followed by the growth of an n-type polysilicon layer. The polysilicon is then patterned (mask 1) using optical lithography to define the polysilicon gate. The polygate is then used as a mask for a deep ion implantation

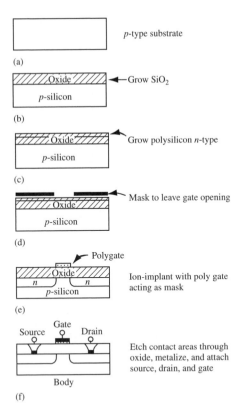

Figure 4.25   Basic steps involved with the fabrication of a small-signal planar (long-channel) enhancement-mode n-channel MOSFET

of the two n^+ regions that form the source and the drain on either side of the gate. Contacts to the source and drain regions are then opened up by patterning (mask 2) the oxide layer – again with optical lithography. Next, the metal interconnect is formed by the deposition and patterning (mask 3) of a metal layer, such as aluminum, and a final oxide passivation layer that is deposited and patterned (mask 4) to leave just the wire-bonding pads exposed. A p-channel MOSFET is usually made in a CMOS process by depositing an N-Well in the p-type wafer and so it usually takes a minimum of five masks[9] for a CMOS circuit with additional steps for LOCOS rather than junction isolation.

As in the bipolar process, the properties of the small-signal planar MOSFET are determined by the accuracy of the photolithography. Once again, lateral and vertical transistors can be made, and this is known as a *diffused-channel metal oxide semiconductor (DMOS) process*. The lateral MOSFET can have a smaller channel[10] and so runs at lower power and higher frequencies, whereas the vertical MOSFET is a power device. Figure 4.26 illustrates the DMOS process used to make a lateral n-channel MOSFET.

---

[9] A sixth mask can be used for threshold implant adjustment.
[10] Commonly referred to as a short-channel device.

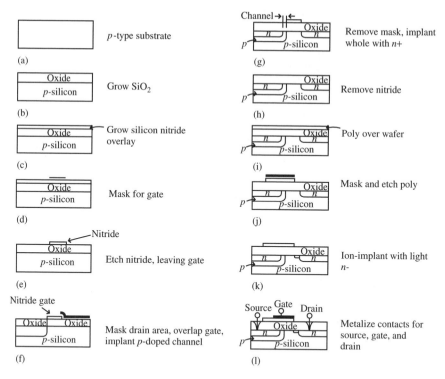

Figure 4.26 Basic steps involved with the fabrication of a lateral (short-channel) enhancement-mode n-channel MOSFET

Worked Example E4.3: A Lateral n-channel MOSFET

The process begins with the p-type single-crystalline substrate onto which is grown a layer of oxide and then nitride. The nitride is then patterned (mask 1) and etched to leave a nitride gate. Next, the drain area is masked off (mask 2) and the narrow p-type channel is formed by ion implantation. The nitride gate and masked-off drain area are then removed, and another mask (4) is used to define the n^+ ion implantation regions of the source and drain. Polysilicon is then grown over the whole wafer and patterned (mask 5) to leave a polygate that extends well over the source n^+ region and towards the drain. The polygate is then used as a mask for a light n^- ion implantation of a thin channel beneath the oxide. Finally, windows are opened up through the oxide by another lithographic process (mask 6) to the source and drain regions and a metal layer that is deposited and patterned (mask 7) to form the connections to the gate, source, and drain regions.

The DMOS process for the lateral MOSFET leads to a reproducible short-channel device with a high transconductance (see Section 4.3.4), and this process is widely used to make high-speed switching circuitry. Power MOSFETs are made using a vertical DMOS process that is slightly more complicated than the lateral DMOS process. Figure 4.27 shows most of the steps required to *make a vertical enhancement-mode n-channel power MOSFET*. The process starts with a heavily n^+-doped n-type substrate rather than a p-type

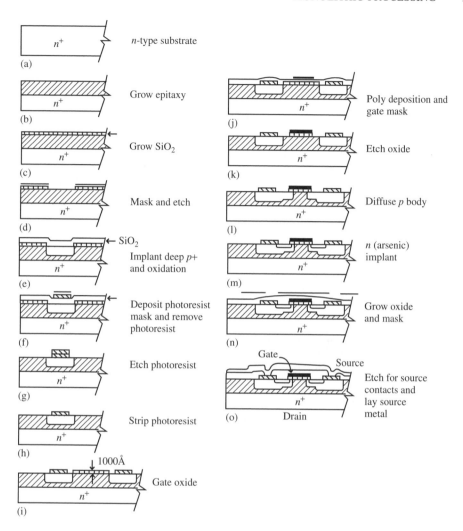

Figure 4.27 Basic steps involved with the fabrication of a vertical (short-channel) enhancement-mode *n*-channel power MOSFET

substrate. A thick lightly doped n^- epi-layer is grown on the wafer and from thereon the steps are shown in Figure 4.27. The vertical configuration produces low channel resistance and hence large currents to flow through the device. Vertical power MOSFETs are used as power drives in actuators, whereas small-signal lateral MOSFETs are used to amplify and condition signals in sensors.

4.3.4 Characteristics of FETs

Figure 4.24 shows the different types of FETs that can be fabricated today. A cross section of an *n*-channel junction FET is shown in Figure 4.28 alongside the symbols used to denote the *n*-channel and *p*-channel depletion types.

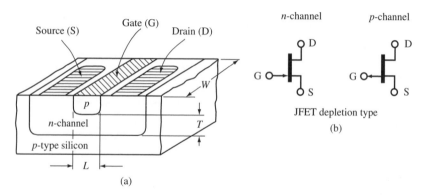

Figure 4.28 (a) Cross section of an *n*-channel JFET and (b) symbols for an *n*-channel and *p*-channel JFET

The more commonly used small-signal *n*-channel MOSFET is shown in Figure 4.29. MOSFETs can not only be *n*-channel or *p*-channel but also be of the depleted type or the enhanced type. Figure 4.29 shows the symbols used to represent the four basic types of MOSFETs.

FETs can be used to make a number of different types of microsensors. For example, a FET can be used to make an ion-selective or gas-sensitive chemical sensor by modifying its gate and exposing it to the local environment (see Section 8.6). For this reason, it is useful to provide the basic properties of an MOSFET, especially for the less experienced reader who can see the transfer characteristics of such a device. Figure 4.30 shows both the output characteristics $I_D–V_{DS}$ and transfer characteristic $I_D–V_{GS}$ of a typical *n*-channel enhancement-mode DMOSFET.

The drain current I_D just starts to flow when the gate-source voltage reaches the device threshold voltage V_T (or off voltage for the depletion-type devices). When operating the

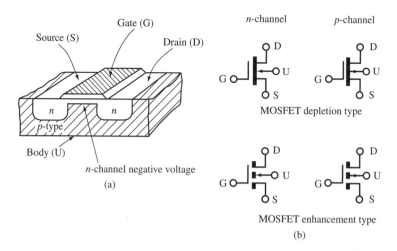

Figure 4.29 (a) Cross section of an *n*-channel MOSFET and (b) symbols for *n*-channel and *p*-channel depletion-type and enhancement-type MOSFETs

device in the linear region (i.e. $V_{DS} < V_{GS} - V_T$, $V_{GS} > V_T$), the drain current is given by

$$I_D = K_n \left[2 \left(V_{GS} - V_T \right) V_{DS} - V_{DS}^2 \right] \qquad (4.24)$$

where K_n is the device constant and, for an n-type MOSFET, is related to the channel length L, width W, electron mobility μ_n, gate oxide capacitance C_o' by

$$K_n = \frac{1}{2} \mu_n C_o' \left(\frac{W}{L} \right) \qquad (4.25)$$

In the saturated region of operation (i.e. $V_{DS} > V_{GS} - V_T$), the device is switched on and the drain current simplifies to

$$I_D = \frac{K_n}{2} \left(V_{GS} - V_T \right)^2 \qquad (4.26)$$

as shown by the transfer characteristic illustrated in Figure 4.30. Pinch-off occurs when V_{GS} is less that V_T, and, ideally, the drain current is zero when the device is switched off.

The basic dynamic properties of an FET device in a common-source configuration can be characterised by the low-frequency equivalent circuit[11] shown in Figure 4.31 in which the main small-signal conductances[12] are shown. The low-frequency gate-source, gate-drain, and drain-source conductances are defined as

$$g_{gs} = \frac{d I_G}{d V_{GS}}, \quad g_{gd} = \frac{d I_G}{d V_{GD}}, \quad \text{and } g_{ds} = \frac{d I_D}{d V_{DS}} \qquad (4.27)$$

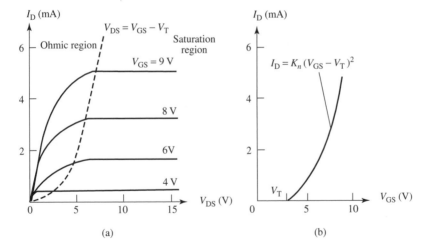

(a)   (b)

Figure 4.30 Typical characteristics of an n-channel MOSFET (enhancement-type): (a) drain (output) characteristic, with dotted line separating ohmic and saturated regions of operation and (b) transfer (input–output) characteristic in the saturated region

---

[11] Leakage current and reactive components are ignored.
[12] The subscripts used are gate g, drain d, source s, and forward f.

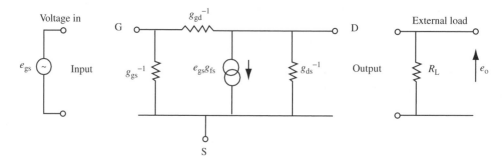

Figure 4.31 Low-frequency, small-signal equivalent circuit of an FET showing the principal conductances. An ideal source voltage (no internal impedance) and ideal load (no reactance) are also shown

An important parameter for both transistors and microsensors is the small-signal forward transconductance g_{fs} that is a measure of the transfer characteristic or sensitivity of a device. The forward transconductance g_{fs} is defined by

$$g_{fs} = \frac{dI_D}{dV_{GS}} \tag{4.28}$$

Therefore, in the case of an MOSFET, the forward transconductance may be found from Equations (4.24) and (4.26) and is

$$g_{fs} = K_n \left[2\left(V_{GS} - V_T\right) - 2V_{DS}\right] \quad \text{(ohmic region)}$$
$$g_{fs} = K_n \left(V_{GS} - V_T\right) \qquad\qquad \text{(saturated region)} \tag{4.29}$$

Clearly, the transconductance is a function of the gate-source voltage and can be determined in the saturation (S) region from Equations (4.27) and (4.26), where

$$g_{fsS} = 2\frac{I_D}{\left(V_{GS} - V_T\right)} \tag{4.30}$$

The low-frequency input conductance g_{is} (when R_L is large) is simply the sum of the gate-source and gate-drain conductances,

$$g_{is} = g_{gs} + g_{gd} \tag{4.31}$$

The output or channel conductance g_{ds} is a function of the gate-source voltage and thus varies with the type of FET. Figure 4.32 shows the variation of channel conductance for n-channel and p-channel FETs.

The channel conductance of an FET that is turned on is low, and this corresponds to V_{DS} being low as well. For an n-channel depletion-type FET, the on-resistance $r_{ds(on)}$ is related to the forward transconductance and is given by, when $V_{GS} > V_T$,

$$r_{ds(on)} = g_{fs}^{-1} = \frac{1}{K_n \left(V_{GS} - V_T\right)} \tag{4.32}$$

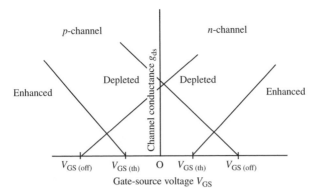

Figure 4.32   Variation of the channel conductance g_{ds} with gate-source voltage for the various types of FETs when the drain-source voltage V_{DS} is set to zero

Generally, the channel conductance is related to the drain-source voltage in the linear (l) region from Equations (4.27) and (4.24), and is given by

$$g_{dsl} = \frac{I_D}{V_{DS}\left[1 - V_{DS}/2\left(V_{GS} - V_T\right)\right]} \qquad (4.33)$$

In the practical use of an FET, the various static and dynamic properties will be affected when a load resistor R_L is applied across the drain and source (see Figure 4.31) to create a common-source voltage amplifier. However, when the output conductance is low, the gain is related simply to the transconductance as follows:

$$A_v \equiv \frac{e_0}{e_{gs}} = -g_{fs}\frac{R_L}{(1 + g_{os}R_L)} \approx -g_{fs}R_L \qquad (4.34)$$

The input capacitance of the FET transistor is an important parameter, and Figure 4.33 shows the principal capacitances within a transistor. A low-input capacitance is desirable because, when coupled with a low on-resistance $r_{ds(on)}$, the switching time is very fast. Short-channel transistors, such as those produced by the DMOS process, have very fast (i.e. nanosecond) switching times and so are used in high-speed circuitry. The output capacitance C_{ds} is mainly determined by the n-p junction capacitance and is inversely proportional to the square root of the drain-source voltage. However, the other capacitances depend on both gate and drain voltages, threshold voltage, and parasitic capacitances. In all these cases, it should be remembered that the device capacitances are in the picofarad range, so care must be taken when designing and interfacing ICs and also while using transistors as either sensing or actuating devices. Any stray capacitance will act as a charge divider and reduce the voltage signals accordingly in a capacitive microtransducer.

4.3.5 SOI CMOS Processing

There are many processes now used for the fabrication of MOS ICs in addition to the standard bulk processes described earlier. One that may have particular relevance to microtransducers and MEMs is the SOI process. Notable successes have been made in

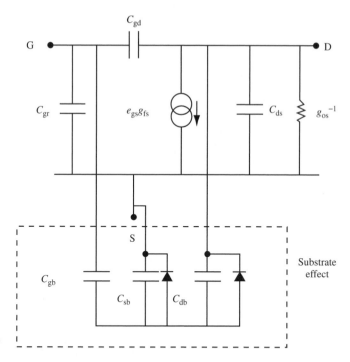

Figure 4.33   Intermediate-frequency, small-signal equivalent circuit of an FET showing the principal capacitances. The additional capacitors and junction diodes to the substrate (or back-gate) are given in the dashed box. A full model would include resistances, other parasitic capacitance, and inductance

depositing thin silicon layers onto insulators. Silicon layers can be deposited by a variety of methods, such as zone-melt recrystallisation and solid-phase epitaxy (Furukawa 1985). Insulating layers can be implanted using an oxygen or nitrogen implant to make SiO_2 or Si_3N_4, and (100)-orientated silicon can be deposited onto glass or sapphire by a process of heteroepitaxy. Some of the most successful methods to produce SOI presently are based on wafer bonding. This involves bonding together two silicon wafers at high temperature, one having a grown oxide on top. Then one of the wafers is thinned down and ends up with the structure of silicon (substrate)-oxide-thin silicon layer. Once the thin layer of silicon has been formed, and it is usually less than 1 micron in thickness, standard bipolar or CMOS can be manufactured on top of this. Figure 4.34 shows the basic structure of a lateral n-channel and p-channel MOSFET fabricated on top of a sapphire substrate. Note that the capacitance of the source and drain to the substrate has been considerably reduced and so SOI CMOS ICs tend to exhibit a high speed at a low power because of the thermal sinking.

In addition, the substrate under an SOI MOSFET can be removed by a bulk-etching process to leave a device mounted in a thin membrane. There are a number of potential uses of SOI technology in microsensor design and Figure 4.35 shows the layout of a gas-sensitive catalytic-gate MOSFET with an integrated FET heater. In this case, the temperature of the CMOS-compatible sensing transistor can be raised to 180 °C, possibly using an MOSFET heater transistor, whereas the surrounding standard integrated circuitry

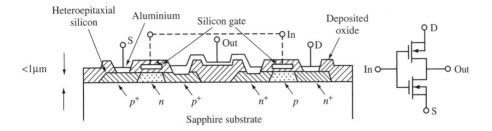

Figure 4.34   Structure of an SOI CMOS device for a high-speed, low-power IC

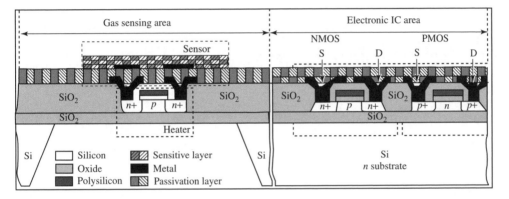

Figure 4.35   Proposed structure of a gas-sensitive MOSFET fabricated using SOI technology. From Udrea and Gardner (1998)

is unaffected (Udrea and Gardner 1998). More details are given in Chapter 15 on the topic of 'smart sensors.'

4.4 MONOLITHIC MOUNTING

There are a number of different technologies that can be used both to package electronic devices and to make circuit interconnections as shown in Figure 4.36. In monolithic mounting, the circuit interconnections have usually been created by the patterning of one or more metallisation layers with, perhaps, some local polysilicon tracks in a CMOS process. Therefore, the mounting process needs both to provide a suitable path for the electrical signals from the single silicon die or chip to the substrate and to attach the die to the substrate. A further consideration is the need to create a suitable path to transfer heat from the chip to the substrate to limit the operating temperature of the IC. Figure 4.36 shows the four main technologies that are used to mount a chip that would normally be standard components, such as MOSFET transistors or TTL logic devices, or, a microtransducer or MEMS device that is of particular interest here, such as a temperature IC or electrostatic microactuator.

The choice of monolithic mounting technique has implications not only to the packaging cost but also to the basic characteristics of the device. Table 4.6 summarises the

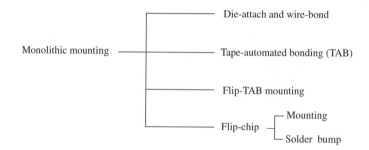

Figure 4.36 The different methods used to mount a monolithic die to a substrate

Table 4.6 Basic features of the four main monolithic mounting technologies. Adapted from Doane and Franzon (1993)

	Die attach and wire bond	TAB	Flip TAB	Flip Chip
Relative cost	1	>2	>2	0.8
Maximum I/O count	300–500	500–700	500–700	>1000
Footprint size, die + (mm)	20–100	3–4	3–4	10
Lead inductance (nH)	2.0–3.5	4.0–5.0	4.0–5.0	<1.0
Peripheral bond pitch (mm)	4–7	3–4	3–4	10
Availability of die	Excellent	Fair	Fair	Poor
Probe test	DC	AC	AC	AC
Reworkability	Poor	Poor/Fair	Fair	Poor

basic features of these technologies and the following sections explain each method in turn. Harper (1997) provides an excellent overview of this field.

In monolithic mounting, the die is bonded onto the substrate that is part of the IC package. The IC package is then connected up to other electronic circuit components, normally through a printed or hybrid circuit board. IC packages come in a variety of sizes and forms, but the two main types are the dual-in-line[13] package (DIP) and surface-mount technology (SMT). DIPs are popular for smaller input–output (I/O) counts, whereas SMT permits higher component densities. Figure 4.37 shows an example of a low-cost plastic DIP, a metal DIP, and a metal SMT package.

Table 4.7 illustrates the characteristics of the common IC packages in terms of their size, electrical characteristics, thermal characteristics, usual gates, and relative cost (Ginsberg 1992).

4.4.1 Die Bonding and Wire Bonding

Die and wire bonding have been used for more than 25 years and involve a two-stage process. First, the die is attached mechanically to the substrate either by an organic adhesive, such as a silver-loaded epoxy, or by a metal solder. This is a low-temperature

---

[13] Sometimes abbreviated to DIL package

Figure 4.37   Examples of (a) a plastic DIP; (b) a metal DIP; and (c) a surface-mount package

process and care must be taken not to stress the die through differences in the thermal expansions of the materials. Next, the electrical contacts between the die and substrate are made through the bonding of a metal wire (see Figure 4.38). Gold or aluminum wires of varying diameter (or ribbons) can be attached by thermocompression, thermosonic, or ultrasonic bonding. Thermocompression bonding is commonly employed and requires both heat ($>300\,^{\circ}$C) and pressure to join the two metals together, usually by forming a ball or stitch. In ultrasonic wedge bonding, the heat is generated by ultrasound and so the substrate remains around room temperature. Finally, thermosonic bonding uses a combination of ultrasound and pressure, and better results are obtained at intermediate substrate temperatures of $125\,^{\circ}$C.

4.4.2 Tape-Automated Bonding

Tape-automated bonding (TAB) has a number of advantages over die- and wire-bonding methods. First, TAB connects the die onto the substrate both electrically and mechanically. The dies are thermocompressively bonded onto tiny beam leads that have been etched in a metal tape (see Figure 4.39).

These inner leads have a smaller pitch than wire bonds and then fan out to a larger pitch that is bonded onto the substrate. The gang bonding of the leads by means of a

Table 4.7 Characteristics of common IC packages. From Ginsberg (1992)

Package type	Range of physical dimensions	Electrical characteristics	Thermal characteristics (°C/W)	Usable gates	Relative cost (per pin)
Through-hole DIP	16 to 64 pins 100 mils pin pitch 0.75 to 2.3 in body length 0.3 to 0.7 in body width	R: medium L: high C: low	Ceramic/plastic 70–40/120–80	Up to 17 000	1
Surface mount SOIC	16 to 28 pins 10 mils pin pitch 50 to 70 mils body length 0.3 to 0.4 in body width	R: medium L: medium C: low	Ceramic/plastic 110–80/130–105	Up to 6500	Ceramic 6 Plastic 2.5
Surface mount OFPT	48 to 260 pins 10 mils pin pitch 0.65 to 1.7 in body width	R: medium L: medium C: low	Plastic 95–60	Up to 17 000	6
Surface mount CLCC	28 to 84 pins 40 to 50 mils pin pitch 0.45 to 0.97 in body width	R: medium L: medium C: medium	Ceramic 70–45	Up to 25 000	30
Surface mount PLCC	28 to 84 pins 50 mils pin pitch 0.49 to 1.19 in body width	R: medium L: medium C: low	Plastic 65–50	Up to 17 000	2
Through-hole PGA	64 to 299 pins 70 mils pin pitch 1.033 to 1.7 in body width	Ceramic/plastic R: low/low L: low/low C: high/low	Ceramic/plastic 40–19/46–38	Up to 75 000	Ceramic 60 Plastic 12

Note: R: Resistance, L: Inductance, C: Capacitance. Assumes 1.5 μm CMOS technology for usable gates.

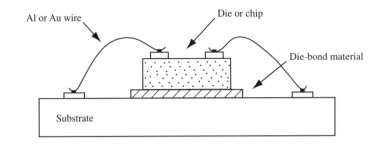

Figure 4.38   Die- and wire-bonding technique

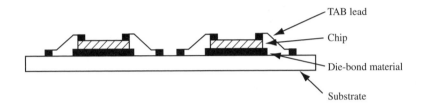

Figure 4.39   Tape-automated bonding technique

hot thermode produces a faster throughput than wire bonding. Moreover, the reduced inductance of a probe means that the devices can be AC-tested.

The disadvantages of TAB include the relatively high cost of the process and the need for a large device footprint. This problem is overcome in flip-chip mounting.

4.4.3 Flip TAB Bonding

In flip TAB bonding, the die is mounted upside down on the substrate, as shown in Figure 4.40. The major advantage of flip TAB over regular TAB mounting is that the die can be subsequently attached to a metal lid for better thermal management.

4.4.4 Flip-Chip Mounting

Finally, flip-chip mounting of the die has a number of key advantages. It provides an excellent contact between the die and substrate by eliminating the wire or beam lead

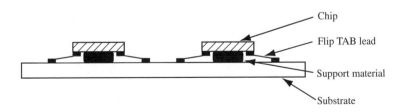

Figure 4.40   Flip TAB technique

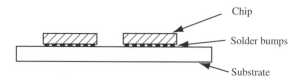

Figure 4.41   Flip-chip mounting technique

entirely (see Figure 4.41). Solder bumps are placed on the substrate and then the die is mounted facedown, and the solder is melted to make the connection. The small footprint and pitch, coupled with short interconnect of about 50 µm, and hence low inductance, make this a very attractive technology at a relatively low cost.

Full details of these bonding methods may be found in textbooks such as Doane and Franzon (1993).

4.5 PRINTED CIRCUIT BOARD TECHNOLOGIES

Once electronic components have been made and packaged, such as the monolithic ICs described in Sections 4.3 and 4.4, they need to be connected with other components to form a circuit board. The most common way to do this is to make a PCB, which is also known as a printed wiring board (PWB). There are a number of different PCB technologies based on different dielectric materials and their fabrication process. Here, we consider the three main kinds of organic PCBs – solid, flexible, and moulded; the ceramic PCB is known as a *thick film hybrid circuit board* and is discussed in Section 4.6.1.

4.5.1 Solid Board

Solid (and flexible) PCBs generally consist of an organic dielectric material on top of which is a thin metal layer – predominantly copper. The copper layer is patterned using a photoresist material and an acid etch to define the tracks between the electronic components. In the case of surface-mount devices, a single-sided organic PCB can be used as illustrated in Figure 4.42(a). Single-sided PCBs are simpler to make and are increasingly used with the greater availability of surface-mount components. However, the majority of organic PCBs are double-sided with multilayer boards used in special cases, such as the need to introduce ground planes and thereby reduce the electrical interference between high-speed switching logic and analogue circuitry (Figure 4.42(b) and (c)). A double-sided PCB has copper tracks patterned on both sides of the dielectric material. Electrical connections between the layers are formed by drilling holes through the board, and this is followed by the plating of the sides of the holes. Clearly, the metal will be thinner here, and passing large currents down through holes can be a problem. Finally, a solder mask is prepared and, if required, a protective layer is patterned, leaving just the solder areas exposed.

In a solid organic PCB, the dielectric material consists of an organic resin reinforced with fibres. The fibres are either chopped or woven into the fabric, and the liquid resin is added and processed using heat and pressure to form a solid sheet. The most

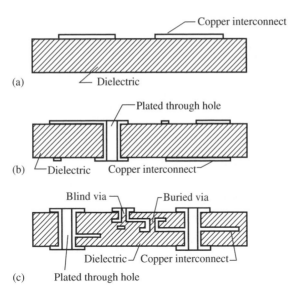

Figure 4.42 Schematic cross section of three types of organic PCBs: (a) single-sided; (b) double-sided; and (c) multilayered

Table 4.8 Material properties of some common fibres used in organic PCBs

	Units	e-glass	s-glass	Quartz	Aramid
Thermal expansion	ppm/°C	5.0	2.8	0.54	−5.0[a]
Dielectric constant at 1 MHz	–	5.8	4.52	3.5	4.0
Dissipation factor at 1 MHz	10^{-3}	1.1	2.6	0.2	1.0
Maximum elongation	%	4.8	5.5	5.0	4.5
Softening temperature	°C	840	975	1420	300
Specific gravity	g/cm^3	2.54	2.49	2.20	1.40
Specific heat capacity	J/g.°C	0.827	0.735	0.966	1.092
Tensile strength	kg/mm	350	475	200	400
Thermal conductivity	W/m.°C	0.89	0.9	1.1	0.5
Young's modulus	kg/mm	7400	8600	7450	13 000

[a] Along axis of fibre; radial is 60 ppm/°C

commonly used fibres are paper, e-glass, s-glass, quartz, and aramid. The precise choice of the dielectric material depends on the technical demands presented by the device and application proposed, and the properties, such as the permittivity and loss factor, are frequently the most important. Table 4.8 gives some of the properties of the fibres that are commonly used in organic PCBs.

4.5.2 Flexible Board

In flexible PCBs, the resin used to make a solid dielectric material is replaced by a thin flexible dielectric material and the metal is replaced by a ductile copper foil. Again, a

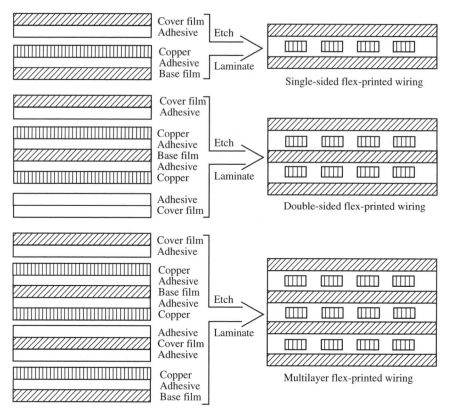

Figure 4.43 Schematic cross section of three types of flexible PCBs: (a) single-sided; (b) double-sided; and (c) multilayered

Table 4.9 Material properties of some resins used in organic PCBs

	Units	Epoxy	Polyimide	Cyanate ester	PTFE
CTE	ppm/°C	58	49	55	99
Dielectric constant at 1 MHz	–	4.5	4.3	3.8	2.6
Poisson's ratio	–	0.35	0.33	0.35	0.46
Temperature	°C	130	260	260	–
Thermal conductivity[a]	W/m.°K	0.3	0.3	0.3	0.3
Young's modulus	GPa	3.4	4.1	3.4	0.03

[a] Approximate values

number of different organic materials can be used to make a flexible wiring board such as polyimide (Kapton), polyester terephthalate (Mylar), random fibre aramid (Nomex), Teflon, and polyvinyl chloride (PVC). The copper foil is processed as before by optical lithography, and layers can be joined together to form multilayer laminates. The layers are usually bonded together using an adhesive such as acrylic, epoxy, polyester, and

Table 4.10 Material properties of some dielectric films used in flexible organic PCBs

	Units	Polyimide	FEP	Polyester	Epoxy polyester	Aramid paper
Density	g/cm^3	1.40	2.15	1.38	1.53	0.65
Dielectric constant at 1 MHz	–	4.00	2.30	3.40	–	3.00
Dielectric strength, min.	kV/mm	79	79	79	5.9	15.4
Dimensional stability, max.	%	0.15	0.3	0.25	0.20	0.30
Dissipation factor at 1 MHz	10^{-3}	12	0.7	7.0	0	10
Elongation, min.	%	40	200	90	15	4
Initial tear strength	g	500	200	800	1700	–
Tensile strength, min.	MPa	165	17	138	34	28
Volume resistivity (damp heat)	Ω·cm	10^6	10^7	–	10^5	10^6

polytetrafluroethylene (PTFE). Figure 4.43 shows the way in which single-sided, double-sided, and multilayer flexible PWBs are constructed.

Table 4.9 gives some typical properties of the resins used in flexible organic PCBs. Care is needed to match these properties with those of the copper layer and the nature of the circuit, for example, high frequency or high power.

Flexible PCB dielectric and adhesive films are now manufactured to a standard, and Table 4.10 shows the Class 3 properties of some dielectric films according to the standard IPC-FC-231. Accordingly, organic PCB laminates can now be constructed with increased confidence in their performance.

4.5.3 Plastic Moulded

The most common forms of PCB – the organic PCB and the ceramic PCB (see next section) – are planar, that is, the metal interconnects are formed in two dimensions with plated through holes joining one layer to another. However, it is possible to make a three-dimensional PCB by the moulding of a suitable plastic. A three-dimensional PCB can be made from extruded or injection-moulded thermoplastic resins with a conductive layer that is selectively applied on its surface. However, high-temperature thermoplastics are required to withstand the soldering process, and commonly used materials are polyethersulfone, polyetherimide, and polysulfone. Plastic moulded PCBs have several advantages over organic PCBs, such as superior electrical and thermal properties and the ability to include in the design, noncircular holes, connectors, spacers, bosses, and so on. More often than not, a moulded PCB is in essence an IC chip carrier package. Plastic moulded PCBs may prove to be advantageous in microtransducers and MEMS applications, in particular, when the assembled microstructure has an irregular structure or needs special clips or connectors. The plastic moulded IC package may also be used as part of a hybrid MEMS before full integration is realised. Future Micro-moulds may be fabricated using microstereolithography (see Chapter 7).

4.6 HYBRID AND MCM TECHNOLOGIES

4.6.1 Thick Film

PCBs can also be formed on a ceramic board, and these may be referred to as *ceramic PCBs*. A ceramic board, such as alumina, offers a number of advantages over organic PCBs, because a ceramic board is much more rigid, tends to be flatter, has a lower dielectric loss, and can withstand higher process temperatures. In addition, alumina is a very inert material and hence is less prone to chemical attack than an organic PCB. Ceramic PCBs can be processed in a number of different ways, such as thick-film, thin-film, co-fired, and direct-bond copper. The most important technology is probably the thick film. Circuit boards have been made for more than twenty years using this technology and are usually referred to as *hybrid circuits*.

In thick-film technology, a number of different pastes have been developed (known as inks), and these pastes can be screen-printed onto a ceramic base to produce interconnects, resistors, inductors, and capacitors.

Example:

1. Artwork is generated to define the screens or stencils for the wiring layers, vias, resistive layers, and dielectric layers.
2. Ceramic substrate is cut to size using laser drilling, and perforations that act as snapping lines are included after the process is complete.
3. Substrate is cleaned using a sandblaster, rinsed in hot isopropyl alcohols, and heated to 800 to 925 °C to drive off organic contaminants.
4. Each layer is then in turn screen-printed to form the multilayer structure. Each paste is first dried at 85 to 150 °C to remove volatiles and then fired at 400 to 1000 °C.
5. The last high-temperature process performed is the resistive layer (800 to 1000 °C).
6. A low-temperature glass can be printed and fired at 425 to 525 °C to form a protective overlayer or solder mask.

Thick-film technology has some useful advantages over other types of PCB manufacture. The process is relatively simple – it does not require expensive vacuum equipment (like thin-film deposition) – and hence is an inexpensive method of making circuit boards.

Figure 4.44 shows a photograph of a thick-film PCB used to mount an ion-selective sensor and the associated discrete electronic circuitry (Atkinson 2001). The thick-film process is useful here not only because it is inexpensive but also because it forms a robust and chemically inert substrate for the chemical sensor. The principal disadvantage of thick-film technology is that the packing density is limited by the masking accuracy – some hundreds of microns. Photolithographically patterned thin-film layers can overcome this problem but require more sophisticated equipment.

4.6.2 Multichip Modules

Increasingly, PCB technologies are being used to make multichip modules (MCMs). A multichip module is a series of monolithic chips (often silicon) that are connected and

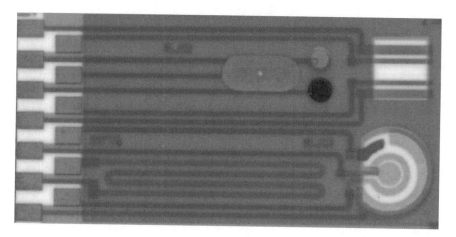

Figure 4.44 ISFET sensor and associated circuitry mounted on a ceramic (hybrid) PCB. From Atkinson (2001)

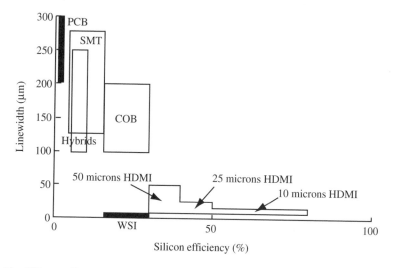

Figure 4.45 Silicon efficiency rating and line width of different interconnection and substrate technologies. After Ginsberg (1992)

packaged to make a self-contained unit. This module can then be either connected directly to peripheral ports for communication or plugged into another PCB. One important reason for using MCM instead of a conventional die-packaging approach is that the active silicon efficiency rating is improved (see Figure 4.45). In other words, the total area of the semiconductor die is comparable to the MCM substrate area. As can be seen from the figure, conventional PCB technologies and even SMT and hybrid are much poorer than the high-density MCM methods.

The ceramic-based technology is referred to as an *MCM-C structure*; other MCM-C technologies include high-temperature co-fired ceramic (HTCC) and low-temperature co-fired ceramic (LTCC). Table 4.11 lists the relative merits of different MCM-C technologies.

Table 4.11 Relative merits of MCM-C technologies. Adapted from Doane and Franzon (1993), with one being the best

Property	Thick-film	HTCC	LTCC	Benefit
Top-layer dimensional stability	1	3	2	Improved wire-bond, assembly yield stability
Low K values	1	3	1	Improved high-frequency performance
High-conductivity metallisation	1	3	1	Smaller line and space designs
High mechanical strength	2	1	3	More rugged package
High thermal conductivity	2	1	3	Good thermal characteristics
CTE matched to alumina or silicon	2	3	1	Capability of assembly
Hermeticity	2	1	1	Development of packages
Excellent dielectric control	3	1	1	More consistent electrical performance
Surface roughness	3	2	1	Better high-frequency performance

Table 4.12 Properties of some commonly used MCM-C materials. Adapted from Doane and Franzon (1993)

Property	Units	Al_2O_3	Al_2O_3	BeO	AlN
Purity	%	99.5	96	99.5	98–99.8
Colour	–	White	White	White	Dark grey
CTE at 25 to 400°C	$10^{-6}/°C$	7.6	7.1	9.0	4.4
Density	g/cm^3	3.87	3.7	3.01	3.255
Dielectric constant at 1 MHz	–	9.9	9.5	6.5	8.8–8.9
Dielectric loss tangent at 1 MHz	10^{-3}	0.1	0.4	0.4	0.7–2.0
Dielectric strength	kV/mm	24	26	9.5	10–14
Flexural strength	GPa	400	250	170–240	280–320
Resistivity	Ω·cm	10^{14}	10^{14}	10^{15}	$>10^{13}$
Specific heat capacity	J/g.°K	–	–	–	0.74
Thermal conductivity	W/m.°K	20–35	20–35	250–260	80–260

The choice of ceramic substrate is important and the >99 percent alumina (Al_2O_3) has a low microwave loss, good strength and thermal conductivity, and good flatness. However, it is expensive and 96 percent alumina can be used in most applications. In cases in which a high thermal conductivity is required (e.g. power devices), beryllia (BeO) or aluminum nitride (AlN) can be used, although these involve a higher cost. Table 4.12 summarises the key properties of the ceramic substrates.

In addition, modules wherein interconnections are made by thin films are classified as MCM-D and those made by plastic (organic) laminate-based technologies are classified as MCM-L. Table 4.13 shows a comparison of the typical properties of the three main types of MCM interconnection technologies.

Table 4.13 Comparison of MCM interconnection technologies. Adapted from Doane and Franzon (1993)

Property	Thick film	HTCC	Thin film	Laminate
MCM class	MCM-C	MCM-C	MCM-D	MCM-L
Dielectric material:	Glass-ceramic	Alumina	Polyimide	Epoxy-glass
Dielectric constant	6–9	9.5	3.5	4.8
Thickness/layer (μm)	35–65	100–750	25	120
Min. via diameter (μm)	200	100–200	25	300
Conductive materials:	Cu (Au)	W (Co)	Cu (Al, Au)	Cu
Thickness (μm)	15	15	5	25
Line width (μm)	100–150	100–125	10–25	75–125
Line pitch (μm)	250–350	250–625	50–125	150–250
Bond pad pitch (μm)	250–350	200–300	100	200
Maximum number of layers	5 to 10+	50+	4–to 10	40+
Electrical properties:				
Line resistance ($\Omega\cdot$cm)	0.2–0.3	0.8–1	1.3–3.4	0.06–0.09
Sheet resistance (mΩ/sq)	3.0	10.0	3.4	0.7
Propagation delay (ps/cm)	90	102	62	72
Stripline capacitance of 50 Ω line (pF/cm)	4.3	2.1	1.25	1.46

MCM technology has several advantages for integrating arrays of microtransducers and even MEMS (Jones and Harsanyi 1995). First, the semiconductor dies can be fabricated by a different process, with some dies being precision analogue (bipolar) components and others being digital (CMOS) logic components. Second, the cost of fabricating the MCM substrate is often less expensive than using a silicon process, and the lower die complexity improves the yield. Finally, the design and fabrication of a custom ASIC chip is a time-consuming and expensive business. For most sensing technologies, there is a need for new silicon microstructures, precision analogue circuitry, and digital readout. Therefore, fabricating a BiCMOS ASIC chip that includes bulk- or surface-micromachining techniques is an expensive option and prohibitive for many applications.

Figure 4.46 shows the layout of a multichip module (MCM-L) with the TAB patterns shown to make the interconnections (Joly *et al.* 1995). This MCM-L has been designed for a high-speed telecommunications automatic teller machine (ATM) switching module, which, with a power budget of 150 W, is a demanding application.

4.6.3 Ball Grid Array

There are a number of other specialised packaging technologies that can be used as an alternative to the conventional PCB or MCM. The main drive for these technologies is to reduce the size of the device and maximise the number of I/Os. For example, there are three types of ball grid array (BGA) packages. Figure 4.47 shows these three types: the plastic BGA, ceramic BGA, and tape BGA. The general advantages of BGA are the smaller package size, low system cost, and ease of assembly. The relative merits of

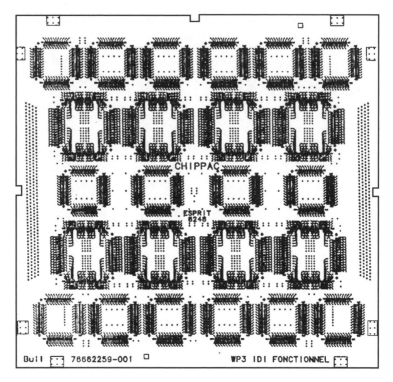

Figure 4.46 Example of a high-density MCM-L substrate with TAB patterns. From Joly *et al.* (1995)

plastic and ceramic PGA packages are similar to those already discussed for PCBs and MCMs. The tape BGA uses a TAB-like frame that connects the die with the next layer board.

4.7 PROGRAMMABLE DEVICES AND ASICs

The microtechnologies described in this chapter are used to make a variety of different microelectronic components. Figure 4.48 shows the sort of devices that can be made today. These are subdivided into two classes – standard components, which are designed for a fixed application or those that can be programmed, and application-specific ICs (ASICs), which are further subdivided. The standard components that may be regarded as having fixed application are discrete devices (e.g. *n*-*p*-*n* transistors), linear devices (e.g. operational amplifiers), and IC logic families of TTL and CMOS (e.g. logic gates and binary counters, random access memory). The other types of standard component may be classified as having the application defined by hardware or software programming.

In hardware programming, the application is defined by masks in the process, and examples of these devices include programmable logic arrays (PLAs) and read-only memory (ROM) chips. There has been a move in recent years to make software programmable components. The most familiar ones are the microprocessors (such as the Motorola 68 000 series or Intel Pentium) that form the heart of a microcomputer and its

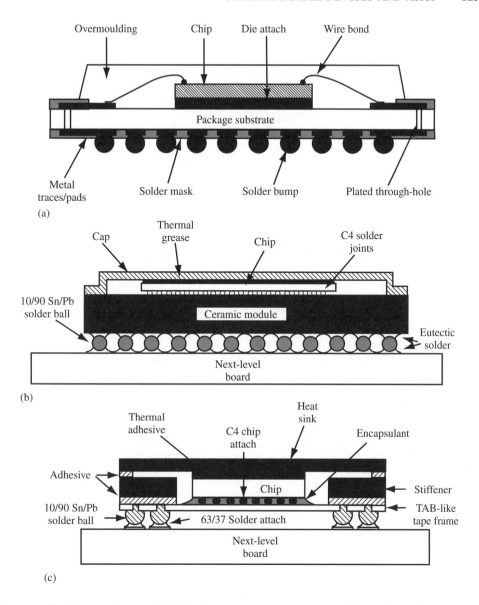

Figure 4.47   Three main types of ball grid array packages: (a) plastic; (b) ceramic; and (c) tape-ball

nonvolatile memory. For example, erasable programmable read-only memory (EPROM) in which the memory is erased by UV light and its easier-to-use successor, electronically erasable programmable read-only memory (EEPROM). In recent years, there has been a strong move toward software programmable array devices; these components do not have the mathematical capability of a microprocessor but are able to perform simple logical actions. As such, they can be high-speed stand-alone chips or glue chips – that is, chips that interface an analogue device with a microcontroller or communication chip. Examples of these are programmable logic devices (PLDs) and PGAs that may have several thousand gates to define. The newest type of component is the programmable analogue array (PAA) device, and these may well become increasingly important in sensing applications in which the design of an operational amplifier circuit can be set and reset through I/O ports.

It should be noted that this classification of devices into hardware and software programming is not universally accepted. Sometimes, it may be more useful to distinguish

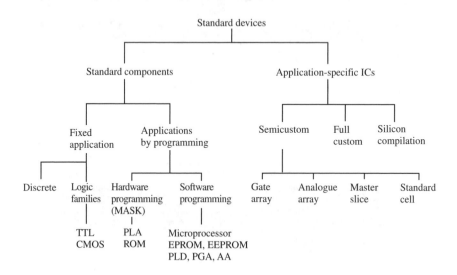

Figure 4.48 Manufacturing methods for common microelectronic devices and ICs

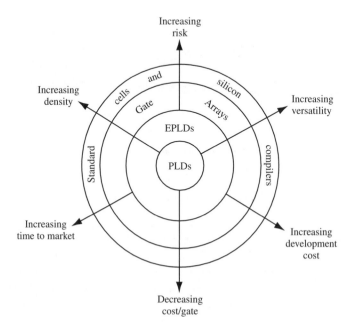

Figure 4.49 Diagram showing the various trade-offs between the different technologies adopted to make an ASIC chip

between devices according to where they were programmed. In this case, devices may all be regarded as 'electrically' programmed and then they can be subdivided into those programmed by the manufacturer (as in mask-programmed parts) and those programmed by the user (as in field-programmable gate array).

The second class of components are those called *application-specific integrated circuit ICs (ASICs)* (see Figure 4.48). There are several types of ASIC and these are referred to as full-custom, semicustom, and silicon compilation. Full-custom ASICs are those that are defined down to the silicon level and, therefore, there is great scope for the optimisation of the device layout, reduction in silicon die area, and speed of operation. However, a full-custom design can be an expensive option and is only useful for large volumes. Silicon compilation is the exact opposite; hence, it is rather wasteful of silicon and pushes up the process costs while minimising the design cost. The more common approach, and more relevant for the manufacturing of microtransducers, is that of a semicustom ASIC chip. This has four subdivisions. In gate arrays, the device has been partly processed and the designer simply defines the interconnection of the digital logic devices by one or two customised masks. Thus, most of the process is common to a number of end users, and hence the costs are greatly reduced. In analogue arrays, the same principle is applied, except that this time a range of analogue components are being connected and an analogue circuit is formed. In master slice, the wafer run can be split at a later stage into different subprocesses. The last type of semicustom approach is the standard cell in which the designer selects standard logic or analogue circuit functions from a software library and then connects them together on the silicon die. The design time is reduced by using standard cells with a standard process.

The various trade-offs of the ASIC technologies are illustrated in Figure 4.49 such as risk, cost, density, and flexibility. Strictly speaking, PLDs are not ASICs but they have been included here because they are often the main competitors to an ASIC chip. Although the number of equivalent gates per chip in PLDs is only 500 to 3000, the cost advantage is often attractive.

When deciding upon which ASIC technology to use, it is important to weigh the relative costs involved, such as the development time and the nonrecurring engineering costs (mask making etc.), and design consideration such as the architecture required and the number of gates. In the final analysis, it is usually the volume that dictates the cost to manufacture the chips; the production charges per 1000 gates are shown against total volume in Table 4.14. For example, modern microprocessor and memory chips are manufactured in enormous volume (millions of chips per year) and so the cost is dominated by the time to process and, hence, the size of wafer processed. Current microelectronic plants use wafers of a diameter of 8″ or more, and companies have to build new plants that cost nearly one billion dollars as larger diameter wafers become available. This situation is usually not applicable to the manufacture of microsensors because of the much reduced volume and higher added value.

However, all of these production costs per kgate are low compared with the cost of fabricating a nonstandard component. For instance, when integrating a microtransducer or MEMS with a standard IC, it is nearly always necessary to develop nonstandard pre- or postprocessing steps, such as surface or bulk micromachining (see next chapter). This cost issue is critical for the eventual success of a component on the market and therefore, we will return to it later on in Chapter 8, having first described the different fabrication methods and technologies associated with microtransducers and MEMS.

Table 4.14 Typical costs of different ASIC (and programmable device) technologies. Adapted from Ginsberg (1992)

Device	Capability	Density (kgates)	Development time	NRE costs (k€)	Production volume (1000s)	Production cost in k€ per kgates
Full custom	RAM, ROM, Analogue	1–100	Long	50	<2.5 2.5–10 >10	N/A 1 2–3 1
Standard and compiled cells	RAM, ROM, Analogue	1–10^a 1–50	Moderate	15^a –50	<2.5 2.5–10 >10	5–10 3–4 2–3
Gate arrays	Logic only	1–50	Moderate	15–100	<2.5 2.5–10 >10	N/A 3–4 2–3
PLDs	Fixed logic	0.5	Short	<5	<2.5 2.5–10 >10	8 7 6
FPGAs	Fixed logic	1–3	Moderate	5–20	<2.5 2.5–10 >10	10–20 7–15 5–12

^{a}Costs shown for PC-based and workstation-based design, respectively

REFERENCES

Atkinson, J. (2001). University of Southampton, UK. Personal communication.

Colclaser, R. A. (1980). *Microelectronics Processing and Device Design*, Wiley & Sons, New York, p. 333.

Doane, D. A. and Franzon, eds. (1993). *Multichip Module Technologies and Alternatives*, Van Nostrand Reinhold, New York, p. 875.

Furakawa, S. (1985). *Silicon-on-insulator: Its Technology and Applications*, D. Reidel Publishing Company, Dordrecht, p. 294.

Ginsberg, G. L. (1992). *Electronic Equipment Packaging Technology*, Van Nostrand Reinhold, New York, p. 279.

Gise, P. and Blanchard, R. (1986). *Modern Semiconductor Fabrication Technology*, Prentice-Hall, New Jersey, p. 264.

Harper, C. A. (1997). *Electronic Packaging and Interconnection Handbook*, McGraw-Hill, USA.

Hart, P. A. H. (1994). *Bipolar and Bipolar-MOS Integration*, Elsevier Science, Amsterdam, p. 468.

Joly J., Kurzweil, K. and Lambert, D. (1995). "MCMs for computers and telecom in CHIPPAC programme," in W. K. Jones and G. Harsanyi, eds., *Multichip Modules with Integrated Sensors*, NATO ASI series, Kluwer Academic Publishers, Dordrecht, p. 324.

Jones, W. K. and Harsanyi, G., eds. (1995). *Multichip Modules with Integrated Sensors,* NATO ASI series, Kluwer Academic Publishers, Dordrecht, p. 324.

Sze, S. M. (1985). *Semiconductor Devices, Physics and Technology*, Wiley & Sons, New York, p. 523.

Udrea, F. and Gardner, J. W. (1998). UK Patent GB 2321336A, "Smart MOSFET gas sensor," Published 22.7.98, Date of filing 15.1.97. International Publication Number: WO 98/32009, 23 July 1998, Gas-sensing semiconductor devices.

5

Silicon Micromachining: Bulk

5.1 INTRODUCTION

The emergence of silicon micromachining has enabled the rapid progress in the field of
microelectromechanical systems (MEMS), as discussed previously in Chapter 1. Silicon
micromachining is the process of fashioning microscopic mechanical parts out of a silicon
substrate or, indeed, on top of a silicon substrate. It is used to fabricate a variety of
mechanical microstructures including beams, diaphragms, grooves, orifices, springs, gears,
suspensions, and a great diversity of other complex mechanical structures. These mechanical structures have been used successfully to realise a wide range of microsensors[1]
and microactuators. Silicon micromachining comprises two technologies: bulk micromachining and surface micromachining. The topic of surface micromachining is covered in
the next chapter. Further details can be found in the two-volume *Handbook of Microlithography, Micromachining, and Microfabrication* (Rai-Choudhury 1997).

Bulk micromachining is the most used of the two principal silicon micromachining
technologies. It emerged in the early 1960s and has been used since then in the fabrication
of many different microstructures. Bulk micromachining is utilised in the manufacture of
the majority of commercial devices – almost all pressure sensors and silicon valves and 90
percent of silicon acceleration sensors. The term *bulk micromachining* expresses the fact
that this type of micromachining is used to realise micromechanical structures within the
bulk of a single-crystal silicon (SCS) wafer by selectively removing the wafer material.
The microstructures fabricated using bulk micromachining may cover the thickness range
from submicrons to the thickness of the full wafer (200 to 500 μm) and the lateral size
ranges from microns to the full diameter of a wafer (75 to 200 mm).

Etching is the key technological step for bulk micromachining. The etch process
employed in bulk micromachining comprises one or several of the following techniques:

1. Wet isotropic etching

2. Wet anisotropic etching

3. Plasma isotropic etching

4. Reactive ion etching (RIE)

5. Etch-stop techniques

---

[1] Chapter 8 is devoted to this topic.

Some of these etch processes have already been used as a standard technology in the microelectronics industry, for example, RIE (Chapter 2).

In addition to an etch process, bulk micromachining often utilises wafer bonding and buried oxide-layer technologies. However, the use of the latter in bulk micromachining is still in its infancy.

This chapter describes the commonly used bulk-micromachining processes and gives a set of worked examples[2] that illustrate the applications of each one, or a combination, of these important processes. The discussion includes the important topics of etch-stops and wafer-to-wafer bonding.

5.2 ISOTROPIC AND ORIENTATION-DEPENDENT WET ETCHING

Wet chemical etching is widely used in semiconductor processing. It is used for lapping and polishing to give an optically flat and damage-free surface and to remove contamination that results from wafer handling and storing. Most importantly, it is used in the fabrication of discrete devices and integrated circuits (ICs) of relatively large dimensions to delineate patterns and to open windows in insulating materials. The basic mechanisms for wet chemical etching of electronic materials were described in Section 2.4. It was also mentioned that most of the wet-etching processes are isotropic, that is, unaffected by crystallographic orientation.

However, some wet etchants are orientation-dependent, that is, they have the property of dissolving a given crystal plane of a semiconductor much faster than other planes (see Table 5.1). In diamond and zinc-blende lattices, the (111) plane is more closely packed than the (100) plane and, hence, for any given etchant, the etch-rate is expected to be slower.

A commonly used orientation-dependent etch for silicon consists of a mixture of potassium hydroxide (KOH) in water and isopropyl alcohol. The etch-rate is about 2.1 μm/min for the (110) plane, 1.4 μm/min for the (100) plane, and only 0.003 μm/min for the (111) plane at 80 °C; therefore, the ratio of the etch rates for the (100) and (110) planes to the (111) plane are very high at 400:1 and 600:1, respectively.

Table 5.1 Anisotropic etching characteristics of different wet etchants for single-crystalline silicon

Etchant	Temperature (°C)	Etch-rate (μm/hour) of		
		Si(100)	Si(110)	Si(111)
KOH:H$_2$O	80	84	126	0.21
KOH	75	25–42	39–66	0.5
EDP	110	51	57	1.25
N$_2$H$_4$H$_2$O	118	176	99	11
NH$_4$OH	75	24	8	1

[2] Appendix M provides a list of all the worked examples provided in this book.

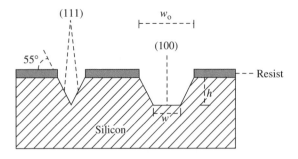

Figure 5.1   Anisotropic etching of (100) crystal silicon

Figure 5.1 shows orientation-dependent etching of (100)-oriented silicon through patterned silicon dioxide (SiO_2), which acts as a mask. Precise V-grooves, in which the edges are (111) planes at an angle of approximately 55° from the (100) surface[3], can be realised by the etching. If the etching time is short, or the window in the mask is sufficiently large, U-shaped grooves could also be realised. The width of the bottom surface, w, is given by

$$w = w_0 - 2h \coth(55°) \quad \text{or} \quad w = w_0 - 1.4h \tag{5.1}$$

where w_0 is the width of the window on the wafer surface and h is the etched depth. If (110)-oriented silicon is used, essentially straight walled grooves with sides of (111) planes can be formed as shown in Figure 5.1.

Worked Example E5.1: Mechanical Velcro

Objective:

The objective is to apply isotropic and anisotropic wet etching to fabricate a dense regular array of microstructures that act as surface adhesives (Han *et al.* 1992). The principle of bonding is that of a button snap, or a zipper, but in a two-dimensional configuration. The bonding principle is shown by a schematic cross section in Figure 5.2. When two surfaces fabricated with identical microstructures are placed in contact, the structures self-align and mate. Under the application of adequate external pressure, the tabs of the structures deform and spring back, resulting in the interlocking of the two surfaces. Thus, the structures behave like the well-known 'Velcro' material.

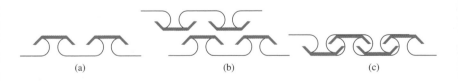

(a)   (b)   (c)

Figure 5.2   Basic steps involved in bonding together silicon 'Velcro'

[3] The value of 55° is important to remember.

Process Flow:

1. A 120-nm SiO_2 layer is grown at $1000\,^\circ C$ in dry oxygen on (100) silicon wafers. The oxide is patterned using optical lithography into an array of $10\,\mu m^2$ rectangular islands, with one edge aligned 45° to the (110) flat (see Figure 5.3(a)).

2. After photoresist stripping, the wafer is immersed in an anisotropic etch bath that consists of aqueous KOH (33–45 percent, $84\,^\circ C$, 4 min) and isopropyl alcohol. The etching results in a truncated pyramid with exposed (212) planes, which are the fastest etching surfaces. The (212) planes intercept the (100) base plane at an angle of 48° (See Figure 5.3(b)).

3. After stripping the masking oxide and cleaning the samples with a conventional chemical sequence, a thick SiO_2 layer (~ 1.0 to $1.5\,\mu m$) is grown at $1000\,^\circ C$ in wet oxygen. The oxide is patterned by a second mask that consists of an array of Greek crosses, each approximately 18-μm wide, aligned to the original array (see Figure 5.3(c)).

4. The oxide crosses act as a mask for a second etch in KOH (~ 3 min), which removes some of the underlying silicon. Finally, the microstructures are completed by etching the wafer for two minutes in an isotropic etching bath ($15:5:1$ $HNO_3:CH_3CO_2H:HF$). This step provides the vertical clearance for the interlocking mating structures and the lateral undercut necessary to produce the four overhanging arms. Although the isotropic silicon etch also attacks the oxide, the selectivity is sufficiently large so as not to cause a significant problem (see Figure 5.4(d)).

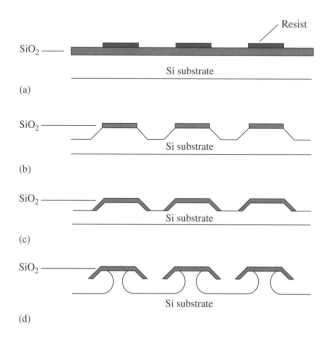

Figure 5.3 Process flow for the fabrication of silicon microvelcro

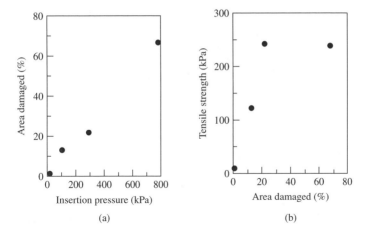

Figure 5.4   (a) Damaged area against insertion pressure and (b) tensile strength against area damaged (Han *et al.* 1992)

Mechanical Testing:

Patterned samples, nominally (8×8) mm^2, were interlocked by applying a load to the upper substrate; the insertion pressure is monitored by placing the entire assembly on an electronic force scale. The bond strength of the mating structures is then characterised by direct measurements of the tensile load needed to induce failure. Bond strength is determined by applying a tensile load through a pulley and measuring the force necessary for separation. Separation of the samples (failure) is always accompanied by damaged areas only on some regions of the mating surfaces, implying that the samples are only interlocked over these damaged regions. The fraction of the damaged area is found to be proportional to the insertion pressure (Figure 5.4(a)). Also, the tensile load necessary to induce failure is proportional to the fraction of the damaged area (Figure 5.4(b)). Extrapolation of the straight line plot of the area damaged against the tensile strength to 100 percent interlocking yields a tensile strength of approximately 1.0 MPa.

Failure Analysis:

The analysis assumes a simple cantilever model as shown in Figure 5.5.

In the figure, F_n is the interaction force between the tabs and l is the length of the tab. The bending stress, σ, is given by

$$\sigma(x) = \frac{M(x)y}{I_z} \tag{5.2}$$

where x is measured from the edge of the tab that is attached to the substrate, the bending moment $M(x)$ is given by $F_n(l - x)$, I_z is the moment of inertia ($bh^3/12$) of the rectangular cross-sectional area of width b and thickness h about the centroidal axis (z-axis), and y is the distance from the neutral plane. The maximum bending stress σ_{max} occurs when $x = 0$ and $y = \pm h/2$ and is given by

$$\sigma_{\mathrm{max}} = \frac{6F_n l}{bh^2} \tag{5.3}$$

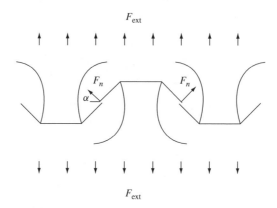

Figure 5.5 Simple cantilever model of the failure mode of silicon microvelcro (Han *et al.* 1992)

Similarly, the maximum shearing stress τ_{max} occurs at the neutral plane $y = 0$:

$$\tau_{max} = \frac{3F_n}{2bh} \tag{5.4}$$

σ_{max} is higher than τ_{max} with the ratio of τ_{max}/σ_{max} equal to $h/4l$ for the design used.

Visual examination of the tested samples indicates that failure is accompanied by damage to the edge of the tab and is consistent with the failure occurring when σ_{max} exceeds the yield point σ_{yp}. From Figure 5.5, we have

$$F_n = \frac{F_{ext}d^2}{4\sin\alpha} \tag{5.5}$$

where F_{ext} is the tensile load (i.e. force per unit surface area) applied to the sample. Using Equation (5.3), we get

$$\sigma_{max} = \frac{3F_{ext}d^2l}{2bh^2\sin\alpha} \quad \text{or } F_{ext} = \frac{2\sigma_{max}bh^2\sin\alpha}{3d^2l} \tag{5.6}$$

which, when we include friction with a static coefficient μ, becomes

$$F_{ext} = \frac{2\sigma_{max}bh^2\sin\alpha}{3d^2l(1 + \mu\cot\alpha)} \tag{5.7}$$

The tensile strength (failure load) of the structure can be found by substituting design values for b, h, l, α, and μ (0.5) and by substituting σ_{yp} (6×10^5 kPa for the oxide) for σ_{max} in Equation (5.7) to obtain a value of F_{ext} equal to 1.1 MPa, which is in agreement with the value obtained from the extrapolation of data in Figure 5.4(b). This finding confirms that the failure mechanism is that of bending stress, which exceeds the oxide yield point at the tab edge.

Worked Example E5.2: Undoped Silicon Cantilever Beams

Objective:

To fabricate a cantilever beam oriented in the (100) direction on (100) silicon wafers (Choit and Smits 1993).

Process Flow:

1. A layer of SiO_2 that is 0.5 µm thick is grown on a (100) n-type silicon wafer. The wafer is spin-coated with a layer of positive photoresist. The masks needed to fabricate the cantilevers are shown in Figure 5.6(a, b). Cross-hatched areas represent opaque regions of the masks. For mask 1 (Figure 5.6 (a)),

$$w_1 = 2(w_b + t_b) \quad \text{and} \quad l_1 = l_b \tag{5.8}$$

where l_b, w_b, and t_b are the length, width, and thickness of the beam, respectively. l_1 and w_1 are shown in Figure 5.6(a). For mask 2 (Figure 5.6(b)), we have

$$w_2 = w_b + 2d \quad \text{and} \quad l_2 = l_b \tag{5.9}$$

where d is a small parameter that corrects design errors and mask misalignment. l_2 and w_2 are shown in Figure 5.6(b). The two masks have essentially the same pattern, except that mask 2 has a smaller beam width than mask 1. The wafer is patterned with mask 1. The wafer is oriented in such a way that the length of the cantilever beam is in the (010) direction of the wafer, as shown in Figure 5.6(c).

2. The wafer is then immersed in a bath of buffered oxide etch (BOE) to remove the SiO_2 in the areas that are not covered by photoresist, and this is followed by dissolving the resist in an acetone bath. A transverse cross section of the beam region after the resist has been removed is shown in Figure 5.7(a). The wafer is now ready to be bulk-etched in sodium hydroxide (NaOH) at 55 °C.

3. Etching will take place in regions where the (100) planes of silicon are exposed. Lateral etching of silicon directly underneath the SiO_2 passivation layer will also occur; the lateral planes that are etched are the (100) equivalent planes. These planes are normal to the substrates. The rate of downward etching is the same as that of lateral etching; this will result in walls that are almost completely vertical (see Figure 5.7(b)). Planes are formed at the clamped end of the cantilever beam (111).

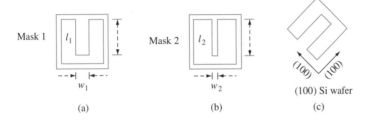

Figure 5.6 Masks required to fabricate the cantilever

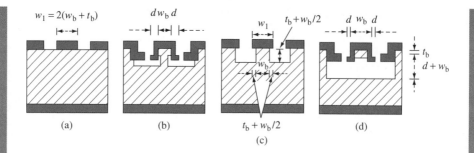

Figure 5.7   Four process steps to make and release the cantilever

4. The wafer is then etched in BOE to remove all the SiO_2 and is then cleaned and oxidised to grow a fresh layer of SiO_2 that is 1 μm thick. The wafer is spin-coated with a layer of positive photoresist and patterned with mask 2. After the unprotected oxide is etched away in BOE, the resist is removed in acetone. The wafer is then etched in NaOH at 55 °C until the bulk silicon is completely under-etched in the areas that are directly underneath the beam. Figure 5.7(c, d) shows the evolution of the silicon cantilevers etched in this way at different stages of the final etching in NaOH.

5.3 ETCH-STOP TECHNIQUES

Many different chemical etchants for silicon are known. The properties that make some of these etchants indispensable to micromachining of three-dimensional structures are selectivity and directionality. As etching processes in polar solvents are fundamentally charge transport phenomena, it is not surprising that the etch-rate may be dopant-type-dependent, dopant-concentration-dependent, and bias-dependent. Etch processes can be made selective by the use of dopants – heavily doped regions etch more slowly – or even halted electrochemically when observing the sudden rise in current through an etched n-p junction.

A region where wet (or dry) etching tends to slow down (or halt) is called an *etch-stop*. There are several ways in which an etch-stop region can be created. In the following subsections, two such methods by which etch-stops are created are discussed. These methods are:

- doping-selective etching (DSE)
- bias-dependent etching BSE

5.3.1 Doping-Selective Etching (DSE)

Silicon membranes are generally fabricated using the etch-stop phenomenon of a thin, heavily boron-doped layer, which can be epitaxially grown or formed by the diffusion or implantation of boron into a lightly doped substrate. This stopping effect is a general property of basic etching solutions such as KOH, NaOH, ethylenediamine pyrocatechol

(EDP), and hydrazine (see Table 5.2). Because of the heavy boron-doping, the lattice constant of silicon decreases slightly, leading to highly strained membranes that often show slip planes. They are, however, taut and fairly rugged even in a few microns of thickness and are approximately 1 cm in diameter. The technique is not suited to stress-sensitive microstructures that could lead to the movement of the structures without an external load. In this case, other etch-stop methods should be employed.

Early studies (Greenwood 1969; Bohg 1971) on the influence of boron doping on the etch rates of EDP for (100) silicon at room temperature have shown a constant etch rate of approximately 50 μm/h for the resistivity range between 0.1 and 200 Ω·cm corresponding to boron concentration from 2×10^{14} to $5 \times 10^{17} cm^{-3}$. As the boron concentration is raised to about a critical value of $7 \times 10^{19} cm^{-3}$, corresponding to a resistivity of approximately 0.002 Ω·cm, the silicon remains virtually unattacked by the etching solution (see Table 5.2). Figure 5.8 shows the boron-doping etch-stop properties for both KOH and EDP.

The dependence of the etch rate on the dopant concentration is typically exploited for undercutting microstructures that are defined by a masked heavy boron diffusion

Table 5.2 Dopant-dependent etch rates of selected silicon wet etchants

Etchant (Diluent)	Temperature (°C)	(100) Etch rate (μm/min) for boron doping $\ll 10^{19}$ cm^{-3}	Etch rate (μm/min) for boron-doping $\sim 10^{20} cm^{-3}$
EDP (H_2O)	115	0.75	0.015
KOH (H_2O)	85	1.4	0.07
NaOH (H_2O)	65	0.25–1.0	0.025–0.1

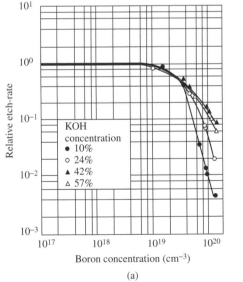
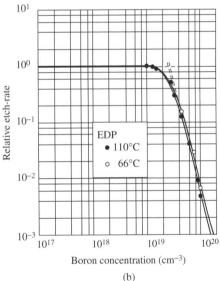

Figure 5.8 Boron etch-stop properties for (a) KOH and (b) EDP etchants

in a lightly doped n- or p-type substrate. If the etch-stop concentration threshold lies in between the substrate and diffusion concentrations, the p-substrate (or n-substrate for that matter) is etched out from underneath the high boron diffusion. A silicon microstructure with a geometry defined by the diffusion mask and a thickness close to the diffusion depth is hence left freely suspended.

The main benefits of the high boron etch-stop are the independence of crystal orientation, the smooth surface finish, and the possibilities it offers for fabricating released structures with arbitrary lateral geometry in a single etch step. On the other hand, the high levels of boron required are known to introduce considerable mechanical stress into the material; this may even cause buckling or even fracture in a diaphragm or other double-clamped structures. Moreover, the introduction of electrical components for sensing purposes into these microstructures, such as the implantation of piezoresistors, is inhibited by the excessive background doping. The latter consideration constitutes an important limitation to the applicability of the high boron dose etch-stop. Consequently, bias-dependent BSE, commonly referred to as *an electrochemical etch-stop*, is currently the most widely used etch-stop technique.

5.3.2 Conventional Bias-Dependent BSE or Electrochemical Etch-Stop

In electrochemical etching of silicon, a voltage is applied to the silicon wafer (anode), a counter electrode (cathode) in the etching solution. The fundamental steps of the etching mechanism are as follows:

1. Injection of holes into the semiconductor to raise it to a higher oxidation state Si^+

2. Attachment of negatively charged hydroxyl groups, OH^-, to the positively charged Si

3. Reaction of the hydrated silicon with the complexing agent in the solution

4. Dissolution of the reaction products into the etchant solution

In bias-dependent etching, oxidation is promoted by a positive voltage applied to the silicon wafer, causing an accumulation of holes at the Si−solution interface. Under these conditions, oxidation at the surface proceeds rapidly while the oxide is readily dissolved by the solution. Holes such as H^+ ions are transported to the cathode and released there as hydrogen gas bubbles. Excess hole-electron pairs can, in addition, be created at the silicon surface, for example, by optical excitation, thereby increasing the etch rate.

Figure 5.9 shows an electrochemical cell that is used to etch Si in a 5 percent hydrofluoric (HF) solution. The cathode plate used is made of platinum. In the etching situation shown in Figure 5.9, holes are injected into the Si electrode and they tend to reside at the Si surface where they oxidise Si at the surface to Si^+. The oxidised silicon interacts with incoming OH^- that are produced by dissociation of water in the solution to form the unstable Si(OH), which dissociates into SiO_2 and H_2 gas. The SiO_2 is then dissolved by HF and removed from the silicon surface.

The current density−voltage characteristics for different silicon types and resistivities are shown in Figure 5.10. It is apparent from Figure 5.10 that the current density is very much dependent on the type and the resistivity (doping level) of Si. This dependence on the type and resistivity is the property that is utilised in the electrochemical etch-stop phenomenon.

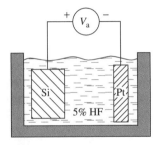

Figure 5.9 Electrochemical cell with 5 percent HF solution to etch silicon. The voltage V_a applied to the silicon is relative to a platinum reference electrode

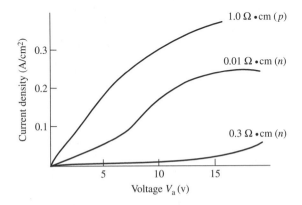

Figure 5.10 Plot of electrochemical current density against voltage for silicon doped to different resistivities

To understand the mechanisms of electrochemical etch-stop, it is important to explore in more detail the current–voltage $(I-V)$ characteristics in etching solutions that exhibit strong electrochemical etch-stop effects. The curves in Figure 5.11 are typically $(I-V)$ characteristics for n- and p-type silicon in KOH. We can easily see the similarity of Figure 5.11 to the well-known curve of a diode, except that at a certain voltage the current suddenly and sharply drops. Let us define the open circuit potential (OCP) as the potential at which the current I is zero, and the passivating potential (PP) as the potential at which the current suddenly drops from its maximum value. The two regions of interest are the ones separated by the PP. Only cathodic to the PP is the sample etched, whereas just anodic to it, an oxide grows and the surface is passivated. The insulating oxide layer that is formed during the etching process brings about the drastic fall in current at the PP. The difference between this etch and the HF etch described in Figure 5.9 is the fact that in the latter etch the oxide is dissolved by the HF solution, whereas in the former etch the oxide is not readily dissolved in the KOH solution. Another important feature of Figure 5.11 is the different behaviour of the two dopant types. When applying a voltage between the two passivating potentials of n- and p-type, one expects, in accordance with the characteristics shown in Figure 5.11, that only the p-type sample, and not the n-type sample, would be etched. This is the doping-selective effect that is used as an etch-stop.

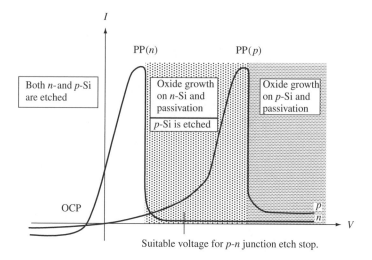

Suitable voltage for *p-n* junction etch stop.

Figure 5.11   Current–voltage characteristics of *n*-Si and *p*-Si in KOH. No current *I* flows at the OCP and the current stops above the passivating potential (PP) (Linden *et al.* 1989)

The growth of an anodic oxide is believed to result from the progressing competition between the oxidation of the silicon and the dissolution of the oxide products at the silicon–solution interface. Ellipsometric measurements have given evidence of such an oxide layer (Palik *et al.* 1985a). The results have been fitted with a multilayer model that suggests a graded connective layer of SiO_x.

Measurements on etch rates in KOH as a function of applied potential (Palik *et al.* 1985a) show that at the OCP there is very little difference in the etch rates between the *n*- and *p*-type substrates, whereas there is a marked difference at other potentials. Furthermore, the etch rate is not proportional to the current, and in fact, the etch rate attains a maximum at the OCP (current is zero) and slows down as the PP (current is a maximum) is approached; eventually, the etch stops when the current drops. The etch rates at the OCP, therefore, seem to be independent of free-carriers concentration, and it seems reasonable to suggest that the chemical mechanism is dominant. However, at the other potentials, in which the etch rates for the *n*- and *p*-type dopants differ, this is probably not the case. At these potentials, it is possible that a combined chemical and electro-chemical mechanism is responsible for the etching process. A chemical–electrochemical mechanism is proposed (Palik *et al.* 1985b) in which a chemical reaction takes place at the silicon surface with sequential attacks of Si–Si bonds by H_2O and OH, resulting in the discharge of the OH into the etching layer giving soluble silicate of the form $Si(OH)_2O_2$.

It is then followed by the more rapid electrochemical reaction that oxidises underlying bonds. The reduction–oxidation couple OH^-/H_2O is assumed to supply the species for etching, electrolysis, and oxidation.

The hydrogen gas produced per Si atom has been measured (Palik *et al.* 1985b) and was found to be $2H_2$ per Si; this indicates a chemical reaction for a Si–Si bond at the OCP as follows:

$$Si - Si + H_2O \longrightarrow Si - H + Si - OH \tag{5.10}$$

and

$$Si - Si + Si - OH + OH^- \longrightarrow Si - O^- + Si - OH + H_2 \tag{5.11}$$

The net reaction for dissolution of a silicon atom would, therefore, be

$$Si + 2HO + 2OH^- \longrightarrow H_2 + Si(OH)_2(O^-)_2 \tag{5.12}$$

The nature of the band bending at the surface will play a critical role in the effect that an applied bias gives. A model relating the DSE and the p^+-Si etch-stop with band bending and charge transfers has been suggested but is beyond the scope of this book.

The conventional electrochemical etch-stop technique is an attractive method for fabricating both microsensors and microactuators because it has the potential for allowing reproducible fabrication of moderately doped n-type silicon microstructures with good thickness control. However, a major limiting factor in the use of the conventional electrochemical etch-stop process is the effect of reverse-bias leakage current in the junction. Because the selectivity between n- and p-type silicon in this process is achieved through the current-blocking action of the diode, any leakage in this diode will affect the selectivity. In particular, if the leakage current is very large, it is possible for etching to terminate well before the junction is reached. In some situations, the etching process may fail completely because of this leakage. This effect is well known, and alternative biasing schemes that employ three and sometimes four electrodes have been proposed to minimise this problem. In the three-electrode setup (see Figure 5.12), a reference electrode is introduced for more accurate control over the potential of the etchant.

In the four-electrode setup (Figure 5.13), a fourth electrode is used to contact the p-type substrate to gain direct control over the p-n forward-bias voltage. The four-electrode approach allows etch-stopping on lower quality epitaxial layers (larger leakage currents) and should also enable etch-stopping of p-type epitaxial layers on an n-type substrate.

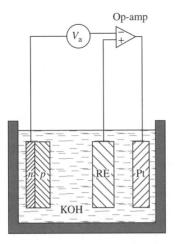

Figure 5.12   Basic arrangement of a three-electrode electrochemical cell for silicon etch-stop

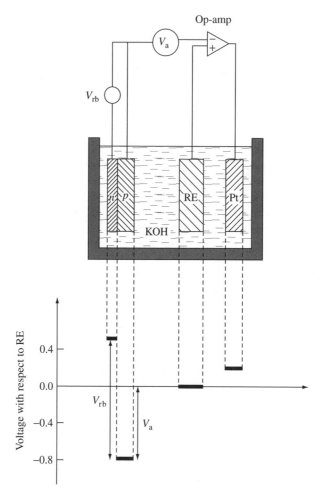

Figure 5.13   Basic arrangement of a four-electrode electrochemical cell for silicon etch-stop. Voltages are indicated relative to the reference electrode RE

The conventional junction etch-stop permits the formation of n-type structures. If the doping types of the silicon in Figures 5.12 and 5.13 are reversed such that the bias is applied to a p-type layer, and the n-type is exposed to solution, the diode is now forward-biased by the applied potential. If the forward-diode current density is low enough, it is possible that the n-type silicon will not passivate, even if the potential applied to the p-type layer is greater than the passivation potential for silicon. If this is the case, the n-type silicon will etch until it reaches the p-type layer, thus permitting the formation of a p-type membrane. In practice, this process can be very difficult to implement because it requires a very tight control of the potential applied to the p-type layer and it is very sensitive to the diode-forward characteristics. To circumvent this weakness of the conventional electrochemical etch-stop technique, an alternative dopant-selective technique that uses pulsed anodising voltages applied to silicon samples immersed in etching solutions was developed (Wang *et al.* 1992). This alternative technique is called *selective etching by pulsed potential anodisation* and is described in the following section.

5.3.3 Selective Etching of *n*-Type Silicon by Pulsed Potential Anodisation

The pulsed potential anodisation technique selectively etches *n*-type silicon (Wang *et al.* 1992). The difference in the dissolution time of anodic oxide formed on *n*- and *p*-type silicon samples under identical conditions is used to create an etch selectivity. The mechanism responsible for this dissolution time difference is not fully understood at present. However, it is believed to be due to a difference in oxidation rates caused by the limited supply of holes in *n*-type samples (Wang *et al.* 1992). This technique is applicable in a wide range of anodising voltages, etchant compositions, and temperatures. It differs from the conventional *p-n* junction etch-stop in that the performance of the etch-stop does not depend on the rectifying characteristics or quality of a diode. Using this technique, *p*-type microstructures of both low and moderate doping can be fabricated. Hence, the pulsed potential anodisation technique opens up the possibility for the creation of fragile microstructures in *p*-type silicon.

The main problems with the conventional electrochemical etch-stop and the pulsed potential anodisation technique are related to the etch holders required for contacting the epitaxial layer (and the substrate for two, three, or four electrodes) and for protecting the epitaxial side of the wafer from the etchant. Any leakage in these holders interferes with the correct operation of the etch-stop. Moreover, mechanical stress introduced by the holder is known to reduce substantially the production yield in many cases. Therefore, development of a reliable wafer holder for anisotropic etching with electrochemical etch-stop is not straightforward. The process of making contact with the wafer itself can also be critical and difficult to implement. Therefore, single-step fabrication of released structures with either the conventional electrochemical etch-stop or the pulsed potential anodisation techniques may be troublesome. An alternative etch-stop technique that does not require any external electrodes (or connections to be made to the wafer) has been recently developed. This new technique is what is referred to as *the photovoltaic electrochemical etch-stop technique* (PHET) (Peeters *et al.* 1994).

5.3.4 Photovoltaic Electrochemical Etch-Stop Technique (PHET)

The PHET approach is able to produce the majority of structures that can be produced by either the high boron or the electrochemical etch-stop (Peeters *et al.* 1994). PHET does not require the high impurity concentrations of the boron etch-stop and does not require external electrodes or an etch holder as is required in conventional electrochemical etch-stop or in pulsed anodisation technique. Free-standing *p*-type structures with arbitrary lateral geometry can be formed in a single etch step. In principle, PHET is to be seen as a two-electrode electrochemical etch-stop in which the potential and current required for anodic growth of a passivating oxide is not applied externally but is generated within the silicon itself. The potential essentially consists of two components, namely, the photovoltage across an illuminated *p-n* junction and the 'Nernst' potential of an *n*-Si/metal/etchant solution electrochemical cell.

Worked Example E5.3: Formation of an Array of Thin Membranes

Objective:

The objective is to use electrochemical etching to fabricate an array of membranes with thickness in the range 3 to 8 μm and sides that are between 0.5 and 5 mm. The array is to be fabricated on a *p*-type silicon substrate (Linden *et al.* 1989).

Process Flow:

1. The silicon wafers used are standard commercial, 280 μm thick, (100)-oriented silicon wafers. The wafers are boron-doped to a resistivity of 7 to 10 Ω·cm, which corresponds to a doping concentration of approximately 1.5×10^{15} cm^{-3}. To produce a *p-n* junction on the front surface of the wafer, a phosphorus-doped *n*-layer is diffused on the *p*-type silicon wafer. The diffusion is performed by a predeposition, in which oxygen is bubbled through a flask containing POCl$_3$, followed by a driven-in diffusion with a mixed gas ambient of nitrogen and oxygen. Typical *n*-doping concentrations are around 10^{17} cm^{-3}.

2. Standard photolithographic methods are used to form a SiO$_2$ mask on the backside (*p*-type side) of the wafer. The sides of the membranes range between 5 mm and 0.1 mm and the thickness (the thickness of the *n*-layers) range from 8 μm down to 3 μm (see Figure 5.14(a)).

3. The wafer is mounted on the etch apparatus as shown in Figure 5.14(b). The voltage between the two passivating potentials is chosen as 1.9 V. The etch, using a KOH solution, is performed through the *p*-type material and stopped when it reaches the *n*-type material. Etching through the *p*-type silicon usually takes around four to five hours.

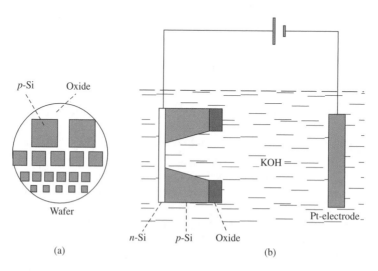

(a)

(b)

Figure 5.14 Method to fabricate an array of thin membranes: (a) design of an oxide mask and (b) the electrochemical cell providing a back-etch

Worked Example E5.4: **Fabrication of Cantilever Beams**

Objective:

The objective is to fabricate a cantilever beam that is a few microns thick on a *p*-type silicon that supports wafers using electrochemical etching (Linden *et al.* 1989).

Process Flow:

The starting silicon wafers used are again standard commercial, 280 μm thick[4], (100)-oriented silicon wafers. The wafers are boron-doped with a resistivity of 7 to 10 Ω·cm, which corresponds to a doping concentration approximately 1.50×10^{15} cm^{-3}. Two different techniques can be used to define the beam.

Technique A: The diffused-pattern technique

In the diffused-pattern technique, an *n*-type (a diffused *n*-layer of *n*-doping concentration $\sim 10^{17}$ cm^{-3}, as described in Worked Example 5.3) pattern that describes the beam is diffused into the wafer. The wafer is masked with SiO$_2$ grown in wet O$_2$ at 1000 °C. Then, by stripping the patterning oxide and by performing the electrochemical etch, the beam will be totally defined by the diffusion process. The steps for the beam fabrication using this method are shown in Figure 5.15. The etching is performed in an apparatus similar to the one described in Worked Example 5.3 (see Figure 5.14(b)).

In this technique, the pattern is diffused, thus resulting in an *n*-type pattern defining the beam. Because of lateral diffusion during the drive-in, this technique will result in beams with 'rounded' corners (Figure 5.15 (c)); this means that we will not end up with the sharp corners that are usually associated with anisotropic etching.

Technique B: The etched-pattern technique

In the etched pattern technique, an *n*-type layer (dopant concentration $\sim 10^{17}$ cm^{-3}) is diffused over the entire wafer. Oxidation is performed simultaneously with the drive-in of the dopant. The oxide is patterned to cover the surfaces of the forming beam. The wafer is then immersed in the KOH etching solution as in Worked Example 5.3 (Figure 5.14(b)) without any bias, until the *n*-type layer is etched through and the *p*-type layer is exposed. Following this nonbias etch, a voltage bias is applied and the beam will be etched correctly. The processing sequence for the beam fabrication using the etched-pattern technique is outlined in Figure 5.16(a) through to (d).

This method has all the advantages of the anisotropic KOH etch, that is, giving perfectly sharp corners defined by the (111) crystal planes. On the other hand, there will be under-etching alongside the beam under the oxide mask when etching the top layer without bias (if the beam is parallel to the (100) planes). In addition, the tip of the beam will be 'polygonal' instead of being square in shape (Figure 5.16(e)) because the etchant finds faster etching planes at the convex corners. Another possible advantage is the fact that the beam surfaces are protected by the SiO$_2$ throughout the whole etching

---

[4] Wafer thickness normally increases with diameter, so 280 μm-thick wafers (3" diameter) are now hard to source; however, larger (e.g. 4") wafers can be thinned down.

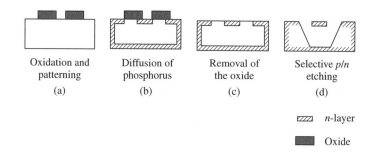

Oxidation and patterning	Diffusion of phosphorus	Removal of the oxide	Selective *p/n* etching
(a)	(b)	(c)	(d)

▨▨ *n*-layer

■ Oxide

Figure 5.15   Process flow of diffused pattern technique

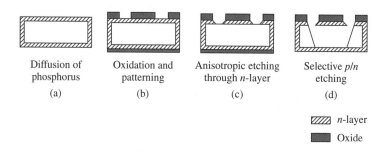

Diffusion of phosphorus	Oxidation and patterning	Anisotropic etching through *n*-layer	Selective *p/n* etching
(a)	(b)	(c)	(d)

▨▨ *n*-layer

■ Oxide

Figure 5.16   Process flow of etched-pattern technique

process and are not exposed to the etchant, which helps in stopping microdamage on the surfaces.

5.4 DRY ETCHING

As discussed in Sections 5.2 and 5.3, bulk-micromachining processes can yield SCS microstructures using crystal-orientation-dependent and dopant-concentration-dependent wet chemical etchants, such as EDP, KOH, and hydrazine to undercut the SCS structures from a silicon wafer. However, the type, shape, and size of the SCS structures that can be fabricated with the wet chemical etch techniques are severely limited. A dry-etch-based process sequence to produce suspended SCS mechanical structures and actuators has been developed (Zhang and McDonald 1992). The process is called *single-crystal reactive etching and metallisation* (SCREAM). SCREAM uses RIE processes to fabricate released SCS structures with lateral feature sizes down to 250 nm and with arbitrary structure orientations on a silicon wafer. SCREAM includes process options to make integrated, side-drive capacitor actuators. A compatible high step-coverage metallisation process using metal sputter deposition and isotropic metal dry etch is used to form side-drive electrodes. The metallisation process complements the silicon RIE processes that are used to form the movable SCS structures.

For the SCREAM process, mechanical structures are defined with one mask and are produced from a silicon wafer. The process steps used in SCREAM are illustrated in

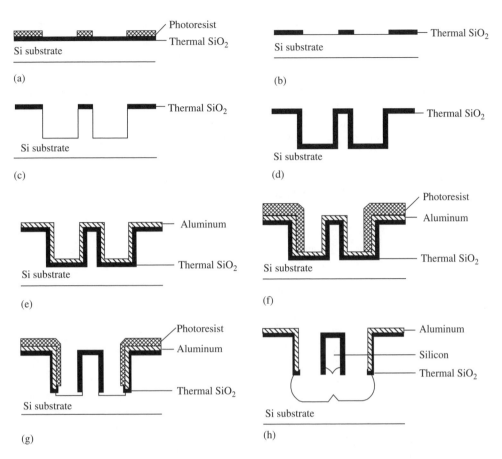

Figure 5.17   SCREAM process flow for a straight cantilever beam with integrated electrodes (Zhang and McDonald 1992)

Figure 5.17. This figure shows the SCREAM process sequence for the fabrication of a straight cantilever beam including the integrated metal electrodes adjacent to each side of the beam. A layer of SiO_2, used as an etch mask, is thermally grown on the silicon substrate. The pattern to produce free-standing SCS structures is created using photolithography. The photoresist pattern on the SiO_2 is transferred to the silicon dioxide using fluorocarbon-based oxide etching plasma. The photoresist is then stripped by an O_2 plasma etch and the SiO_2 pattern is subsequently transferred to the silicon substrate using a Cl^--based or a HBr-based RIE. Following the silicon etch, a sidewall silicon dioxide layer is thermally grown in wet O_2 at high temperatures. The thermal oxidation process reduces possible damage on the sidewalls of the silicon steps created during the RIE process. The lateral dimensions of the SCS structures are reduced during the thermal oxidation. A metal layer is conformally deposited on top of the thermal oxide using a sputter deposition system. Before the metal deposition, contact windows are opened to allow electrical contact to both the silicon substrate and the movable silicon structures. After the metal sputter deposition, photoresist is spun on the metal layer to refill the few-microns-deep trenches. A metal side-electrode pattern is created in the photoresist using a wafer stepper to expose the photoresist. The metal side-electrode is then transferred to the metal by a metal RIE process. After the metal electrodes are patterned, the SiO_2 is etched back with fluorocarbon-based RIE. Finally, the silicon mechanical structures are released from the silicon substrate using an RIE process.

The SCREAM process described in the preceding text can be used to fabricate complex, circular and triangular structures in SCS. These structures can include integrated, high-aspect-ratio, and conformable capacitor actuators. The capacitor actuators are used to generate electrostatic forces and so produce micromechanical motion.

Worked Example E5.5: **Fabrication of Straight Cantilever Beam with Integrated Aluminum Electrodes Using SCREAM**

Objective:

To fabricate a free-standing cantilever beam 200 μm long, 0.8 μm wide, and 3.5 μm thick and coated with a 150 nm-thick SiO_2 layer (Zhang and McDonald 1992).

Process Flow:

The steps are based on the process shown in Figure 5.17, with the metal being aluminum.

1. The starting substrate is an arsenic-doped, 0.005 Ω·cm, n-type (100) silicon wafer. A layer of SiO_2 that is 400 nm thick is thermally grown on the substrate in a conventional furnace in a steam O_2 ambient at 1100 °C. The oxide is used as an etch mask. The pattern to produce free-standing SCS structures is created using photolithography in positive photoresist spun on the oxide layer on silicon. A wafer stepper is used to expose the photoresist. The minimum feature size of the SCS beam structures is 400 nm (see Figure 5.17(a)).

2. The photoresist pattern on the oxide is transferred to the oxide in a CHF_3/O_2 plasma etch at flow rates of 30 sccm[5] and 1 sccm, a chamber pressure of 30 mTorr and a DC self-bias of 470 V in a conventional parallel-plate RIE tool. The etch rate of the oxide is 23 nm/min. The photoresist on top of the oxide is then stripped off by an O_2 plasma etch (Figure 5.17(b)).

3. The oxide is then transferred to the silicon substrate using a $Cl_2/BCl_3/H_2$ RIE in a commercial RIE tool. Three etch steps are required to accomplish a deep vertical silicon etch (Figure 5.17(c)). The parameters for these etch steps are given in Table 5.3.

Table 5.3   Silicon etch parameters (Zhang and McDonald 1992)

Step	Cl_2 flow-rate (sccm)	BCl_2 flow-rate (sccm)	H_2 flow-rate (sccm)	Pressure (mTorr)	Time (min)
1	0	28	14	20	2
2	4	28	14	20	3
3	28	4	0	30	28

[5] sccm is standard cubic centimeter per minute.

4. Following the silicon etch, a sidewall SiO_2 layer of 150 nm is thermally grown in wet O_2 at 1000 °C. The thermal oxidation process reduces possible damage on the sidewalls of the silicon steps created during the $Cl_2/BCl_3/H_2$ RIE. The lateral dimensions of the SCS structures are reduced during the thermal oxidation (Figure 5.17(d)).

5. A 400 nm layer of Al with a step coverage of 60 percent on the sidewalls is conformably deposited using DC magnetron sputter deposition in a commercial deposition tool. The sputtering is performed at a pressure of 9 mTorr and at a temperature of 20 °C in argon gas and a beam current of 5 A. Before the Al deposition, contact windows are opened to allow electrical contact to both the silicon substrate and the movable silicon structure (see Figure 5.17(e)).

6. After the Al deposition, positive photoresist is spun on the Al layer at 2.5 krpm for 45 seconds (3.6 µm thick) to fill the 3.8 µm-deep trenches. An Al side-electrode pattern is created in the resist using a wafer stepper to expose the resist (Figure 5.17(f)).

7. The Al side-electrode pattern is then transferred to the Al by means of an aluminum dry-etch process; this step is required to clear the conformal Al layer deposited on the high steps where the resist has been removed. The etch process is an anisotropic RIE process that utilises flow rates of 20 sccm of Cl_2 and 40 sccm of barium trichloride (BCl_3) at a chamber pressure of 50 mTorr and a DC self-bias voltage of 250 V. The photolithography and the Al RIE steps produce smooth edges on the Al pattern over a topography with 3.8 µm steps (Figure 5.17(g)).

8. After the Al electrodes are patterned, the SiO_2 is etched back with a CF_4 plasma in a standard parallel-plate RIE tool. Anisotropic etch profiles are preferred to the oxide etch-back to remove the oxide from the bottom of the trenches but to retain the SiO_2 on the top and sidewalls of the SCS structures. Therefore, a low chamber pressure of 10 mTorr and a high DC self-bias of 600 V are selected in the CF_4 RIE process (Figure 5.17(h)).

5.5 BURIED OXIDE PROCESS

This process is still under development and is based on the separation by the implanted oxygen (SIMOX) technique that was developed originally to enable the manufacture of silicon-on-insulator (SOI) wafers for ICs[6]. The buried oxide process generates microstructures by means of exploiting the etching characteristics of a buried layer of SiO_2. After oxygen has been implanted into a silicon substrate using suitable ion-implantation techniques, high-temperature annealing causes the oxygen ions to interact with the silicon to form a buried layer of SiO_2. The remaining thin layer of SCS can still support the growth of an epitaxial layer of thickness ranging from a few microns to tens of microns.

In micromachining, the buried SiO_2 layer is used as an etch-stop. For example, the etch rate of an etchant such as KOH slows down markedly as the etchant reaches the SiO_2 layer. However, this process has the potential for generating patterned SiO_2 buried layers by appropriately implanting oxygen.

[6] An SOI field-effect transistor (FET) microheater for high-temperature gas sensors is described in Chapter 15.

5.6 SILICON FUSION BONDING

The construction of any complicated mechanical device requires not only the machining of individual components but also the assembly of the components to form a complete set. In micromachining, bonding techniques are used to assemble individually micromachined parts to form a complete structure. Wafer bonding, when used in conjunction with micromachining techniques, allows the fabrication of three-dimensional structures that are thicker than a single wafer. Several processes have been developed for bonding silicon wafers. The most common bonding process is fusion bonding.

In the last decade, several groups (Lasky 1986; Ohashi *et al.* 1986; Apel *et al.* 1991) have demonstrated that the fusion of hydrophilic silicon wafers is possible for obtaining SOI materials. Since then, wafer-bonding techniques have found different applications in the field of microelectronics; several static random access memory (SRAM), complementary metal oxide semiconductor (CMOS), and power devices have been fabricated on bonded SOI material. For micromechanical applications, fusion bonding rendered possible the fabrication of complex structures by combining two or more patterned wafers. This section describes the principles of wafer fusion bonding and presents fusion-bonding processes for MEMS device fabrication.

5.6.1 Wafer Fusion

In its simplest form, the process of wafer fusion bonding is the mating together of a pair of wafers at room temperature, followed by thermal annealing at temperatures between 700 and 1100 °C. At room temperature, the wafers adhere via hydrogen bridge bonds of chemisorbed water molecules that subsequently react during the annealing process to form $Si-O-Si$ bonds. Consequently, wafer pretreatment procedures that include hydrophilisation steps (wet cleaning processes and plasma hydrophilisation) support the process.

A major concern of all bonding processes is the presence of noncontacting areas, which are generally called *voids*. Voids are mainly caused by particles, organic residues, surface defects, and inadequate mating. Therefore, both the surfaces that are fusion-bonded have to be perfectly smooth and clean because the smallest of particles could cause large voids. Optimised processing includes wafer surface inspection, surface pretreatment (hydrophilisation and cleaning), and mechanically controlled, aligned mating in a particle-free environment.

5.6.2 Annealing Treatment

As discussed in the previous subsection, wafer fusion bonding involves a high-temperature annealing step that is to be performed after the room-temperature contacting of the surfaces. This annealing step is necessary to increase the strength of the bond. However, the high-temperature annealing step (usually at a temperature above 800 °C) may introduce problems, such as doping profile broadening, thermal stresses, defect generation, and contamination. Annealing also prevents the use of bonding technology for compound semiconductor materials because their dissociation temperature is often low. In addition, postmetallisation bonding also requires bonding temperatures that are less than 450 °C as most of the common metals that are used in device fabrication melt below this temperature. Therefore, to make full use of the potential provided by wafer bonding for microstructures, low-temperature bonding methods have to be developed. Attempts to lower bonding temperatures and still achieve reasonable bond strength are currently under way.

Three annealing temperature ranges are of interest in wafer bonding:

1. Temperature less than 450 °C for postmetallisation wafers.

2. Temperature less than 800 °C for wafers with diffusion dopant layers (e.g. p^+ etch-stop layers).

3. Temperature greater than 1000 °C for wafer bonding before processing. According to the reaction mechanism, annealing at temperatures above 1000 °C for several hours should result in an almost complete reaction of the interface. A 1000 °C anneal for about two hours gives sufficiently high bond strength for all subsequent treatments (Harendt *et al.* 1991); it is not possible to separate the two bonded Si wafers without breaking the silicon.

An 800 °C anneal results in sufficient bond strength for subsequent processes such as grinding, polishing, or etching. However, the bonding is incomplete, as suggested by partial delamination of thinned films after stress treatment (Harendt *et al.* 1991). The low-temperature anneal ($T < 450$ °C) is inadequate for full wafer bonding. Although a significant increase in bond strength is already measurable after annealing at 200 °C (Kissinger and Kissinger 1991), additional voids develop during annealing in the temperature range 200 to 700 °C and disappear at higher temperatures. If annealing is interrupted in this temperature range, these voids remain after cooling; they probably originate from the interfacial water, which dissolves and reacts at temperatures above 800 °C. Patterned wafers, however, have been successfully annealed at 450 °C without the development of additional voids. In this case, the cavities probably act as buffers for the water. Table 5.4 is taken from Harendt *et al.* (1991) and it gives the bond quality for different annealing temperatures.

5.6.3 Fusion of Silicon-Based Materials

Fusion bonding of polysilicon, SiO_2, or silicon nitride to silicon proceeds in a manner similar to silicon-to-silicon bonding. In the case of polysilicon bonding to silicon, a polishing step for the two surfaces to be bonded is necessary. This polishing step produces

Table 5.4 Bond quality data taken from Harendt *et al.* (1991)

Structure	Annealing temperature (°C)	Bond strength (Jm^{-2})	Voids (% nonbonding)
Si/Si	450	0.5	–
Si/Si	800	0.6	0.3
Si/Si	1000	2.6	0.3
Si/Si$_3$N$_4$(140 nm)	800	0.9	0.2
Si/Si$_3$N$_4$(140 nm)	1000	Cleavage	0.2
Si/Si$_3$N$_4$(300 nm)	1000	Cleavage	25

two smooth defect-free surfaces. The bonding mechanism is most probably identical to silicon-to-silicon fusion bonding in that in both cases, Si−OH groups are present at the surface. Thus, pretreatment (hydrophilisation) and annealing conditions are similar.

Because of the dissimilar mechanical characteristics of the different bonded materials, the yield of void-free wafers can be significantly reduced by wafer bow or defects caused by stress during thermal treatment. Bonding of wafers covered with a thin thermal oxide or a thin silicon nitride results in homogenous bonded wafers, whereas oxides with thicker oxide (or nitride) films develop voids during annealing (see Table 5.4).

5.7 ANODIC BONDING

Silicon-to-silicon anodic bonding is a bonding technique used to seal silicon together by use of a thin sputter-deposited glass layer. The equipment used for anodic bonding is shown in Figure 5.18. The equipment is basically a heat-chuck element with an electrode that is capable of supplying high voltage across the structure to be bonded. The system may automatically control the temperature and power supply during the bonding process.

After surface cleaning and polishing, one of the wafers (referred to here as the top wafer) is initially given a glass film that is a few microns thick. This glass film is sputtered on the wafer surface. The top wafer is placed on top of a second silicon wafer, which is usually referred to as the support wafer (Figure 5.19); these two wafers are to be bonded. The support wafer rests on the aluminum chuck shown in Figure 5.18. The two wafers are usually sealed together by anodic bonding at temperatures less than 400 °C with an electrostatic DC voltage of 50 to 200 V. The negative electrode is connected to the top sputter-coated wafer. The voltage should be applied over a time that is long enough to allow the current to settle at the steady-state minimised level. Typically, the bonding process is terminated within 10 to 20 minutes. The bond process usually takes place in air at atmospheric pressure.

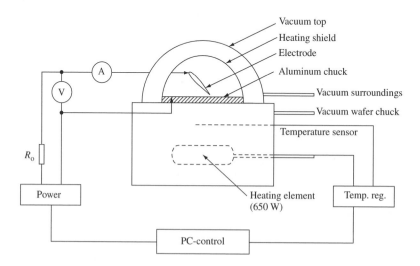

Figure 5.18   Schematic cross section of anodic bonding apparatus (Hanneborg *et al.* 1992)

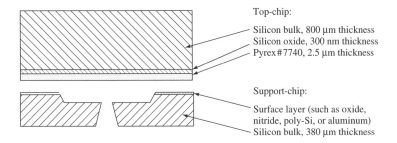

Figure 5.19 Top and support wafers used in a typical anodic bonding process (Hanneborg *et al.* 1992)

Worked Example E5.6: Floating Element Shear Sensor

Objective:

The objective is to fabricate a floating element sensor that consists of a plate (120 μm ×140 μm) and four tethers (30 μm × 10 μm) as shown in Figure 5.20(a) (Shajii and Schmidt 1992). The tethers work as mechanical supports for the plate and the resistors in the transduction scheme. The plate and the tethers are made from a 5 μm thick lightly doped *n*-type silicon layer and are suspended 1.4 μm above another silicon surface as shown in Figure 5.20(b). The entire structure is attached to the lower silicon wafer at the ends of the tethers using a 1.4 μm SiO_2 layer.

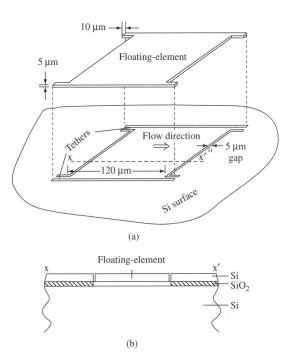

Figure 5.20 Angled view (a) and cross section (b) of a floating sensor based on a rectangular plate with four tethers (Shajii and Schmidt 1992)

Process Flow: (Figure 5.21)

The fabrication of the sensor involves the processing of two wafers (handle wafer and device wafer) that are bonded together. The process proceeds as follows:

1. A 1.4 μm thick layer of SiO_2 is grown on the handle wafer by wet oxidation at 950 °C. The handle wafer is typically a 4, 10 to 20 Ω·cm n-type (100) silicon. Using photolithography, the oxide on top of the handle wafer is patterned as shown in Figure 5.21(a). The device wafer has a 5 μm lightly doped n-type ($\sim 10^{15}$ cm^{-3}) epitaxial silicon layer grown on top of a highly boron-doped ($\sim 10^{20}$ cm^{-3}) p^+ region (Figure 5.21(a)).

2. The front of the handle wafer is bonded to the device wafer. The bonding sequence includes a preoxidation cleaning of the two wafers, hydration of the bonding surfaces using a 3:1 H_2SO_4:H_2O_2 solution for 10 minutes, a deionised water rinse, spin dry, physical contact of the two bonding surfaces, and a high-temperature anneal (1000 °C in dry O_2) for 70 minutes. The device wafer is then thinned in a KOH:H_2O (20 percent KOH by weight at 57 °C) solution until approximately 40 μm of silicon remains. A solution of CsOH:H_2O (60 percent CsOH by weight at 60 °C) is used to stop on the p^+ layer because of its better etch-stop characteristics for heavily boron-doped

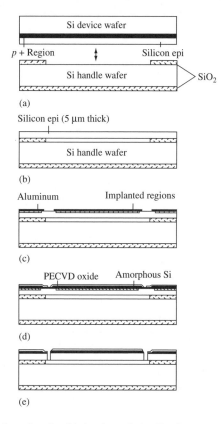

Figure 5.21 Process flow for the fabrication of the floating sensor shown in Figure 5.20 (Shajii and Schmidt 1992)

silicon in comparison with KOH. The $p+$ layer is subsequently removed using a 8:3:1 $CH_3COOH:HNO_3:HF$ liquid etch. A brown stain, indicating porous silicon, resulting from the etch is removed using a mixture of 97:3 $HNO_3:HF$ solution for about 15 seconds leaving a 5 μm-thick silicon epilayer bonded to the patterned oxide (Figure 5.21(b)).

3. A masked implant (as at 80 keV and 7×10^{15} cm^{-2} dose) followed by a one-hour anneal at 850 °C in dry O_2 ambient is performed for better ohmic contact. Sensing tethers are not implanted and their concentration remains at approximately 10^{15} cm^{-3}. Following the implant anneal, 500 nm of Al is deposited using electron-beam deposition, patterned and sintered in N_2 ambient at 375 °C for 30 minutes (Figure 5.21(c)).

4. A thin layer of plasma-enhanced chemical vapour deposition (PECVD) amorphous silicon (500 nm) is deposited over the wafer to protect the aluminum in the subsequent processing steps. A 1 μm-thick PECVD oxide is then deposited and patterned using a $CF_4/CHF_3/He$ plasma etch. This oxide will act as an etch mask for the silicon trench etch that is used to define the floating element. The oxide is plasma-etched and the resist is subsequently removed (Figure 5.21(d)).

5. A CCl_4-based plasma etch is used to trench-etch the silicon. A BOE (7:1 $H_2O:HF$) is used to etch the PECVD oxide that remains after the silicon trench etch, thus completing the fabrication of the floating element. Electrical contacts are made through the amorphous silicon layer to the Al bonding pads either by mechanical probing or by ultrasonic wire bonding. The completed sensor is shown in Figure 5.21(e).

5.8 CONCLUDING REMARKS

In this chapter, we have described the process involved in bulk silicon micromachining that is used to fashion microstructures out of silicon wafers, including a discussion of etching and etch-stops, followed by a discussion of the bonding together of two wafers to make the complete microdevices. Several worked examples are considered to illustrate the role of the various processes.

The alternative approach is that of surface rather than bulk silicon micromachining. This topic is covered in the next chapter, again with worked examples of microstructures. Together, bulk and surface micromachining represent the two key technologies that are essential for the manufacture of many different types of silicon microsensors and microactuators today.

REFERENCES

Apel, U. *et al.* (1991). "A 100-V lateral DMOS transistor with a 0.3 micron channel in a 1 micron silicon-film-on-insulator-on silicon," *IEEE Trans. Electron Devices*, **38**, 1655–1659.

Bohg, A. (1971). "Ethylene diamine-pyrocatechol-water mixture shows etching anomaly in boron-doped silicon," *J. Electrochem. Soc.*, **118**, 401–402.

Choi, W-S. and Smits, J. G. (1993). "A method to etch undoped silicon cantilever beam," *J. Microelectromech. Syst.*, **2**, 82–86.

Greenwood, J. C. (1969). *J. Electrochem. Soc.*, **116**, 1325–1330.

Han, H., Weiss, L. E. and Reed, M. L. (1992). "Micromechanical velero," *J. Microelectromech. Syst.*, **1**, 37–43.

Hanneborg, A., Nese, M., Jakobsen, H., and Holm, R. (1992). "Silicon-to-thin film anodic bonding," *J. micromech. Microeng.*, **2**, 117–121.

Harendt, C., Graf, H-G., Hofflinger, B. and Penteker, E. (1991). "Silicon direct bonding for sensor applications – characterisation of the bond quality," *Sensors and Actuators A*, **25**, 87–92.

Kissinger, W. and Kissinger, G. (1991). *Proc. 1st Symp. on Wafer Bonding*, Phoenix, USA, abstract no. 461, p. 681.

Lasky, J. B. (1986). "Wafer bonding for silicon-on-insulator technologies," *Appl. Phys. Lett.*, **48**, 78–80.

Linden, Y., Tenerz, L., Tiren, J. and Hok, B. (1989). "Fabrication of three dimensional silicon structures by means of doping-selective etching," *Sensors and Actuators*, **16**, 67–82.

Ohashi *et al.* (1986). *Proc. IEDM*, **86**, 210–213.

Palik, E. D., Bermudez, V. M. and Glembocki, O. J. (1985a). "Ellipsometric study of the etch-stop mechanism in heavily doped silicon," *J. Electrochem. Soc.*, **132**, 135–141.

Palik, E. D., Bermudez, V. M. and Glembocki, O. J. (1985b). In C. Fung *et al.*, eds., *Micromachining and Micropackaging of Transducers*, Elsevier, Amsterdam, pp. 135–149.

Peeters, E., Lapadatu, D., Puers, R. and Sansen, W. (1994). "PHET, an electrodeless photovoltaic electrochemical etch-stop technique," *J. Microelectromech. Syst.*, **3**, 113–123.

Rai-Choudrey, P. ed. (1997). *Handbook of Microlithography, Micromachining and Microfabrication*, 2 vols., SPIE Press, Washington, USA.

Shajii, J., Ng, K-Y. and Schmidt M. A. (1992). "A microfabrication floating-element shear stress sensor using wafer-bonding technology," *J. Microelectromech. Syst.*, **1**, 89–94.

Wang, S.S. *et al.* (1992). "An etch-stop utilising selective etching of *n*-type silicon by pulsed potential anodisation," *J. Microelectromech. Syst.*, **1**, 187–192.

Zhang, Z. L. and McDonald, N. C. (1992). "A RIE process for submicron, silicon electromechanical structures," *J. Micromech. Microeng.*, **2**, 3138.

6
Silicon Micromachining: Surface

6.1 INTRODUCTION

Since the beginning of the 1980s, much interest has been directed toward micromechanical structures fabricated by a technique called *surface micromachining*. The resulting two-and-a-half-dimensional structures[1] are mainly located on the surface of a silicon wafer and exist as a thin film – hence the half dimension. The dimensions of these surface-micromachined structures can be an order of magnitude smaller than the bulk-micromachined structures. The main advantage of surface-micromachined structures is their easy integration with integrated circuit (IC) components, as the same wafer surface can also be processed for IC elements. However, as miniaturisation is immensely increased by silicon surface micromachining, the small sizes or masses that are created are often insufficient for viable sensors and, particularly, for actuators. The problem is most acute in capacitive mechanical microsensors (Section 8.4) and especially in capacitively driven microactuators because of the low coupling capacitances. Deep etching techniques, such as lithography, electroplating, and moulding process (LIGA), have been developed to address this problem but are difficult to realise in silicon.

There are several common approaches to the making of microelectromechanical system (MEMS) devices using surface micromachining. The first of these approaches is the sacrificial layer technology for the realisation of mechanical microstructures. The second approach incorporates IC technology and wet anisotropic etching and the third approach uses plasma etching to fabricate microstructures at the silicon wafer surface. These approaches are illustrated in this chapter by a set of ten worked examples that cover a range of microsensor and MEMS devices from cantilever beams, resonant comb structures through to micromotors.

6.2 SACRIFICIAL LAYER TECHNOLOGY

Sacrificial layer technology uses, in most situations, polycrystalline rather than single-crystal silicon as the structural material for the fabrication of microstructures. Low-pressure chemical vapour deposition (LPCVD) of polysilicon (poly-Si) is well known

---

[1] Full three-dimensional structures can now be made using microstereolithography; this important technology is discussed in Chapter 7.

in standard IC technologies (see Chapter 4) and has excellent mechanical properties that are similar to those of single-crystalline silicon[2]. When polycrystalline silicon is used as the structural layer, sacrificial layer technology normally employs silicon dioxide (SiO_2) as the sacrificial material, which is employed during the fabrication process to realise some microstructure but does not constitute any part of the final miniature device.

In sacrificial layer technology, the key processing steps are as follows:

1. Deposition and patterning of a sacrificial SiO_2 layer on the substrate

2. Deposition and definition of a poly-Si film

3. Removal of the sacrificial oxide by lateral etching in hydrofluoric acid (HF), that is, etching away of the oxide underneath the poly-Si structure

Here, we refer to poly-Si and SiO_2 as the structural and sacrificial materials, respectively. The reason for doing this is that in almost all practical situations this is the preferred choice of material combination. However, other material combinations are also being used in surface micromachining, some of which are discussed in Section 6.3.

6.2.1 Simple Process

The simplest of surface-micromachining processes involves just one poly-Si layer and one oxide layer. This process is a one-mask process and is illustrated in Figure 6.1 in which it is designed to form a poly-Si cantilever anchored to a Si substrate by means of an oxide layer. The oxide sacrificial layer is deposited first (Figure 6.1(a)). The poly-Si structural layer is then deposited on top of the oxide. Next, the poly-Si layer is patterned, forming both the cantilever beam and the anchor region (Figure 6.1(b)). Following the poly-Si patterning step, the cantilever beam is released by laterally etching the oxide in an HF solution. The oxide etch needs to be timed so that the anchor region is not etched away (Figure 6.1(c)).

To implement successfully the process described in the preceding paragraph, the release etch must be very carefully controlled. If the release etch is continued for too long a period, the anchor region will be completely cut, resulting in device failure. However, to avoid such a failure, the process may be extended to a two-mask process in which the poly-Si cantilever is directly anchored to the substrate. This two-mask process is shown in Figure 6.2. In this process, the deposited oxide (Figure 6.2(a)) is patterned for an anchor opening by the first mask (Figure 6.2(b)). This is followed by a conformal deposition of poly-Si and subsequent patterning of the poly-Si cantilever beam using the second mask (Figure 6.2(c)). The cantilever is then released by a lateral oxide etch in HF solution (Figure 6.2(d)). Because the anchor region in this case is made out of poly-Si, the oxide release etch poses no threat of device failure.

Modifications of these simple one-mask and two-mask processes are the addition of bushings (or dimples) and/or an insulating layer between the cantilever and the substrate. The process with an added bushing is shown in Figure 6.3. An additional mask is needed to pattern the bushing mould in either the one-mask or in the two-mask process (Figure 6.3(b)). Because of the difference in depth between the bushing mould and the

---

[2] See Appendix G for tabulated properties of silicon.

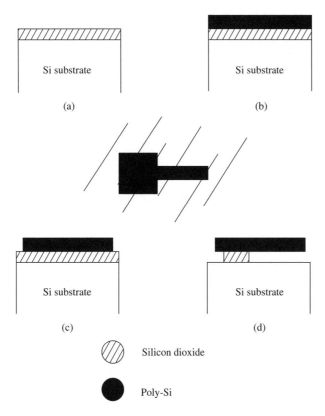

Figure 6.1 Process flow for a polysilicon cantilever anchored to a silicon substrate by means of an oxide layer

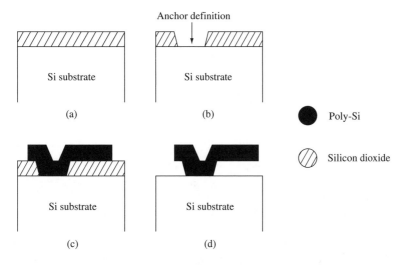

Figure 6.2 Process flow for a polysilicon cantilever anchored directly to a silicon substrate

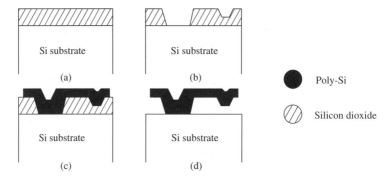

Figure 6.3   Process flow for a polysilicon cantilever with a dimple anchored directly to a silicon substrate

anchor hole, it is not possible to pattern both openings simultaneously using the same mask. The preferable mask sequence for the patterning of the two openings is first the patterning of the bushing mould followed by the anchor region definition since the latter opening is the deepest of the openings. The rest of the process proceeds as described in the one-mask and two-mask processes.

The process in which an insulating layer is incorporated between the cantilever and substrate is illustrated in Worked Example 6.1. The insulating layer that works very well with the poly-Si–oxide combination is silicon nitride (see also Section 6.3). This process and three other worked examples are now presented in turn:

- Freestanding polysilicon beam

- Linear motion microactuator

- Rotor on a centre-pin bearing

- Rotor on a flange bearing

Worked Example E6.1A: Freestanding Poly-Si Beam

Objective (A):

The objective is to fabricate a poly-Si freestanding beam that rests on the surface of a silicon wafer but is raised above it by a silicon nitride–insulating step.

Process Flow (A):

A layer of silicon nitride is first deposited by LPCVD on the surface of a silicon wafer (Figure 6.4(a)). The thickness of the nitride film corresponds to the height of the step on which the freestanding beam base is to rest. This nitride film also acts as a protective layer for the silicon substrate. A layer of sacrificial SiO_2 is then deposited by chemical vapour deposition (CVD) on top of the nitride layer (Figure 6.4(b)) and patterned as shown in Figure 6.4(c). This patterned oxide island is of a thickness that is equal to the height above the nitride-layer surface of the freestanding beam. Poly-Si is then deposited by LPCVD on the patterned oxide as shown in Figure 6.4(d). When the sacrificial SiO_2 island is laterally etched, the freestanding beam shown in Figure 6.4(e) is finally created.

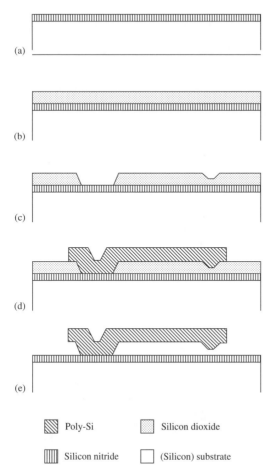

(a)

(b)

(c)

(d)

(e)

⊠ Poly-Si   ▦ Silicon dioxide

▥ Silicon nitride   ☐ (Silicon) substrate

Figure 6.4   Process flow for a freestanding polysilicon cantilever beam anchored to a silicon substrate via an insulating nitride layer

Worked Example E6.1B: Freestanding Poly-Si Beam[3]

Objective (B):

The objective is to fabricate a poly-Si freestanding beam that rests on the surface of a silicon wafer and is on the same level as a nitride layer that has been deposited on top of the silicon wafer.

Process Flow (B):

This second procedure for beam fabrication is based on the localised oxide isolation of silicon (LOCOS) process in which windows are opened in silicon nitride on the silicon substrate and, subsequently, thermal SiO_2 is grown in the openings (Figure 6.5(a)). Using this patterned SiO_2 as a sacrificial layer allows the construction of planar poly-Si beams,

---

[3] For details see Linder *et al.* (1992).

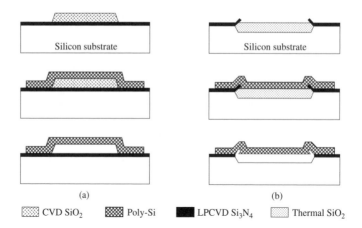

(a)   (b)

▨ CVD SiO$_2$ ▨ Poly-Si   ■ LPCVD Si$_3$N$_4$ ▨ Thermal SiO$_2$

Figure 6.5   Process flow for freestanding polysilicon beams using (a) CVD and (b) thermal silicon dioxide as the sacrificial layer

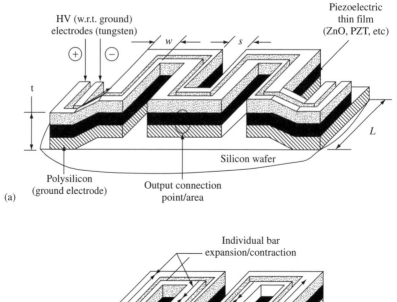

Figure 6.6   Linear motion microactuator (a) perspective view and (b) expansion/contraction and net displacement (Robbins *et al.* 1991)

that is, the beam is on the same level as the surrounding nitride layer (Figure 6.5(b)). However, oxide growth also occurs underneath the nitride at the edges of the windows and thereby pushes up the nitride mask – this is called the 'bird's beak effect.' As a consequence of this effect, steps in the form of spikes are created at the edges of the LOCOS poly-Si beams Figure 6.5(b).

Worked Example E6.2: Linear-Motion Microactuator[4]

Objective:

The use of piezoelectric materials for microactuators is receiving increasing attention as an alternative to electrostatic-based and thermal-based actuation. Perceived advantages of piezoelectric materials include greater energy densities, lower operating voltages, and greater force-generating capabilities than electrostatic actuators. Piezoelectric materials also have faster response times and greater efficiency than thermal actuators. The objective in this example is to fabricate the linear actuator shown in Figure 6.6(a) and (b). The linear-motion actuator uses folded-path geometry as seen in the figure. When a voltage is applied to the dual electrodes on the top surface of a piezoelectric thin film of lead zirconium titanate (PZT), the PZT either expands or contracts along its length, depending on the polarity of its voltage with respect to the poly-Si layer. The alternating expansion and contraction from one bar to the next and the mechanical series connection of the bars cause the net change in the length of each bar to add to that of the other bars (Figure 6.6(b)). This cumulative effect permits a substantial increase in the actuation range of this type of device.

Process Flow:

The process flow is shown in Figure 6.7.

1. The process starts with the deposition and patterning of the sacrificial material (SiO_2) as shown in Figure 6.7(a).

2. This is followed by the deposition of a poly-Si layer as the structural layer. The poly-Si layer is then patterned as shown in Figure 6.7(b).

3. The poly-Si deposition and patterning is followed by a deposition and patterning of PZT (see Figure 6.7(c)).

4. The fourth step is to deposit and pattern the metal electrodes (Figure 6.7(d)), followed by an etch in HF solution to remove the sacrificial oxide (Figure 6.7(e)) and release the mechanical microstructure.

6.2.2 Sacrificial Layer Processes Utilising more than One Structural Layer

The worked examples described in Section 6.2.1 use only one structural (poly-Si) and one sacrificial layer (SiO_2). However, in principle, a surface-micromachining process may comprise more than one structural layer and more than one sacrificial layer. Descriptions of processes with more than one structural or sacrificial layer are given in the following

---

[4] For details see Robbins *et al.* (1991).

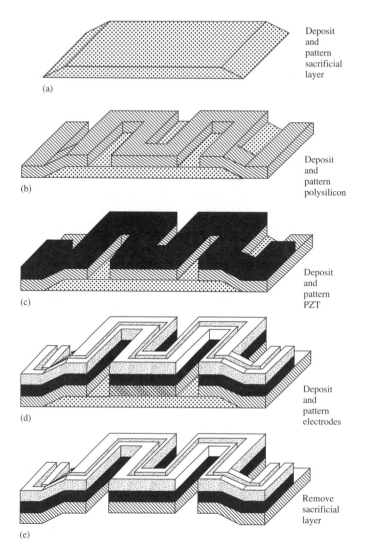

Figure 6.7 Process flow for the linear motion actuator shown in Figure 6.6 (Robbins *et al.* 1991)

two worked examples that are designed to produce rotors with either a centre-pin bearing or a flange bearing:

Worked Example E6.3: Rotor on a Centre-Pin Bearing[5]

Objective:

The objective is to fabricate a disk-shaped rotor made of poly-Si that is free to rotate about a poly-Si centre bearing.

---

[5] For details see Mehregany and Tai (1991).

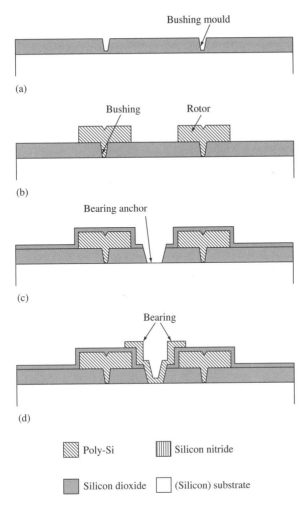

Figure 6.8 Process flow for a rotor on a centre-pin bearing (Mehregany and Tai 1991)

Process Flow:

Cross-sectional schematics of the process are shown in Figure 6.8:

1. The process starts with the deposition of an oxide layer as the first sacrificial layer on which the bushing moulds are patterned (Figure 6.8(a)).

2. Following the patterning of the first oxide layer, the first poly-Si structural layer is conformably deposited and patterned using the second mask as shown in Figure 6.8(b). The bushings are formed automatically on the deposition of the poly-Si layer. These bushings are often necessary to prevent stiction to the substrate when the structure undergoes the sacrificial oxide wet-etching process.

3. Another oxide layer is deposited. This second oxide layer is also used as a sacrificial layer. The second and the first oxide layers are patterned using a third mask that

Figure 6.9   Top view of the rotor structure in Figure 6.8

carries the bearing anchor opening (Figure 6.8(c)). Note that at this stage of the process, the poly-Si rotor is totally encased within the two oxide layers.

4. The second poly-Si structural layer is then deposited and patterned using the fourth mask. This step defines the centre bearing as shown in Figure 6.8(d).

5. The rotor is finally released by etching the two sacrificial oxide layers in HF solution. A top view of the rotor structure is shown in Figure 6.9.

Worked Example E6.4: Rotor on a Flange Bearing[6]

Objective:

The objective is to fabricate a poly-Si disk-shaped rotor that is free to rotate about a poly-Si flange bearing.

Process Flow:

The process flow here is very similar to that in Worked Example 6.3 and uses only minor modifications to those processes already described. A cross-sectional view of the process is shown in Figure 6.10.

1. As in Worked Example 6.3, the process starts with the deposition of an oxide as the first sacrificial layer. This step is followed by the deposition and patterning of the first poly-Si structural layer as shown in Figure 6.10(a).

2. Using the second mask, the first sacrificial oxide is under-etched for flange formation; this is schematically shown in Figure 6.10(b).

3. The second oxide sacrificial layer is then deposited conformably and the bearing anchor region is patterned using the third mask (Figure 6.10(c)).

4. The poly-Si second structural layer is then deposited and the bearing is patterned. The flange in the bearing forms automatically upon deposition of the second poly-Si layer (Figure 6.10(d)). Following this step, the two sacrificial oxide layers are dissolved in an HF solution. When released, the rotor rests on the bearing flange and does not come in contact with the substrate. The rotor now slides on the flange as it rotates.

---

[6] For details, see Mehregany and Tai (1991).

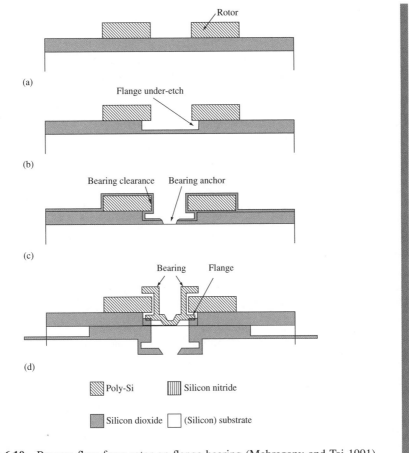

Figure 6.10   Process flow for a rotor on flange bearing (Mehregany and Tai 1991)

6.3 MATERIAL SYSTEMS IN SACRIFICIAL LAYER TECHNOLOGY

An important consideration in the fabrication of an ideal mechanical microstructure is that it is without any residual mechanical stress, so that the films deposited have no significant residual strain. In particular, doubly supported freestanding structures will buckle in the presence of a relatively modest residual compressive strain in the structural material. By choosing the appropriate deposition conditions and by optimising the annealing step, an almost strain-free structural material layer can be obtained[7].

Surface micromachining requires a compatible set of structural materials, sacrificial materials, and chemical etchants. The structural materials must possess the physical and chemical properties that are suitable for the desired application. In addition, the structural materials must have appropriate mechanical properties, such as high yield and fracture strengths, minimal creep and fatigue, and good wear-resistance. The sacrificial materials

---

[7] Test structures are normally placed on the wafers to measure the actual strain.

must also have good mechanical properties to avoid device failure during the fabrication process. These properties include good adhesion and a low residual stress to eliminate device failure by delamination and/or cracking. The etchants must have excellent etch selectivity and must be able to etch off the sacrificial materials without affecting the structural ones. In addition, the etchants must also have appropriate viscosity and surface tension characteristics.

Some of the common material systems that are used in surface micromachining are discussed in the following three sections, followed by one worked example (a silicon condenser microphone).

6.3.1 Polycrystalline Silicon and Silicon Dioxide

This material system has already been mentioned in Section 6.1. The poly-Si/oxide material system is the most common one and uses the poly-Si deposited by LPCVD as the structural material and the thermally grown (or LPCVD) oxide as the sacrificial material. The oxide is readily dissolved in HF solution without the poly-Si being affected. Silicon nitride is often used together with this material system for electrical insulation. The advantages of this material system include the following:

1. Both poly-Si and SiO_2 are used in IC processing; therefore, their deposition technologies are readily available.

2. Poly-Si has excellent mechanical properties and can be doped for various electrical applications. Doping not only modifies the electrical properties but can also modify the mechanical properties of poly-Si. For example, the maximum mechanically sound length of a freestanding beam is significantly larger for phosphorus-doped as compared with undoped poly-Si[8]. However, in most cases, the maximum length attainable is limited by the tendency of the beam to stick to the substrate.

3. The oxide can be thermally grown and deposited by CVD over a wide range of temperatures (from about 200 to 1200 °C), which is very useful for various processing requirements. However, the quality of oxide will vary with deposition temperature.

4. The material system is compatible with IC processing. Both poly-Si and SiO_2 are standard materials for IC devices. This commonality makes them highly desirable in sacrificial layer technology applications that demand integrated electronics.

6.3.2 Polyimide and Aluminum

In this second material system, the polymer 'polyimide' is used for the structural material, whereas aluminum is used for the sacrificial material. Acid-based aluminum etchants are used to dissolve the aluminum sacrificial layer. The three main advantages of this material system are as follows:

---

[8] Caution is needed when doping because it can induce a significant amount of stress.

1. Polyimide has a small elastic modulus, which is about 50 times smaller than that of polycrystalline silicon.

2. Polyimide can take large strains before fracture.

3. Both polyimide and aluminum can be prepared at relatively low temperatures (<400 °C).

However, the main disadvantage of this material system lies with polyimide in that it has unfavourable viscoelastic characteristics (i.e. it tends to creep), and so devices may exhibit considerable parametric drift.

6.3.3 Silicon Nitride/Polycrystalline Silicon and Tungsten/Silicon Dioxide

In the third material system of silicon nitride/poly-Si, silicon nitride is used as the structural material and poly-Si as the sacrificial material. For this material system, silicon anisotropic etchants such as potassium hydroxide (KOH) and ethylenediamine pyrocatechol (EDP) are used to dissolve the poly-Si[9].

In the fourth material system of tungsten/oxide, tungsten deposited by CVD is used as the structural material with the oxide as the sacrificial material. Here again, HF solution is used to remove the sacrificial oxide.

Finally, we give a worked example in which silicon nitride is employed as the structural material as before but as a variant and, unusually, aluminum is used as the sacrificial layer instead of poly-Si.

Worked Example E6.5: **Silicon Condenser Microphone[10]**

Objective:

The objective is to fabricate a silicon condenser microphone using silicon nitride as the structural layer and aluminum as the sacrificial layer. In this case, gold is used to make the electrical tracks.

Process Flow:

1. First, the backside and polished front side of a silicon wafer are covered with 1.8 μm and 0.45 μm thermal oxides, respectively. A square window is then etched on the oxide on the backside. A sacrificial layer of aluminum about 1.0 μm thick is then evaporated onto the polished front side of the wafer and patterned, followed by plasma-enhanced chemical vapour deposition (PECVD) of 1.3 μm of silicon nitride (Figure 6.11(a)).

2. The diaphragm-to-be silicon nitride film is provided with a 30 nm titanium adhesion layer and a 30-nm gold electrode. Holes and V-grooves are then etched anisotropically from the backside using KOH, and the KOH etch stops at the SiO_2 etch-stop layer (Figure 6.11(b)).

---

[9] Details of these etches are given in Section 2.4.
[10] For details see Scheeper *et al.* (1992).

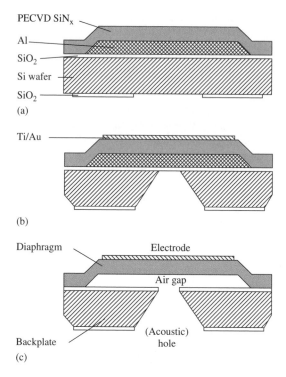

PECVD SiN$_x$

Al

SiO$_2$

Si wafer

SiO$_2$

(a)

Ti/Au

(b)

Diaphragm   Electrode

Air gap

Backplate   (Acoustic) hole

(c)

Figure 6.11   Process flow to fabricate a silicon condenser microphone using silicon nitride as the structural layer and aluminum as the sacrificial layer (Scheeper *et al.* 1992)

3. The SiO$_2$ etch-stop is finally removed in an HF solution. Subsequently, the aluminum sacrificial layer is etched in an H$_3$PO$_4$/HNO$_3$/CH$_3$COOH/H$_2$O mixture at 50 °C to yield the microphone shown in Figure 6.11(c).

6.4 SURFACE MICROMACHINING USING PLASMA ETCHING

Surface micromachining can also be realised using a dry-etching process rather than a wet-etching process. Plasma etching of the silicon substrate, with SF$_6$/O$_2$-based and CF$_4$H$_2$-based gas mixtures, is advantageous because high selectivities for photoresist silicon dioxide and aluminum masks can be achieved. However, when using plasma etching, a large undercut of the mask is generally produced. This is due to the isotropic fluorine atom etching of silicon that is known to be high compared with the vertical etch induced by ion bombardment. In contrast, reactive ion etching (RIE) of poly-Si using a chlorine–fluorine gas combination produces virtually no undercut and produces almost vertical etch profiles with photoresist used as the masking material. Thus, rectangular silicon patterns, which are up to 30 µm deep, can be formed using chlorine–fluorine plasmas out of poly-Si films and the Si wafer surface. A deep etch process is essential for microactuators and, therefore, the deep RIE process is an attractive option. Here, we illustrate its use to make two MEMS devices:

- Centre-pin-bearing side-drive micromotor

- Gap comb-drive resonant actuator

Worked Example E6.6: **Centre-Pin-Bearing Side-Drive Micromotor[11]**

Objective:

The objective is to fabricate a centre-pin, variable-capacitance, and side-drive micromotor, such as the salient-pole and wobble types.

Process Flow:

The flow process in this case adds to what has already been described in the previous Worked Examples 6.3 and 6.4. The rotor is the main component of the micromotor; however, we also need to incorporate stator poles to form the micromotor. A variable-capacitance side-drive micromotor clearly requires electrically conducting materials for both the rotor and the stator. Heavy doping of the poly-Si with phosphorus to form n-type poly-Si satisfies this requirement. In addition, the stator poles need to be electrically isolated from the substrate, the rotor, and one another. This electrical isolation is achieved by LPCVD of an insulating silicon nitride layer. Figures 6.12 and 6.13 show the top and the cross-sectional views, respectively, of the salient-pole and wobble micromotors.
 The process flow is shown in Figure 6.14 and runs as follows:

1. First, an insulation bilayer that consists of 1 μm silicon-rich silicon nitride is deposited by LPCVD over a 1 μm thermally grown oxide and it completely covers it. This insulation bilayer is required to survive the release etching and to withstand high voltages during the operation of the micromotor. The etch rate of silicon-rich silicon nitride in HF solution is negligibly small compared with that of the oxide. Also,

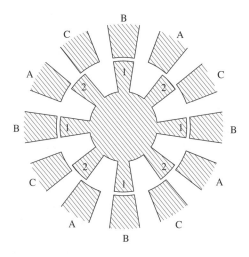

Figure 6.12   Top view of the salient pole micromotor

[11] For details, see Mehregany and Tai (1991).

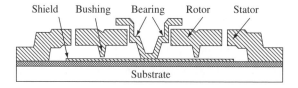

Figure 6.13   Cross-sectional view of the salient pole micromotor

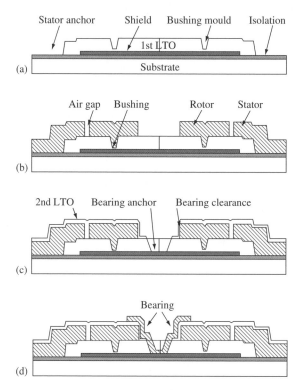

Figure 6.14   Process flow for the centre-bearing side-drive micromotor of Figures 6.12 and 6.13 (Mehregany and Tai 1991)

the bilayer insulation structure proves to be very effective in eliminating electrical breakdown through the substrate and permits operating voltages of up to about 250 V on the stator poles.

2. The second step is to deposit a 0.35 μm-thick heavily doped poly-Si and pattern the poly-Si layer to form the shield. This step is followed by the deposition of the first thermal low-temperature oxide (LTO) sacrificial layer, which is subsequently patterned for the bushings and stator anchors as shown in Figure 6.14(a).

3. A 2.5 μm-thick poly-Si layer is deposited and heavily doped with phosphorus. Phosphorus doping is achieved using POCl$_3$ at 950 °C. The poly-Si is patterned by RIE to form the rotor, stator, and air gaps (Figure 6.14(b)). In the RIE process, a 0.5 μm

thermal oxide is used as an etch mask. The final rotor-stator poly-Si thickness is 2.2 μm because of the thermal oxidation used for the mask formation.

4. A second sacrificial LTO layer is grown; this provides 0.3 μm of LTO coverage on the rotor and stator sidewalls and approximately 0.5 μm of LTO coverage on the top surfaces. The bearing anchor is then defined and etched through the two sacrificial oxide layers down to the electric shield below (Figure 6.14(c)).

5. A 1 μm-thick poly-Si layer is deposited, heavily doped with phosphorus, and then patterned to form the bearing as shown in Figure 6.14(d). At this point, the completed device is immersed in HF solution to dissolve the sacrificial LTOs and release the rotor.

Worked Example E6.7: Gap Comb-Drive Resonant Actuator[12]

Objective:

Comb-drive actuators are widely used, as their output force is easily controlled by the applied voltage, and the output force required to drive passive structures is extracted more easily than that from rotational actuators. A top view of the resonator to be fabricated is shown in Figure 6.15. The drive force of the actuator is obtained by applying a voltage between the stator and the drive electrodes; this force is inversely proportional to the gap width between the electrodes. Therefore, reducing the gap width between the two electrodes is the most effective means of reducing the high drive voltage greater than 25 V that is normally required.

Process Flow:

It is widely acknowledged that masking precisely controlled submicron gaps from thick poly-Si (e.g. ~4 μm as used in this example) using commonly available lithography and etching systems is not an easy process. The process flow described subsequently

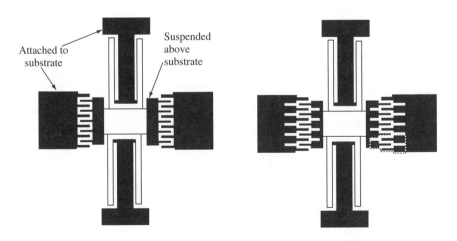

Figure 6.15   Top view of a gap comb-drive resonant actuator (Hirano *et al.* 1992)

---

[12] For details, see Hirano *et al.* (1992).

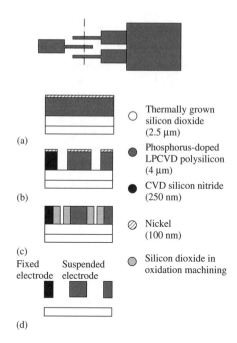

Figure 6.16   Process flow to fabricate the gap comb-drive resonant actuator in Figure 6.15

outlines a method of fabricating submicron gaps for comb-drive actuators called *oxidation machining*. The process flow for the actuator's fabrication is depicted in Figure 6.16.

1. A 4 μm-thick poly-Si (doped by ion implantation and annealed at 1100 °C for 1 hour) is deposited by LPCVD on top of a 2.5 μm thermally grown oxide on the substrate. A 250 nm-thick silicon nitride layer is then deposited by LPCVD over poly-Si, which protects the top surface during the thermal oxidation step. Finally, a 100 nm-thick nickel layer is deposited by vacuum evaporation (Figure 6.16(a)).

2. The shape of the actuator is patterned on the nickel, followed by the wet etching of the nickel film. Using the nickel pattern as a mask, Si_3N_4 and poly-Si are RIE-etched in SF_6 (Figure 6.16(b)).

3. After the removal of the nickel mask, the wafer is cut into 1 cm^2 pieces and the poly-Si is thermally oxidised (Figure 6.16(c)).

4. Figure 6.16(d) shows the actuator's cross section after it is released in HF solution.

6.5 COMBINED IC TECHNOLOGY AND ANISOTROPIC WET ETCHING

Anisotropic wet etching may be combined with an IC process to fabricate freestanding multilayer microstructures without additional masks. Its main merits are low cost and compatibility with standard IC processing. In the first phase, the multilayers are created using IC processing. Usually, the multilayer is composed of the standard insulating and passivating dielectric films, poly-Si layers, and metal layers. The poly-Si and metal layers constitute the active layers and are usually sandwiched between the dielectric films that are necessary for electrical insulation and component passivation. By special design, windows are opened around the multilayer structures for removal of all dielectric layers, thus exposing the silicon surface underneath.

In the second so-called postprocessing phase, the wafers are immersed in anisotropic silicon etchants. Thus, the exposed silicon surface around the multilayer structure is removed, and by under-etching, the microstructures finally become freestanding. Because the active layers are completely contained within the dielectric layers, they are protected against the silicon-etching process.

An alternative approach is to etch anisotropically only the backside of the wafer, that is, use a single-sided etching bath. This technique may be used to make certain structures but tends to be a more time-consuming, and therefore, a more costly process. A worked example is now provided.

Worked Example E6.8: **Integration of Air Gap Capacitor Pressure Sensor and Digital Readout**[13]

Objective:

To fabricate a microstructure that consists of a top plate separated by a small air gap and a bottom plate that has an inlet for pressurised gas (Figure 6.17). The gas pressure moves the top plate upward, increasing the gap, and hence decreasing the capacitance. Metal oxide semiconductor (MOS) electronics[14] are integrated next to the capacitive sensor.

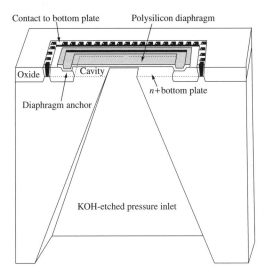

Figure 6.17   Schematic view of an air gap capacitive pressure sensor with integrated digital readout (Kung and Lee 1992)

---

[13] For details, see Kung and Lee (1992).
[14] Standard CMOS process has been described in Section 4.3.

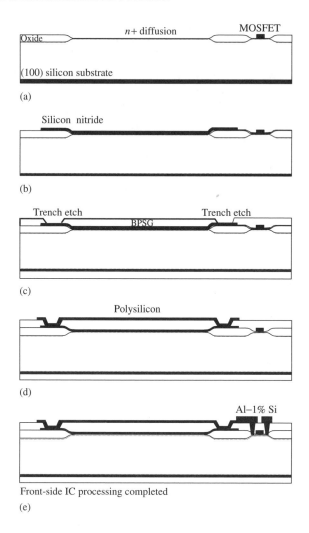

Figure 6.18   Front-side process flow for the resonant actuator given in Figure 6.17

Process Flow (A): Front Side Processing

1. The starting p-type silicon wafer is 500 μm thick, of resistivity 15 to 20 Ω·cm, and of (l00) orientation. An n-channel metal oxide semiconductor (NMOS) process is performed up to the patterning and etching of the poly-Si transistor gates, reoxidation, and junction drive-in of the source–drain regions (Figure 6.18(a)).

2. A 50 nm oxide layer is deposited by LPCVD at 400 °C, followed by a back strip of the oxide and a p^+ boron backside implant (BF_2 at a concentration of $\sim 7 \times 10^{15}$ and energy of ~ 30 keV) to yield a good backside ohmic contact required for the NMOS. Then, 150 nm of Si_3N_4 is deposited by LPCVD at 800 °C, patterned, and plasma-etched using SF_6 (Figure 6.18(b)). The 50 nm oxide is an etch-stop for the Si_3N_4 etch.

3. About 600 nm of borophosphosilicate glass (BPSG) is deposited at 400 °C. BPSG serves as an interlevel dielectric between the poly-Si gate and metal and is also a sacrificial layer that sets the gap. The BPSG is then patterned and plasma-etched using CF_4 that is timed to stop at Si_3N_4 because CF_4 also etches Si_3N_4 (Figure 6.18(c)).

4. A 1 µm layer of poly-Si is deposited by LPCVD at 625 °C, doped with phosphorus using a $POCl_3$ source at 925 °C, and the phosphorus glass wet-etched. A front-coat resist is applied and a backside poly-Si plasma etch is performed using SF_6 stopping on the BPSG layer. The poly-Si layer serves as the top plate of the air gap capacitor and is patterned or plasma-etched in CCl_4 (Figure 6.18(d)).

5. After etching, the process returns to a normal back-end MOS process that consists of patterning or etching contact cuts into the BPSG, followed by a sputter deposition of aluminum (with 1 percent Si) to a thickness of 1.1 µm. The metal is patterned, plasma-etched, and sintered to complete the front-side processing (Figure 6.18(e)). The contact to the n^+ bottom plate is made via n^+ diffusion channels contacted by the metal that surrounds the diaphragm. Contact to the poly-Si is made directly via metal on top of a tab that protrudes from the square structure.

Process Flow (B): Backside Processing

1. Back-to-front infrared alignment is made via two mask-plates in a double-sided mask aligner. After photolithography, the backside is plasma-etched in CH_4, removing both the BPSG and the nitride in one step.

2. The wafer is then immersed in KOH/DI water solution at 80 °C using a custom-designed one-sided etching system. The one-sided etching apparatus is critical because the front circuits would be destroyed without it (Figure 6.l9(a)).

3. Finally, this is followed by a one-sided HF etching of Si_3N_4 and BPSG (Figure 6.l9(b)).

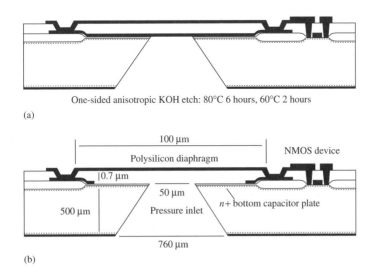

Figure 6.19   Backside process flow for the resonant actuator given in Figure 6.17

6.6 PROCESSES USING BOTH BULK AND SURFACE MICROMACHINING

It is clearly possible to fabricate a variety of microsensor and MEMS devices using either solely bulk-micromachining techniques or surface-micromachining techniques. Some of these devices have been described here via worked examples, and a large number other devices have been described in the literature (Gardner 1994).

However, all such devices suffer from limitations that are inherent in one or the other of these two techniques. Taking advantages of the fabrication possibilities offered by bulk- and surface-micromachining techniques and combining the two techniques in fabricating MEMS devices opens up new opportunities for the fabrication of a new class of MEMS devices that are not possible to fabricate using either of the technique alone. Several devices have been fabricated using a combination of bulk- and surface-micromachining processes. Two of these devices are now described in the Worked Examples 6.9 and 6.10.

Worked Example E6.9: Micronozzles[15]

Objective:

Micronozzles are important in a variety of optical instruments and micromechanical devices, for example, beam-defining elements, high-resolution ink-jet printing heads, microvalves, and flow controllers[16]. The objective in this example is to fabricate a silicon nitride nozzle with a submicron aperture.

Process Flow:

The process flow used in micromachining the micronozzle is depicted in Figure 6.20.

1. The fabrication process starts with a lightly doped (100) Si wafer onto which a composite layer of Si_3N_4 and SiO_2, 160 nm and 500 nm thick, respectively, is deposited. The nitride and oxide are then plasma-etched to form a circular 3 μm-diameter mask (Figure 6.20(a)).

2. Using this mask, the Si is dry-etched to form the Si moulds for the nozzle. The plasma etch is chosen such that the etch is *semianisotropic* for an etch depth of approximately 3.5 μm and a sideways undercutting of approximately 5 μm (Figure 6.20(b)). The oxide mask is first etched in an HF solution and then the nitride mask is removed by further etching Si in a wet isotropic etch that fully undercuts the nitride mask.

3. The next step is to coat the Si moulds in Si_3N_4 to form the nozzles. A thin (few tens of nanometers) layer of padding SiO_2 is deposited, followed by LPCVD of about 350 to 550 nm of Si_3N_4. The deposited layers are highly conformal, replicating even the high-cusped moulds (Figure 6.20(c)). A second layer of SiO_2 is deposited over the nitride. This layer serves as a protective layer for the fine mould tips.

4. To define the nozzle aperture, Si_3N_4 has to be etched back to expose the required portion of the Si mould. This is accomplished by coating the surface with a thick

---

[15] For details, see Farooqui and Evans (1992).
[16] Commercial applications of microstructures are presented in Chapter 1.

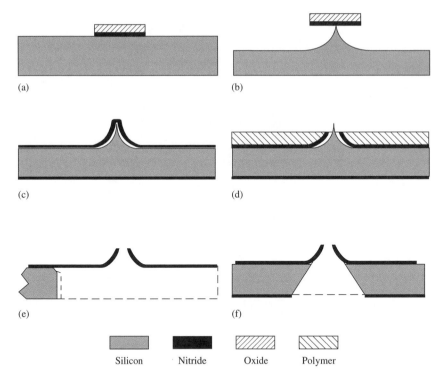

Figure 6.20   Process flow for a micronozzle (Farooqui and Evans 1992)

polymer layer. The coating is then reactively etched back in an oxygen plasma for a predetermined time to expose the required height of the covered tip to give the desired nozzle diameter. This step is followed by the etching of the oxide and the exposed nitride layers (Figure 6.20(d)).

5. The nozzles are finally made freestanding by etching away the Si mould and part of the substrate in KOH followed by a back-sawing technique (Figure 6.20(e)), or a back-masked anisotropic etching (Figure 6.20(f)).

Worked Example E6.10: **Overhanging Microgripper**[17]

Objective:

A top view and a cross-sectional view of the microgripper are shown in Figures 6.21 and 6.22, respectively. The objective is to fabricate a poly-Si microgripper that overhangs from a support cantilever beam that itself protrudes from a silicon die, which serves as the base for the gripper structure. The cantilever is approximately 12 μm thick, 500 μm long, and tapered from a 400 μm width at the base to 100 μm width at the end. This support cantilever accurately locates the overhanging poly-Si microgripper and provides a thin extender for the unit. The poly-Si microgripper is 2.5 μm thick and 400 μm long.

---

[17] For details, see Kim *et al.* (1992).

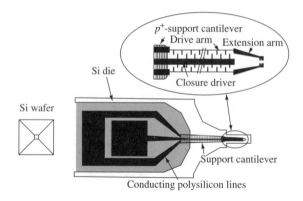

Figure 6.21   Top view of a microgripper

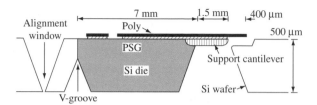

Figure 6.22   Cross-sectional view of a microgripper (Kim *et al.* 1992)

It consists of a closure driver and two drive arms that connect to arms that extend to the gripper jaws. The beam widths for the drive arms and comb teeth are 2 µm, whereas that for the closure drive is 10 µm to provide relative rigidity. When a voltage is applied between the closure driver and drive arms, the drive arms move and close the gripper jaws. Note that the drive arms are at the same electrical potential; this avoids any current flowing between the gripper jaws when they are fully closed and possibly affects the actuation process.

Process Flow:

The process flow for the microgripper's fabrication is shown in Figure 6.23.

1. Using thermally grown oxide as a mask, boron is diffused at 125 °C for 15 hours from a solid dopant source. The oxide mask and the borosilicate glass (BSG) grown during diffusion are subsequently removed (Figure 6.23(a)).

2. A 2 µm thick phosphosilicate glass (PSG) is deposited by LPCVD, followed by LPCVD of 2.5 µm-thick undoped poly-Si deposited at 605 °C. The poly-Si is then patterned by RIE in CCl$_4$ plasma. This step defines the patterns of the gripper and conducting lines. The poly-Si on the wafer backside is subsequently removed (Figure 6.23(b)).

3. Three 2 µm PSG film depositions are made to produce a 6 µm-thick film. These three PSG films and the bottom film are the phosphorus source for diffusion into

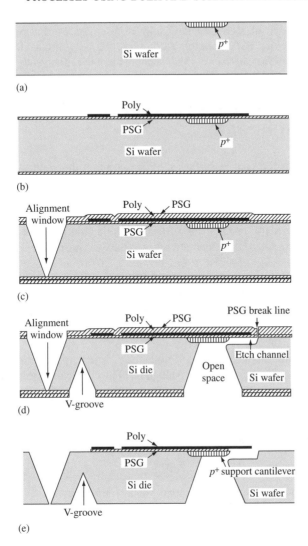

Figure 6.23   Process flow for a microgripper (Kim *et al.* 1992)

the sandwiched poly-Si layer and they also protect poly-Si during subsequent bulk micromachining. Each coating is followed by a one-hour, 1000 °C anneal to drive phosphorus into poly-Si. To make a front-to-back alignment, an alignment window is formed by patterning the PSG on the front side followed by anisotropic etching in EDP (Figure 6.23(c)).

4. Break lines are patterned on the PSG around the poly-Si gripper area. These predetermined break lines effectively prevent any crack in the PSG membrane from propagating to the gripper arms. The PSG on the back wafer can be positioned and patterned in the last masking step. This step is followed by etching in EDP. Convex corners are attached and rounded during the etching. The break lines of the PSG also serve as etching holes on the wafer front side when etched in EDP (Figure 6.23(d)).

5. A final timed etch of sacrificial PSG fully exposes the overhanging poly-Si micro gripper by removing PSG from top and bottom. All poly-Si conducting lines have a PSG layer left underneath them, which anchors them to the substrate (Figure 6.23(e)).

6.7 ADHESION PROBLEMS IN SURFACE MICROMACHINING

After sacrificial layer removal in surface micromachining, the wafers are normally rinsed in deionised (DI) water and dried. The surface tension of the water under the structures pulls them down to the surface of the wafer and, in some cases, causes them to adhere permanently to the wafer surface; this is illustrated in Figure 6.24(a) and (b). This adhesion problem is one of the major problems in surface micromachining and it accounts for almost 90 percent of surface-micromachined structure failures. The problem can be avoided by using thick structural and sacrificial layers; however, this is only possible when the design allows thick layers to be used. In many MEMS applications, the use of very thin layers of structural and sacrificial materials is often necessitated by the design of the device(s).

If longer or thin beams or plates are needed, there are several options to prevent adhesion. One option, as mentioned earlier, is to place small bumps on the bottom surface of the plates to give them a tendency to pull back off the substrate (Figure 6.25). If the spacing of the bumps (also known as dimples) is close enough, the central portion of the plate will in fact never touch the substrate. This critical spacing is a function of the structural and sacrificial layers. For a 2 μm layer, the minimum dimple spacing is approximately estimated to be 50 μm.

There are several other methods of adhesion prevention, and these methods rely on avoiding the problem of surface tension. The following three methods are among the most commonly used methods:

1. Freeze drying (sublimation) of the final rinsing solution; for example, DI water or t-butyl alcohol (Guckel *et al.* 1990).

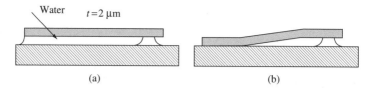

(a)   (b)

Figure 6.24   (a) Wafer that has been washed in water after removal of a sacrificial layer and (b) structural failure as the water tension pulls the structure down and it sticks to the surface

Figure 6.25   A small bump is placed at the bottom of the structure to prevent adhesion

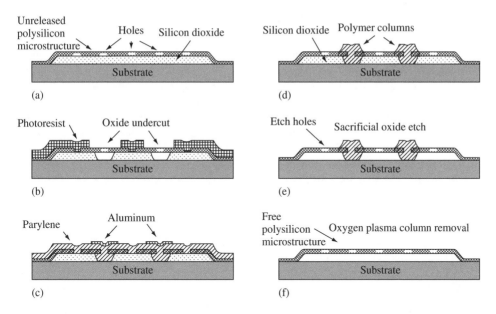

Figure 6.26 Polymer supports are added to the structure *before* the release etch in order to avoid structural failure upon plasma release etch

Table 6.1 Comparison between bulk and surface micromachining technologies

Bulk micromachining		Surface micromachining	
Advantages	Disadvantages	Advantages	Disadvantages
Well established (since 1960)		Uses several materials and allows for new applications	Relatively new (since 1980)
Rugged structures that can withstand vibration and shock	Large die areas that give it high cost	Small die area that makes it cheaper	Less-rugged structures with respect to vibration and shock
Large mass/area (suitable for accelerometers and capacitive sensors)	Not fully integrated with IC processes	Fits well within IC process	Small mass/area, which would typically reduce sensitivity
Well-characterised material (i.e. Si)	Limited structural geometry possible	Wider range of structural geometry	Some of the materials are not very well understood

2. Gradually replacing acetone with photoresist and then spinning and ashing the resist (Hirano *et al.* 1991).

3. Using an integrated polymer support structure during release etching and then ashing in oxygen plasma (Mastrangelo and Saloka 1993). This process is schematically shown in Figure 6.26(a) through to 6.26(f).

6.8 SURFACE VERSUS BULK MICROMACHINING

In this chapter and the previous one, we have described the use of bulk and surface micromachining to fabricate microsensors and MEMS devices. Table 6.1 summarises the relative advantages and disadvantages of these two technologies. Perhaps the most attractive feature is whether the technique may be readily combined with a standard IC process – ideally as simple pre- or postprocessing steps. In this respect, a front side surface micromachining is perhaps the most attractive current option. However, the possibility of a front side bulk etch of a silicon-on-insulator (SOI) CMOS device may become a cost-effective and desirable option in the near future. The use of a micromachined SOI membrane to make a smart gas sensor is discussed later (Section 15.2).

REFERENCES

Farooqui, M. M. and Evans, A. G. R. (1992). "Microfabrication of submicron nozzles in silicon nitride," *J. Microelectromech. Syst.*, **1**, 86–88.

Gardner, J. W. (1994). *Microsensors: Principles and Applications*, John Wiley & Sons, Chichester, p. 331.

Guckel, H., Sniegowski, J. J., Christenson, T. R. and Raissi, F. (1990). *Sensors and Actuators A*, **346**, 21–23.

Hirano, T., Furuhata, T., Gabriel, K. J. and Fujita, H. (1991). Technical Digest of 6th Int. Conf. Solid-State Sensors and Actuators, San Francisco, USA, pp. 63–66.

Hirano, T., Furuhata, T., Gabriel, K. J. and Fujita, H. (1992). "Design, fabrication, and operation of submicron gap comb-drive microactuators," *J. Microelectromech. Syst.*, **1**, 52–59.

Kim, C-J., Pisano, A. P., Muller, R. S. (1992). "Silion-processed overhanging microgripper," *J. Microelectromech. Syst.*, **1**, 31–35.

Kung, J. T. and Lee, H-S. (1992). "An integrated Gir-gap-capacitor pressure sensor and digital read-out with sub-100 attofarad resolution," *J. Microelectromech. Syst.*, **1**, 121–129.

Linder, C. *et al.* (1992). "Surface micromachining," *J. Micromech. Microeng.*, **2**, 122–332.

Mastrangelo, C. H. and Saloka, G. S. (1993). "An investigation of micro structures, sensors actuators machines and systems," *J. Microelectromech. Syst.*, 77–81.

Mehregany, M. and Tai, Y-C. (1991). "Surface micromachined mechanisms and micromotors," *J. Micromech. Microeng.*, **1**, 73–85.

Robbins, W. P. *et al.* (1991). "Design of linear-motion microactuators using piezoelectric thin films," *J. Micromech. Microeng.*, **1**, 247–252.

Scheeper, P. R., van der Donk, A. G. H., Olthuis, W. and Bergveld, P. (1992). "Fabrication of silicon condenser microphones using single wafer technology," *J. Microelectromech. Syst.*, **1**, 147–154.

7

Microstereolithography for MEMS

7.1 INTRODUCTION

Stereolithography (SL) was introduced in 1981 by different teams in the USA (Hull 1984), Europe (Andre *et al.* 1984), and Japan (Kodama 1981). SL is a rapid prototyping, and manufacturing technology that enables the generation of physical objects directly from computer-aided design (CAD) data files.

The stereolithographic process begins with the definition of a CAD model of the desired object, followed by a slicing of the three-dimensional (3-D) model into a series of closely spaced horizontal planes that represent the x-y cross sections of the 3-D object, each with a slightly different z coordinate value. All the 3-D models are next translated into numerical control code and merged together into a final build file to control the ultraviolet (UV) light scanner and z-axis translator. The desired polymer object is then 'written' into the UV-curable resist, layer by layer, until the entire structure has been defined (Figure 7.1).

The first commercially available SL system was produced by 3D Systems in 1987. SL is now widely used in both the automotive and aerospace industries to fabricate industrial products from the basic design to 'show and tell' parts at low cost – before the parts are machined in the conventional manner.

This chapter describes in detail the different stereolithographic techniques and how they can be used to make miniature parts (or microparts). When SL is used to make microparts, it is usually referred to as microstereolithography (MSL). Here, we show the importance of MSL as an enabling technology to make parts for microelectromechanical system (MEMS) devices in materials other than silicon. It is thus a *complementary* technology to the bulk- and surface-micromachining techniques described in Chapters 5 and 6, respectively. Furthermore, MSL permits the fabrication of true 3-D devices, on the micron-to-millimetre scale, including curvilinear and re-entrant microstructures that are difficult to make using conventional silicon micromachining. It is sometimes referred to as the 'poor man's LIGA process'!

The next section describes the fundamental concepts of photopolymerisation and SL to produce an MSL system. The concept of photopolymerisation will already be familiar to those who have read Chapters 2, 4, and 6, because it is also used in the processing of conventional electronic materials and silicon microtechnology (e.g. see Sections 2.3, 4.3, and 6.2).

7.1.1 Photopolymerisation

SL is a photopolymerisation process, that is, a process that joins together a number of small molecules (monomers) in a resin or resist to make larger molecules (polymers), which usually use UV radiation to polymerise (or cure) the resist material. Different types of photopolymers are used in SL prototyping and these are commonly based on either free-radical photopolymerisation or cationic photopolymerisation. The generalised molecular structure of monofunctional acrylate, epoxy, and vinyl ether, which are the three main photopolymer systems, are shown in Figure 7.2 (Jacobs 1996).

In general, the photopolymerisation process is initiated by the incidence of photons generated by an UV light source. The breaking of the C−C double bond (acrylate and vinyl ether) or ring (epoxy) in the monomer enables monomer units to link up and form a chain-like structure. The cross-linked polymer chain finally forms when the chain propagation is terminated. To illustrate this process, the various steps in a free-radical polymerisation sequence are given in Figure 7.3. The selection of the photopolymer depends on the required dimensional accuracy and mechanical properties of each individual photopolymer formulation (Jacobs 1996).

The curing depth and line-width of the photopolymerisation process are the two most critical parameters and these need to be carefully controlled in the SL process. In principle, the curing depth and line-width can be determined from the beam distribution and the absorption of radiation in the resist (Figure 7.4).

Figure 7.1   Basic principle of stereolithography: the writing of 3-D patterns into a series of layers of UV-curable resist at different heights

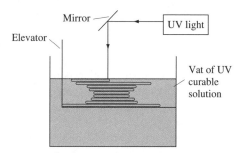

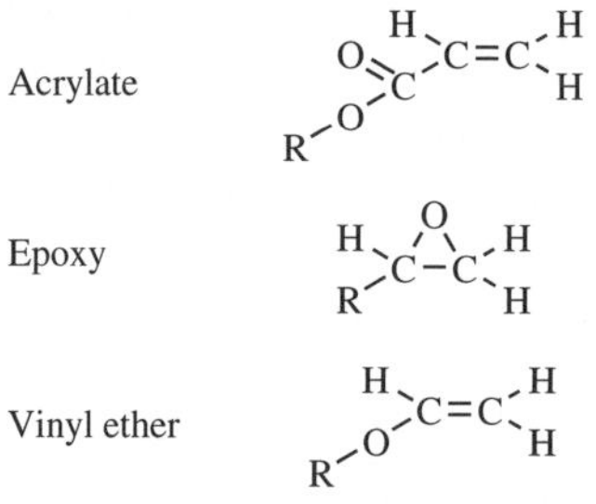

Figure 7.2   Generalised molecular structure of three polymers used in MSL: acrylate, epoxy, and vinyl ether. From Jacobs (1996)

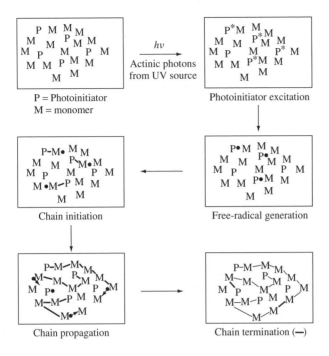

Figure 7.3 Example of a simplified free-radical photopolymerisation sequence used in MSL

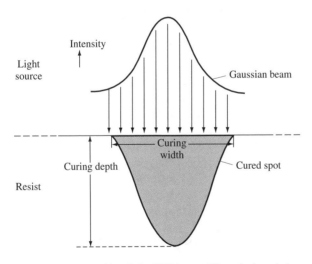

Figure 7.4 Theoretical intensity profile of the UV beam (Gaussian) and the resultant spot cured within the photoresist

The fraction of light transmitted through an absorbing system is given by the Beer–Lambert law (see for example, Wayne (1998)):

$$I_t/I_0 = \exp(-\alpha C d) \tag{7.1}$$

where I_t and I_0 are the transmitted and incident light intensities, C is the concentration of the absorber, d is the distance the light has passed into the absorber, and α is a constant proportional to the absorption coefficient of the absorbing material at the wavelength used.

In the SL process, we consider the UV beam to have a Gaussian profile, which is scanned in a straight line at constant velocity v_s along the x-axis, which is in the surface of the photopolymer, as shown in Figure 7.5 (Jacobs 1992).

The irradiance (radiant power per unit area), $I(x, y, z)$ at any point within the resin can be related to the irradiance incident on the resin surface, $I(x, y, 0)$ using the Beer–Lambert law

$$I(x, y, z) = I(x, y, 0) \exp\left(-z/d_p\right) \tag{7.2}$$

where d_p is the penetration depth of the beam, which depends on the wavelength, the absorption coefficient, and the initiator concentration (from Equation (7.1)). It is defined as the depth of resin that results in a reduction in the incident irradiance by a factor of $1/e$.

Here, we assume a Gaussian radial distribution of the light intensity in the plane orthogonal to the optical axis $O\text{-}X$ shown in Figure 7.6. The relationship between the power P and irradiance I of an axisymmetric Gaussian light source (i.e. laser beam) is

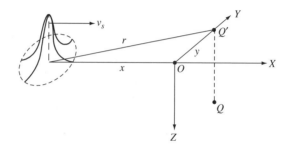

Figure 7.5   Mathematical representation of the line scan of a Gaussian beam. From jacobs (1992)

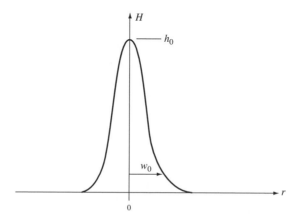

Figure 7.6   Definition of the half-width w_0 of a Gaussian light beam

thus simply

$$P = \int_{r=0}^{r=\infty} I(x, y, 0) 2\pi r \, dr \text{ with } I(x, y, 0) = I(0, 0, 0) \exp(-2r^2/w_0^2) \quad (7.3)$$

where r is the radial distance and r^2 is given by $x^2 + y^2$ according to the Pythagoras theorem and w_0 is defined as the half-width of the Gaussian beam. Because the power of the laser beam is normally known, we can integrate Equation (7.3) to find that $I(0,0,0)$ is equal to $2P/(\pi w_0^2)$. Substituting this result back into Equations (7.2) and (7.3) allows us to define the irradiance at an arbitrary point inside the resin as

$$I(x, y, z) = (2P/\pi w_0^2) \exp(-2r^2/w_0^2) \exp(-z/d_p) \quad (7.4)$$

To derive the working curve of the SL process, the energy per unit area in the beam must be determined and is defined as

$$E(x, y, z) = \int_{t=-\infty}^{t=+\infty} I(x, y, z) \, dt \text{ where } dt = dx/v_s \quad (7.5)$$

where v_s is the scanning speed along the x-axis. The energy along the x-axis is thus given by

$$E(y, z) = \sqrt{\frac{2}{\pi}} \frac{P}{w_0 v_s} \exp\left(-2y^2/w_0^2\right) \exp\left(-z/d_p\right) \quad (7.6)$$

The maximum curing depth is obtained at the point when $y = 0$; the exposure at this point is taken as the critical exposure E_c; thus, we obtain the working curve equation for curing depth C_d

$$C_d = d_p \ln(E_{max}/E_c) \text{ and } E_{max} = E(0, 0) = \sqrt{\frac{2}{\pi}} \frac{P}{w_0 v_s} \quad (7.7)$$

The maximum cured line-width l_w ($2y_{max}$) is obtained at the point of $z = 0$, and is written as

$$l_w = w_0 \sqrt{2 \ln(E_{max}/E_c)} \quad (7.8)$$

It is assumed here that the laser exposure at the point with maximum depth is equal to that at the point with maximum line-width and is taken as critical exposure E_c. It is known that the exposure level at which the gel point (i.e. polymerisation) is reached should be slightly higher than the threshold exposure known as the *critical exposure*.

The relationship between the curing depth and line-width is finally obtained and is given by

$$l_w = 2w_0 \sqrt{C_d/2d_p} \quad (7.9)$$

It is essential that the precise working curves of the curing depth and line-width are known in an SL process. More sophisticated models that take account of other factors, such as dark polymerisation, diffusion of the light source, with calibration performed for each SL system, and photocurable material may be constructed.

7.1.2 Stereolithographic System

All SL systems share the same basic elements or subsystems: the CAD design, layer-preparation functions, and a laser-scanning or imaging system. A typical SL system is schematically shown in Figure 7.7.

The CAD model is a 3-D representation of the object and is exported, most commonly, into a file with a format called *STL*. Slice cross sections are produced from this file and converging parts placed into a 'build' file. Process control software is then used to operate the SL writing process according to the build file. Detailed information on CAD design, STL file formats, part preparation convergence, contour slice, and so on can be found in the references (Jacobs 1992, 1996).

A uniform, flat resin layer of the desired thickness is prepared for curing through standard layer preparation. The liquid resin surface will be the foundation of each layer of the SL model; therefore, the quality of the resin surface has a direct impact on the accuracy of the final structure. Recoating blades, fluidic pumps, surface-finding lasers, and robotic systems have all been considered to achieve a highly uniform resin surface. The general requirements of the SL process are simple in concept, but their implementation is quite challenging. The SL system should satisfy all the following requirements (Jacobs 1996):

- The resin surface must be maintained precisely at the focal plane of the imaging system

- The resin surface must be uniformly flat, leveled, and free of extraneous features

- The resin surface must be a controlled distance above the previously built cross section of the part

The recoating and leveling systems work together and aim to achieve the desired goal of providing a flat layer of liquid resin of proper thickness. A leveling system is used to sense the resin height and permit adjustments. A level-finding system can vary from a

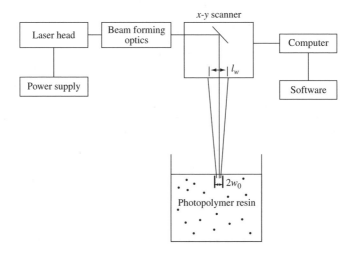

Figure 7.7   Block diagram showing the basic elements of an SL system. From Jacobs (1992)

simple mechanical float mechanism through to an optical proximity sensor system, which monitors the current resin height and allows this to be fed back into a computer-based closed-loop control system.

A number of other steps are required in the coating process, such as the lowering of the part in the vat full of resin, the wiping away of any excess materials, and the smoothing of the remaining materials to provide the desired thickness of liquid resin above the previously solidified layer.

The thickness of the layer typically ranges from 100 to 500 μm. The recoater must be controlled very precisely to achieve thinner layers, because a recoating error of 25 μm becomes very significant when building with thickness below 100 μm (Jacobs 1996). When a smooth, uniform, and accurate coating of the liquid resin has been achieved over the previously solidified polymer layer, the process is not finished until sufficient resin remains within the vat to compensate for any shrinkage that can occur during curing.

An imaging system for SL includes a light source (laser or lamp), beam delivery, and focusing elements (Figure 7.7). The laser (or lamp) chosen for the system must be appropriate for the resin to be used. Wavelength, output beam shape, and power available are all important characteristics (see equations in Section 7.1.1). Beam delivery elements are employed to fold the path of the laser beam and therefore make the SL system as compact as possible.

A typical SL system employs two orthogonally mounted, servo-controlled, galvanometer-driven mirrors to direct the laser beams onto the surface of the vat. The beam is passed though a focusing objective and then hits the resin surface. The beam exposure is controlled by a shutter according to an on–off command generated by the build file. A mechanical shutter requires typically about 1 ms to actuate, and so has now been replaced by an acoustic optical modulator that has a much lower actuation time of about 1 μs, thus allowing a much faster fabrication process. The time needed to write a layer is always a critical parameter of any beam writer because it relates both to throughput and cost.

Applications of SL vary considerably from the quick-cast tooling through to structural analysis. Consequently, SL can speed up product development and improve product quality through superior design and prototyping.

7.2 MICROSTEREOLITHOGRAPHY

The principle of MSL is basically the same as that of SL (Section 7.1), except that the resolution of the process is lower. In MSL, a UV laser beam is focused down to a 1 to 2 μm-diameter spot that solidifies a resin layer of 1 to 10 μm in thickness, whereas in conventional SL, the laser beam spot size and layer thickness are both on the order of 100 to 1000 μm. Submicron control of both the x-y-z translation stages and the UV beam spot enables the precise fabrication of complex 3-D microstructures.

MSL is also called *microphotoforming* and was first introduced to fabricate high aspect ratio and complex 3-D microstructure in 1993 (Ikuta and Hirowatari 1993). In contrast to conventional subtractive micromachining, MSL is an additive process, and therefore, it enables the fabrication of high aspect ratio microstructures with novel smart materials. The MSL process is, in principle, compatible with silicon microtechnology and therefore post-CMOS batch fabrication is also feasible (Ikuta *et al.* 1996; Zissi 1996).

Different MSL systems have been developed in recent years to improve upon their precision and speed. Basically, scanning MSL (Ikuta and Hirowatari 1993; Zissi *et al.* 1996; Katagi and Nakajima 1993; Zhang *et al.* 1999; Maruo and Kawata 1997) and projection MSL (Bertsch *et al.* 1997; Nakamoto and Yamaguchi 1996; Monneret *et al.* 1999) are the two major approaches that have been taken. Scanning MSL builds the solid microparts in a point-by-point and line-by-line fashion (Figure 7.8), whereas projection MSL builds one layer with each exposure (Figure 7.9), thus speeding up the building process by a significant factor (Beluze *et al.* 1999). The details of the two approaches are presented in Sections 7.3 and 7.4.

Another research effort in MSL is the incorporation of a broad spectrum of materials to create MEMS with new special functions. The MSL fabrication of polymer

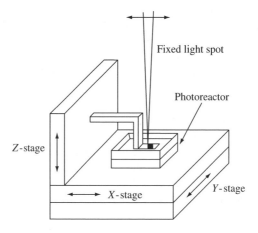

Figure 7.8 Principle of scanning MSL, that is, a vector-by-vector approach. From Beluze *et al.* (1999)

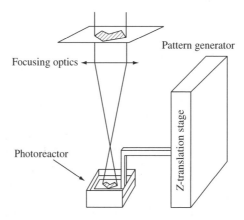

Figure 7.9 Principle of projection MSL, that is, a plane-by-plane approach. From Beluze *et al.* (1999)

microparts and their subsequent electroplating to form metallic microparts have been reported (Ikuta and Hirowatari 1993; Ikuta *et al*. 1996; Zissi *et al*. 1996; Katagi and Nakajima 1993; Maruo and Kawata 1997; Bertsch *et al*. 1997; Nakamoto and Yamaguchi 1996; Monneret *et al*. 1999). Functional polymer (e.g. conducting polymer) microparts possess the unusual characteristics of high flexibility, low density, and high electric conductivity (Ikuta and Hirowatari 1993). Ceramic microstructures have also been fabricated by MSL using both structural and functional ceramic materials (Zhang *et al*. 1999; Jiang *et al*. 1999). The use of MSL to make both ceramic and metallic microparts is discussed in Section 7.7.

7.3 SCANNING METHOD

Most MSL equipment developed today are based on the *scanning* method (Figure 7.8), which is the method employed in conventional SL and is widely used in the industry. With the scanning method, a well-focussed laser beam with beam spot size around 1 micron is directed onto the resin surface to initiate the polymerisation process. A 3-D microstructure is built up by the repeated scanning of either the light beam or the work piece layer by layer.

7.3.1 Classical MSL

The classical setup for SL is shown in Figure 7.10, in which the laser beam is deflected by two low-inertia galvanometric *X-Y* mirrors and is focused by a dynamic lens onto the surface of the workpiece in a photoreactor (vat) that contains a UV photoinitiator (Bertsch *et al*. 1997). An acousto-optical shutter switches the laser beam on and off between the polymerised segments. Small objects can be made with this type of apparatus, but improvements in the beam focus are necessary to obtain the higher resolution needed for microfabrication (less than 100 µm).

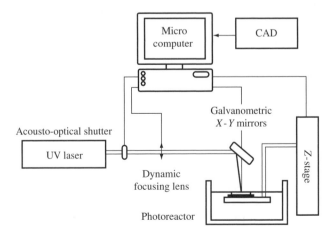

Figure 7.10 Classical apparatus used to perform SL. From Bertsch *et al*. (1997)

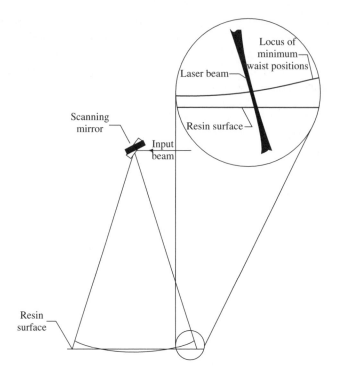

Figure 7.11 Spherical beam swept over a flat resin plane in classical SL. From Jacobs (1996)

It is believed that, in classical MSL, too many mobile optical elements lead to poor focusing. As shown in Figure 7.11, the free liquid resin surface defines a horizontal plane, whereas the motion of the focused laser spot describes a portion of the surface of a sphere. Thus, theoretically, the spot size varies during the scanning process. In classical SL, this defocussing may not be critical as a larger beam size (>100 μm) and longer focal depth are used and an acceptable beam size and shape can be maintained across the flat resin surface (Jacobs 1996). In MSL, however, the depth of focus is relatively short when striving for high-resolution fabrication; in other words, to maintain the focus point at the resin surface is highly critical, and this is the reason classical MSL apparatus employs a dynamic lens for focusing. Even so, focusing still remains a major concern from recent reports (Bertsch *et al.* 1997).

Although this classical MSL system possesses some focusing problems that prevent high-resolution fabrications, it has a fast fabrication speed. Therefore, classical MSL is still an attractive option, as fabrication speed is always the first consideration of production.

7.3.2 IH Process

A series of integrated harden (IH) polymer SL processes have been developed by Ikuta and these are based on the classical scanning MSL method. The IH processes are designed to overcome the beam-focussing problem present in a classical MSL system.

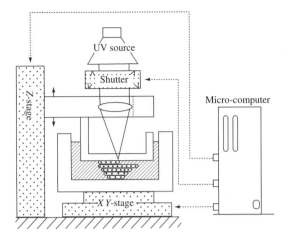

Figure 7.12 Schematic diagram of the apparatus used for an IH process. From Ikuta and Hirowatari (1993)

The apparatus for an IH process is shown in Figure 7.12, where the light source is a UV lamp (Xenon lamp) and where the beam is focused onto the resin surface through a glass window. The focal point of the apparatus remains fixed during the fabrication and the work piece is in a container that is attached to an X-Y stage, which provides the scanning that was realised previously by the galvanometric mirrors. Using an X-Y stage to move the work piece rather than the galvanometric mirrors to deflect beam leads to a smaller focal spot and hence higher resolution. Now, there is no need for a dynamic focus lens because the focal point is fixed. The glass window is attached to the Z-stage so that the layers of precise thickness can be prepared.

The IH process can be used to fabricate polymeric microstructures, whereas metal microstructures can be obtained by first making a polymer micromould, metal-plating, and finally removing the polymer (Figure 7.13, Ikuta and Hirowatari 1993).

The specifications of a typical IH process are listed below:

- 5 μm spot size of the UV beam

- Positional accuracy is 0.25 μm (in the x-y directions) and 1.0 μm in the z-direction

- Minimum size of the unit of harden polymer is 5 μm × 5 μm × 3 μm (in x, y, z)

- Maximum size of fabrication structure is 10 mm × 10 mm × 10 mm

With this IH process, some high aspect ratio and truly 3-D polymer microstructures, such as micropipes and microsprings (Figure 7.14), have been successfully fabricated. In addition, a metallic 3-D micropart was obtained after metal-plating of a polymer cast, as shown in Figure 7.15 (Ikuta and Hirowatari 1993).

The characteristics of an IH process can thus be summarised as follows:

- It is capable of making true 3-D and high aspect ratio microstructures

- It works with different materials

- It requires no mask plates and is thus a cost-effective process

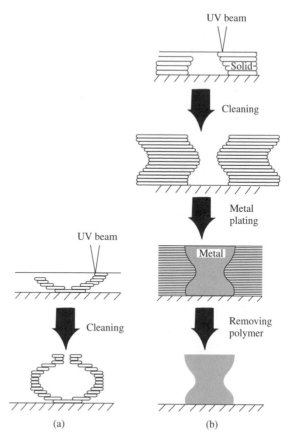

Figure 7.13 Basic steps in the IH process used to make a metallic micropart. From Ikuta and Hirowatari (1993)

- It has a medium range of accuracy (3 to 5 μm)
- It permits desktop microfabrication of parts

It should be pointed out here that the fabrication speed of the IH process is slower than classic MSL because the scanning speed of the *X-Y* stage and container is less than that of the galvanometric scanner.

7.3.3 Mass-IH Process

The slow fabrication speed of the IH process is a concern, and so the so-called mass-IH process was proposed by Ikuta in 1996 to demonstrate the possibility of realising the mass production of 3-D microfabrications by MSL (Ikuta *et al.* 1996).

The mass-IH process is schematically shown in Figure 7.16 and utilises fibre-optic multibeam scanning. The basic idea is that an array of single-mode optical fibres can be used to enhance the speed of device production.

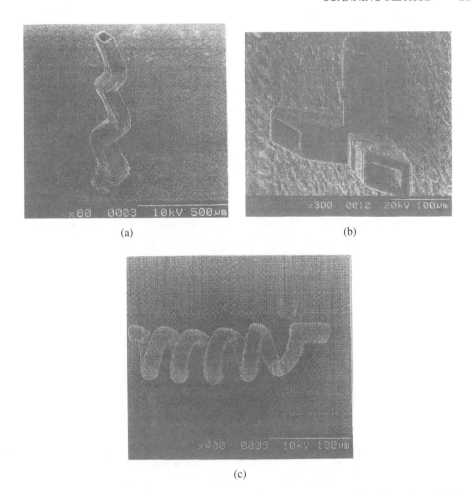

(a)   (b)

(c)

Figure 7.14   Some 3-D polymeric microparts made by an IH process: (a) bending pipe of 100 μm by 100 μm cross section and length 1000 μm; (b) 3-D connected pipes with an inner diameter of 30 μm; and (c) microcoil spring of diameter 50 μm and height 250 μm

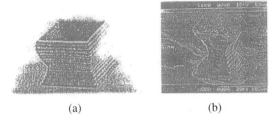

(a)   (b)

Figure 7.15   A polymer mould (a) used to make a metal micropillar of dimensions 60 μm by 60 μm by 100 μm (b) in an IH process. From Ikuta and Hirowatari (1993)

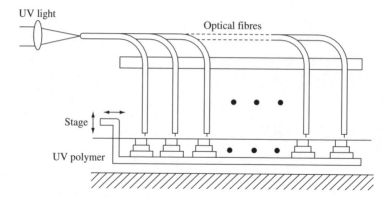

Figure 7.16   The mass-IH process designed to increase the speed of MSL through the use of an array of optical fibres. From Ikuta *et al.* (1996)

Figure 7.17   Prototype of a fibre-optic array (4 mm by 4 mm). From Ikuta *et al.* (1996)

An array with five single-mode optical fibres (4 μm in core diameter) was used to demonstrate the mass-IH process by Ikuta (Ikuta *et al.* 1996) (Figure 7.17). Other specifications of the apparatus are the same as those of the IH process. As an example, five micropipes with lateral windows were made within a period of 40 minutes (Figure 7.18), in which each pipe had a cross section of 250 μm by 250 μm and is 900 μm high. The thickness of each stack layer is 30 μm. Thus, the fabrication speed was significantly improved by this arrangement of 'parallel-processing.' However, this mass-IH process needs to be developed further using optical-fibre arrays and by improving the resolution.

7.3.4 Super-IH Process

Both the IH and mass-IH processes are based on a scanning method with layer preparation, which shares the same basic principle as conventional SL. Two of the problems associated with this kind of layer-by-layer fabrication are as follows:

- The depth resolution is limited by the thickness of the layer that is stacked up

- Viscous UV-curable monomers can deform and hence damage the solidified microstructures

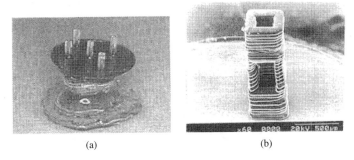

(a)   (b)

Figure 7.18   Some microparts fabricated using the mass-IH process of Ikuta: (a) five 3-D micro-structures on a table of 4.5 mm diameter and (b) a micropipe with a lateral window

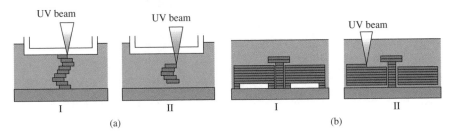

Figure 7.19   (a) Comparison of the solidification processes of conventional MSL (I) and the new super-IH process (II) and (b) processes needed to make movable gear and shaft (I). Conventional MSL needs support structure, whereas the new super-IH process (II) does not need a support

It has been shown by Ikuta *et al.* (1998) that a significant surface tension of the liquid monomer decreases the precision of the fabrication process. The super-IH process has been developed to address this problem and can be used to solidify the monomer at a specific point in 3-D space by focusing a laser beam into a *liquid* UV-curable monomer. The 3-D microstructure is now fabricated by scanning the focused spot in all three dimensions inside the liquid, thus obviating the need for any supports or sacrificial layers. Figure 7.19 illustrates the difference between the conventional MSL processes and the super-IH process (Ikuta *et al.* 1998) in that it does not need a support (a) and therefore can be used to make, for example, a movable gear and shaft in one step. Because the beam is writing directly into the resist, the effects of the monomer viscosity and surface tension are greatly diminished.

A schematic diagram of the experimental setup of the super-IH process is shown in Figure 7.20 and consists of a He–Cd laser of 442 nm wavelength, an optical shutter, a galvano-scanner set, an X-Y-Z stage, an objective lens, and a computer (Ikuta *et al.* 1998). The laser beam is focused inside the monomer volume by coordinating the beam scanning and Z-stage movements; thus, the 3-D structures are formed inside the liquid.

The properties of this system must be precisely tuned to ensure that polymerisation only takes place at the point of focus. The UV monomer system used in the super-IH process is a mixture of urethane acrylate oligomers, monomers, and photoinitiators.

Some interesting polymer microparts with free-moving elements have been fabricated using the super-IH process (Ikuta *et al.* 1998). Figure 7.21 shows a scanning electron

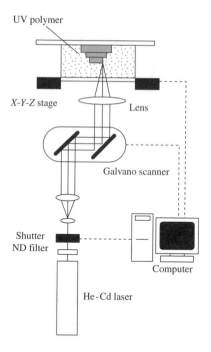

Figure 7.20   Experimental setup of the super-IH process

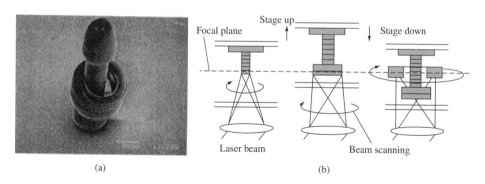

Figure 7.21   (a) SEM image of a microquoits made of solidified polymer using the super-IH process and (b) basic fabrication steps required to make a microquoits. From Ikuta *et al.* (1998)

microscopy (SEM) image of a microquoit (a), and the fabrication steps to make it (b). The focal point is scanned in a circular motion and the radius steadily increases as the stage moves up or down with a constant interval along the optical axis. The shaft was made first from the bottom to the top with a diameter of 10 μm and then the freely movable ring was formed at the middle of the shaft with an inner diameter of 24 μm. The ring is able to remain suspended freely in the liquid because of its high viscosity. Clearly, this process can make microparts with freely moving elements in a single step with no postprocess assembly necessary – this permits the fabrication of more complex MEMS device than that with silicon micromachining.

The resolution of the super-IH process is excellent and is typically less than 1 μm. The fabrication speed can be increased by operating the galvano scanning mirror and X-Y-Z stages together. The disadvantages are that the optics system is more expensive than both the IH and mass-IH processes, and specialised monomer systems need to be developed for each application.

7.4 TWO-PHOTON MSL

As mentioned earlier, conventional MSL is limited in terms of the minimum thickness of the resin layers possible because of viscosity and surface tension effects. In contrast, the two-photon MSL process (like the super-IH) does not have this problem because the resin does not need to be layered.

When a laser beam is focused on a point with a microscope objective lens as shown in Figure 7.22 (Maruo and Kawata 1998), the density of photons decreases with the distance away from the focal plane, but the total number of photons in the beam at every cross section remains the same (see Figure 7.22(b)). Thereafter, the resin is solidified completely in the illuminated region even beyond focal point, leading to a poor resolution; this means that the linear response of the materials to the light intensity based on a single-photon absorption does not have optical sectioning capability. On the other hand, if the material response is proportional to the square of the photon density, the integrated material response is enhanced greatly at the focal point (see Figure 7.22(c)), and, therefore, the polymerisation based on two-photon absorption occurs only in a small volume within the focal depth. Normally, the beam power of the laser has to be extremely high (several kilowatts) to obtain two-photon absorption.

A two-photon MSL apparatus is shown in Figure 7.23 (Maruo and Kawata 1998). The beam is generated by a mode-locked titanium sapphire laser and is directed by two galvanic scanning mirrors. The beam is then focused with an objective lens into the resin. A charge-coupled device (CCD) camera is used to aid focusing and monitor the forming of the microstructure. A Z-stage moves the resin container along the optical axis for multilayer fabrication. The objective lens used by Maruo had a numerical aperture of 0.85 (magnification of 40). The accuracy of the galvano-scanner set (General scanning)

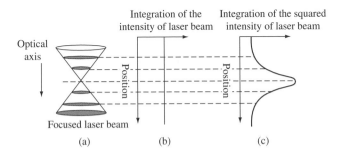

Figure 7.22 Two-photon absorption and one-photon absorption generated by a focused laser: (a) schematic diagram of a focused laser beam; (b) total one-photon absorption per transversal plane, which is calculated by integrating the intensity over the plane versus the optical axis; and (c) total two-photon absorption per transversal plane, which is calculated by integrating the intensity squared over the plane

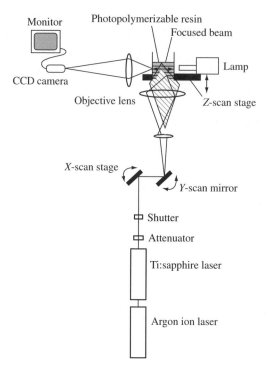

Figure 7.23   Optical setup for the two-photon MSL system. From Maruo and Kawata (1998)

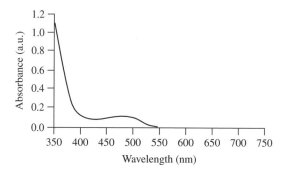

Figure 7.24   Absorption spectrum of the two-photon resin

and the Z-stage (Sigma Optics) were 0.3 and 0.5 μm, respectively. The peak beam power in the resin was about 3 kW with a repetition of 76 MHz and pulse width of 130 fs at a wavelength of 770 nm. The resin used was SCR-500, which is a mixture of urethane acrylate oligomers or monomers and photoinitiators. The absorption spectrum of this resin is shown in Figure 7.24. The resin is transparent at 770 nm, meaning that polymerisation cannot take place by one-photon absorption.

Some high aspect ratio and extremely fine 3-D microstructures have been fabricated using two-photon MSL, as shown in Figure 7.25 (Maruo and Kawata 1998). The lateral

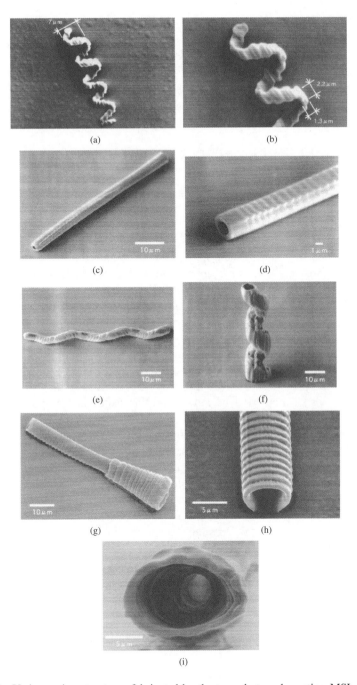

Figure 7.25 Various microstructures fabricated by the two-photon absorption MSL (from Maruo and Kawata (1998)) (a) SEM image of a spiral structure and (b) its magnified view. SEM images of microtubes made of solidified resin (c) length 55.6 μm; (d) inside diameter 1.8 μm and thickness 0.62 μm; (e) microspiral tube of external diameter 4 μm and length 74.3 μm; (f) external diameter 4 μm. SEM image of microfunnels made of solidified polymer of length 63 μm (g) entire part; (h) top part; and (i) bottom view

and depth resolution obtained with two-photon MSL were 0.62 and 2.2 μm, respectively. It is important to note that a depth resolution of only 2.2 μm is difficult to achieve with conventional MSL. However, the longest total length of a structure in the direction of the optical axis was up to 74.3 μm, which is a limitation of two-photon MSL. In addition, the two-photon MSL system is more expensive than most of the other types of MSL equipment.

7.5 OTHER MSL APPROACHES

There have been a number of other approaches to MSL. For example, Bertsch *et al.* (1997) and others (Zissi *et al.* 1996; Zhang *et al.* 1999) report on a system that employs scanning by an X-Y stage and a free surface (Figure 7.26). In this system, all the optics for the beam delivery remain fixed, whereas an X-Y stage simultaneously moves the resin tank and the vertical axis onto which the plate that supports the fabrications is attached. The scanning method is similar to the one Ikuta used in the IH process (Section 7.3.2), in which galvano-scanning is replaced with X-Y stage scanning so that the system is simplified and the precision of the focusing is enhanced. The difference between this MSL method and that of the IH process is that it adopts the free-surface method for layer preparation, rather than the constrained surface (with a window) used in the IH process. The reason is that many fabrications are damaged because of the microparts that adhere to the window in the constrained surface method, and these can be avoided by using a free surface. In addition, the time required in the free-surface method to obtain a fresh layer of resin on top of a cured layer depends on the rheological properties of the resin, so a resin with a low viscosity is preferred.

Free-surface MSL is a single-photon-based photopolymerisation process; the curing volume is relatively large compared with that of two-photon MSL. However, by adding a light-absorbing medium into the resin, the line-width and depth can be decreased (Zissi *et al.* 1996; Zhang *et al.* 1999). If the beam-delivery system is optimised further to obtain the finest beam spot size, the line-width and depth of 1–2 μm and 10 μm, respectively, can be obtained with surface MSL (Zhang *et al.* 1999). Some interesting microfabrications with high aspect ratios and true 3-D features are shown in Figure 7.27 (Zissi *et al.* 1996; Zhang *et al.* 1999).

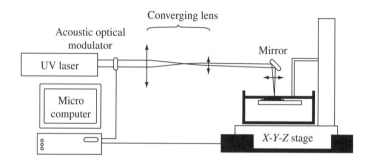

Figure 7.26 Apparatus used for surface MSL

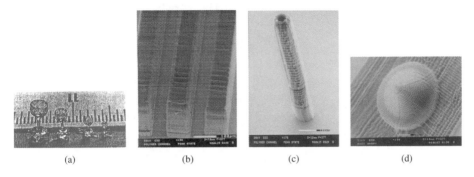

(a)   (b)   (c)   (d)

Figure 7.27   Some high-aspect-ratio microparts fabricated using surface MSL: (a) microcups; (b) a microtube of inner diameter 50 μm and height 800 μm; (c) 100 μm by 300 μm microchannels in HDDA (1.6 hexanediol diacrylate); and (d) a microcone with a bottom diameter of 500 μm and a height of 250 μm

7.6 PROJECTION METHOD

As described in the preceding sections, the scanning MSL can be used for very fine, high aspect ratio 3-D microstructure fabrication, but the fabrication speed is always a major concern – even with the galvano-scanning method. Scanning MSL builds up the objects layer by layer, but each layer is itself built up line by line. Thus, projection MSL has been proposed for the more rapid building of 3-D microstructures. Although it is still building layer by layer, each layer is now written by just one UV exposure through a mask. The reintroduction of a photographic mask plate produces significant savings in time but does add the extra expense of preparing masks.

Basically, there are two types of projection MSL: one is the use of a real photographic mask to project the UV pattern for curing (Nakamoto and Yamaguchi 1996) and the other is to use a dynamic mask referred to here as the *liquid crystal display* (LCD) *projection method* (Bertsch *et al.* 1997).

7.6.1 Mask-Projection MSL

As in standard photolithography, an image is transferred to the liquid photopolymer by shining an UV beam through a patterned mask plate as shown in Figure 7.28 (Suzumori 1994). Then, another fresh layer of liquid photopolymer is prepared on top of the patterned solid polymer. By repeating the above process, a multilayered 3-D microstructure can be built by this mask projection MSL (Katagi and Nakajima 1993; Nakamoto and Yamaguchi 1996; Suzumori *et al.* 1994).

Let us now consider the equations that govern the optics of mask-projection MSL (Nakamoto and Yamaguchi 1996):

When a beam of uniform intensity I_0 passes through a square mask, with the centre of the mask at $x = 0$ and $y = 0$, the depth of the polymer along the z-axis, and onto the surface of the liquid polymer (see Figure 7.29(a)), diffraction occurs, and the intensity I_d may be expressed as follows:

$$l_d(x, y, z) = 0.25 I_0 \left(C_x^2 C_y^2 + C_x^2 S_y^2 + S_x^2 C_y^2 + S_x^2 S_y^2 \right) \qquad (7.10)$$

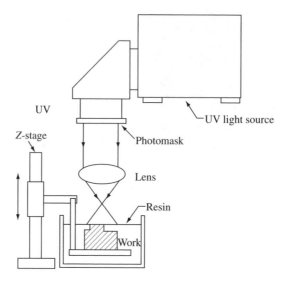

Figure 7.28   Apparatus for mask-projection method of microfabrication

where the various cosine and sine coefficients are defined by

$$
\begin{aligned}
C_x &= \int_{p_1}^{p_2} \cos\left(\pi u^2/2\right) du \\
S_x &= \int_{p_1}^{p_2} \sin\left(\pi u^2/2\right) du \\
C_y &= \int_{q_1}^{q_2} \cos\left(\pi u^2/2\right) du \\
S_y &= \int_{q_1}^{q_2} \sin\left(\pi u^2/2\right) du
\end{aligned}
\tag{7.11}
$$

and the limits on the integrals are defined by,

$$
\begin{aligned}
p_1 &= 2\left(x/a - 0.5\right)/\sqrt{m} \\
p_2 &= 2\left(x/a + 0.5\right)/\sqrt{m} \\
q_1 &= 2\left(y/a - 0.5\right)/\sqrt{m} \\
q_2 &= 2\left(y/a + 0.5\right)/\sqrt{m} \\
m &= 2\lambda(h + z)/a^2
\end{aligned}
\tag{7.12}
$$

where λ is the wavelength of the UV light, a is the length of each side of the square, and h is the distance between the mask and the surface of the resin.

The light intensity inside the resin, $I(x, y, z)$, can be calculated according to the Beer–Lambert law expressed in a slightly different form from that of Equation (7.13)

$$
I(x, y, z) = I_d(x, y, 0)\exp\left(-\alpha z\right)
\tag{7.13}
$$

where α is the absorption coefficient of the resin. The irradiation exposure $E(x, y, z) = I(x, y, z)t$, where t is the irradiation time period. The resin solidifies once the irradiation exposure reaches the threshold value E_c. The parameters α and E_0 determine the solidification properties of the resin and must be determined experimentally. Letting

$E(x, y, z) = E_0$ allows the theoretical shape of the solidified polymer after projection to be determined.

Similar to scanning MSL, the fabrication precision is related to the exposure. In particular, the curing depth strongly depends on the laser exposure as shown in Figure 7.29(b) (Nakamoto and Yamaguchi 1996). The lateral dimension is slightly influenced by the exposure and is determined mainly by the mask pattern in the case of a fixed distance between the mask and resin surface.

Another important parameter in mask-projection MSL is the distance between the mask and the resin surface. A large distance between the mask and resin surface results in a larger lateral dimension because of diffraction (Nakamoto and Yamaguchi 1996) (Figure 7.29(c)). Therefore, to minimise this effect and obtain the highest precision in mask-projection MSL, the mask should be located as close as possible to the resin surface.

An example of a micropart fabricated by the mask-projection MSL process is shown in Figure 7.30 (Nakamoto and Yamaguchi 1996).

The fabrication speed of mask-projection MSL is much faster than scanning MSL as stated earlier. But for complex 3-D microstructures, a large number of masks are needed, and this process is not only time-consuming but also expensive. This disadvantage can be overcome by using dynamic mask-projection MSL.

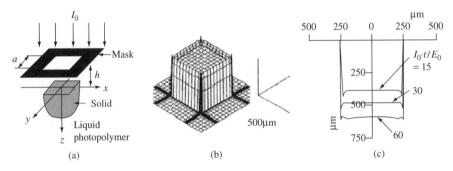

Figure 7.29 Model of mask-projection MSL and the simulation results (a) theoretical model of the mask-based method; (b) simulated cross section of a solidified polymer (a is 500 μm and h is 1000 μm); and (c) cross section of the solidified polymer

Figure 7.30 Example of a polymer microstructure using mask-projection MSL. From Nakamoto and Yamaguchi (1996)

7.6.2 Dynamic Mask-Projection MSL

Dynamic mask-projection MSL utilises a dynamic mask generator rather than a 'static' photographic mask plate and therefore permits the rapid fabrication of complex 3-D micro-objects. A schematic view of a dynamic mask-projection MSL is shown in Figure 7.31 (Bertsch *et al.* 1997). In dynamic MSL, the mask pattern is produced by a computer-controlled LCD rather than by a chrome mask. Once the CAD file has been translated into a numerical control code that is sent to the LCD device via a computer, the LCD can function as a dynamic mask to control the pattern of the layer. The light beam then passes through the LCD mask before it is focused on the resin surface to allow the selective polymerisation of the exposed areas that correspond to the transparent pixels of the LCD.

The rest of the apparatus used in dynamic mask-projection MSL is the same as standard MSL, namely, layer preparation, beam on–off control, and so on. It should be noted that the Z-stage is the only moving element in the system and therefore it is simpler.

In dynamic mask MSL, the liquid–solid phototransformation can once again be described by the Beer–Lambert law. Because time t is now the most critical parameter in the process, the curing depth is usually given in the form of

$$d_c = \ln(t/t_0)/\alpha C \quad \text{with} \quad t_c = Q/\alpha C F_0 \tag{7.14}$$

where, d_c is the curing depth, α is the absorption coefficient (1 mol^{-1}cm^{-1}), C is the concentration of the photoinitiator (mol l^{-1}), t is the exposure time (s), t_c is the exposure time (s) necessary to make the exposure reach the polymerisation threshold energy (s), Q is the number of absorbed photons per unit volume (photon m^{-3}), which is determined experimentally, and F_0 is the incident flux (photon m^{-3}cm^{-1}).

The liquid crystal matrix is inserted between four glass windows that are opaque to UV light; therefore, for this system, it is necessary to use a visible light source and a different set of chemical mixtures (Bertch *et al.* 1997). The lateral dimension is now determined by

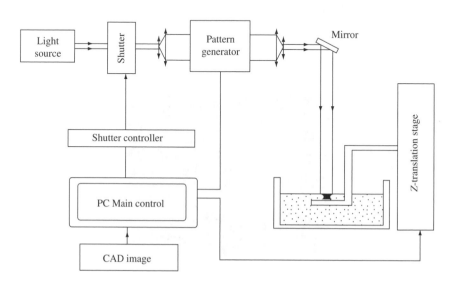

Figure 7.31   Principle of the integral MSL apparatus

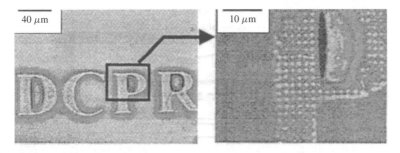

Figure 7.32   Example of a single-layer object manufactured by the MSL process. From Monneret *et al.* (1999)

the LCD mask and so is limited by the resolution and contrast of current LCD displays. By using a LCD composed of a 640 × 480 pixel array, a lateral resolution around 2 μm × 2 μm was obtained (Figure 7.32) by Monneret *et al.* (1999). Thus, dynamic projection MSL gives a reasonable accuracy, which, no doubt, will improve as higher-resolution LCDs are developed.

Some interesting microparts have been fabricated by a number of research groups using dynamic projection MSL, as shown in Figure 7.33 (Nakamoto and Yamaguchi 1996; Monneret *et al.* 1999; Beluze *et al.* 1999).

Dynamic projection MSL has some problems associated with it, such as the limited lateral resolution mentioned earlier and the size of the object that is currently limited to a few millimeters. Nevertheless, dynamic projection MSL is an attractive way of making 3-D microparts in a reasonable time.

7.7 POLYMERIC MEMS ARCHITECTURE WITH SILICON, METALS, AND CERAMICS

The fabrication of new MEMS devices requires the integration of various new functional materials such as polymers, ceramics, metals, and metal alloys. This section describes how the MSL process could be used to fabricate MEMS devices on the basis of these different materials.

7.7.1 Ceramic MSL

Functional and structural ceramic materials possess useful properties such as high temperature or chemical resistance, high hardness, low thermal conductivity, ferroelectricity, and piezoelectricity. The application of ceramic materials in MEMS has attracted a great deal of interest recently (English and Allen 1999; Epstein *et al.* 1997; Bau *et al.* 1998; Polla and Francis 1996; Varadan *et al.* 1996). Three-dimensional ceramic microstructures are of special interest in applications such as microengines (Epstein *et al.* 1997) and microfluidics (Bau *et al.* 1998). Various novel approaches to ceramic microfabrication have been developed. Unlike conventional silicon micromachining, MSL can be used to build the complex ceramic 3-D microstructures in a rapid free-form fashion without the need for high pressures or high temperatures (Jiang *et al.* 1999).

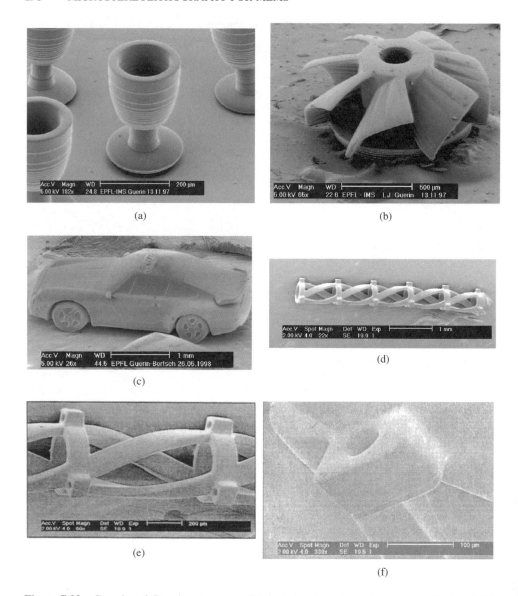

Figure 7.33 Complex 3-D microstructures fabricated using dynamic mask projection MSL: (a) microcup made up of 80 layers of 5 μm thickness; (b) microturbine made of 110 layers of 4.5 μm thickness; (c) microcars made of 673 layers of 5 μm thickness; (d) microsprings made of 1000 layers of 5 μm thickness; (e) close-up of spring; (f) lateral hole in the structure made of imbricated spring; (g) a side mirror of a car made of about 20 layers; (h) wheel of a car; (i) detail of car microprints on the roof of the car; (j) double-end connector made of 700 layers of 5 μm thickness; (k) two channels corresponding to each tube connector with external diameter of 200 μm. From Beluze *et al.* (1998, 1999)

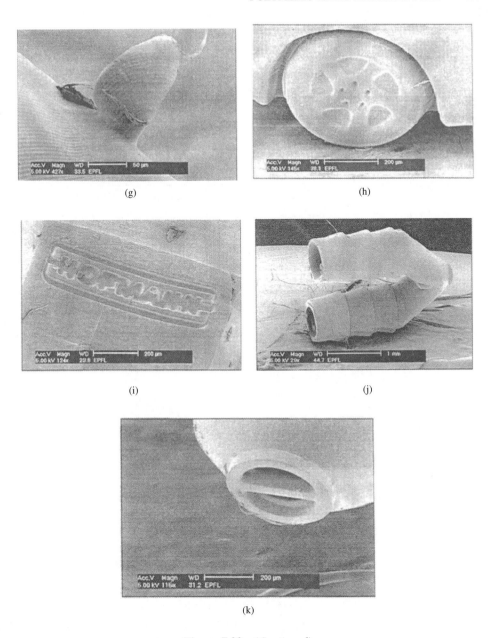

(g)

(h)

(i)

(j)

(k)

Figure 7.33   (*Continued*)

Ceramic MSL differs from polymeric MSL in the following aspects. First, the resin system for ceramic MSL is composed not only of the monomers and photoinitiators that are used in polymer MSL but also of ceramic powders, dispersants, and diluents (Zhang *et al.* 1999; Varadan *et al.* 1996). Dispersants and diluents are used to obtain a homogeneous ceramic suspension with a relatively low viscosity. Upon UV polymerisation, the ceramic particles are bonded together by the polymer and the ceramic body is formed.

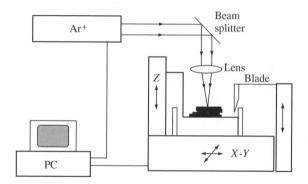

Figure 7.34 MSL apparatus for ceramics. From Jiang *et al.* (1999)

Generally, the viscosity of ceramic suspensions used for MSL is higher than the viscosity of most liquid polymers, leading to slow layer preparation. A precision blade has been designed for the layer preparation to solve this problem (Figure 7.34, Jiang *et al.* 1999). Second, light transportation during MSL is more complicated in the solid–liquid two-phase medium, and this is caused by light scattering off the solid ceramic particles and affecting both the curing depth and the line-width.

The curing depth has been determined from previous macroscale experiments by Griffith and Halloran (1995) to be given by

$$\phi_c = \phi \ln (E/E_c)/Q\xi \quad \text{and} \quad Q = (\Delta n/n_0)^2/(\lambda/\phi)^2 \qquad (7.15)$$

where ϕ is the mean particle size of the ceramic powder, ξ is the volume fraction of the ceramic material in the suspension, n_0 is the refractive index of the monomer solution, Δn is the difference in refractive index of the ceramic solution and the monomer solution, and λ is the wavelength of the UV light.

Therefore, MSL production of ceramic microparts is much more difficult than that of polymer microparts. Furthermore, there is no easy way to estimate the line-width, although a Monte-Carlo simulation approach has been proposed for ceramic MSL (Sun *et al.* 1999).

The fabrication of ceramic microstructures using MSL typically follows steps shown in Figure 7.35. First, the homogeneous ceramic suspension is prepared. Submicron ceramic powders are mixed with monomer, photoinitiator, dispersant, diluent, and so on by ball-milling for several hours. The prepared ceramic suspension is then put into the vat and is ready for exposure defined by the CAD file, after which a (green) body ceramic micropart is obtained. Finally, the green body is put into a furnace first to burn off the polymer binders and then sintered at higher temperatures to obtain the dense ceramic microparts. The temperatures of the binder burnout and sintering vary according to the choice of polymeric and ceramic materials. After sintering, the ceramic microstructures are ready for assembly and use.

The ceramic microparts shown in Figure 7.36 were fabricated from an alumina (50–59.5 percent volume) and lead zinc titanate (PZT) suspension (33 percent volume) (Jiang *et al.* 1999). Even though the resolution of ceramic MSL is poorer than that of polymer MSL, a minimum line-width of around 6 µm is achievable.

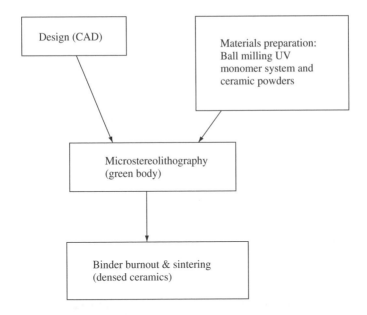

Figure 7.35 Ceramic MSL and postprocessing

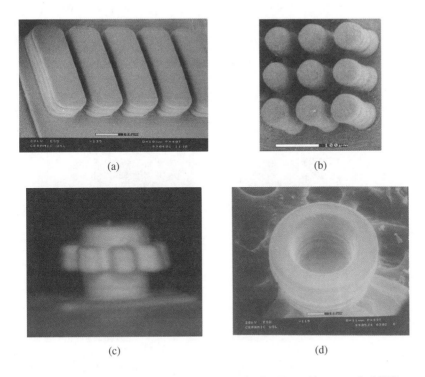

Figure 7.36 Ceramic microparts made by MSL. From Jiang *et al.* (1999)

7.7.2 Metallic Microstructures

Metallic microstructures have been fabricated extensively for MEMS. 3-D metallic microstructures have been built by spatial forming, electrochemical fabrication (EFAB), selective laser sintering, and laser-cladding processes (Taylor *et al.* 1994; Kathuria 1996; Cohen *et al.* 1999). The EFAB process and spatial-forming processes are introduced in this section because of their relatively higher resolution and, therefore, their likelihood to succeed commercially.

7.7.2.1 Spatial forming

Spatial forming combines several different technologies to generate solid metallic microstructures from fine powders (Taylor *et al.* 1994). Similar to projection MSL, cross-section data from solid CAD models are used to define the patterns on a chrome mask. A custom-built offset printing press prints negative materials (space around the solid parts) on a ceramic substrate in multiple registered layers of pigmented organic ink averaging 0.5 μm thick, each layer being cured with UV light. After forming a number of layers of negative materials (e.g. 30), the positive ink, heavily loaded with metal powders (e.g. 50 percent volume), is knifed onto the assembly filling the nonimage voids (Figure 7.37; Thornell and Johansson 1998), followed by curing the filled material with UV light. By repeating the above steps until the desired thickness (e.g. 500 μm) is reached, the green body metallic micropart is created. The green body part is then baked to remove the organic binders and is sintered in a controlled atmosphere in a furnace. The finished pure metallic microparts are thus finally obtained.

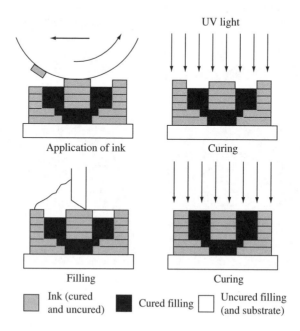

Figure 7.37 Schematics of spatial-forming process. From Thornell and Johansson (1998)

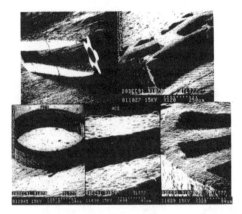

Figure 7.38 A set of demonstration microstructures realised in 17-4 PH stainless steel. About 600 layers of negative material and 20 layers of positive material were printed

This spatial-forming process can be used to mass-produce 3-D microparts (Taylor *et al.* 1994). For example, stainless steel microparts have been fabricated as shown in Figure 7.38, and typical dimensions ranging from $100 \times 150 \times 150$ μm^3 to $0.25 \times 1 \times 20$ mm^3 with a typical minimum feature size of 10 μm (Taylor *et al.* 1994).

7.7.2.2 Electrochemical fabrication process

EFAB is a new micromachining process, which is based on solid free-form fabrication (SFF) principles, to make high aspect ratio and 3-D metallic microsystems (Cohen *et al.* 1999). The major fabrication steps in EFAB include instant masking and selective electroplating, blanket deposition, and planarisation as illustrated in Figure 7.39.

Instant masking uses photolithographically patterned masks on the anode for selective electroplating. The instant mask consists of a conformable insulator because the pattern may be topologically complex. Instant masking patterns a substrate by simply pressing the insulator mask against the substrate. Electroplating materials are then deposited onto the substrate through apertures in the insulator mask, and the insulator mask is removed after the plating of each layer. The mask shown in Figure 7.40 consists of a layer of insulator patterned on a flat Cu disk (Cohen *et al.* 1999). In selective electroplating, pressure is applied between the Cu anode (with mask) and the Ni (cathode) substrate.

Blanket deposition is also based on electroplating but without the use of a mask. Basically, the blanket-deposited material (e.g. Ni) is different from the selectively plated material (Cu), so that one metal acts as a sacrificial material and is removed as the final step.

Planarisation is the next step and involves lapping the surplus materials to achieve a layer of precise thickness and flatness before the deposition of the next layer. By repeating the above steps, a 3-D micrometallic structure can be formed (Figure 7.40).

A schematic view of the apparatus fulfilling the selective-plating, blanket-deposition, and planarisation processes is shown in Figure 7.41 (Cohen *et al.* 1999). An example of a 3-D metallic microstructure fabricated by EFAB is shown in Figure 7.42 and it demonstrates the potential of using EFAB to make MEMS devices.

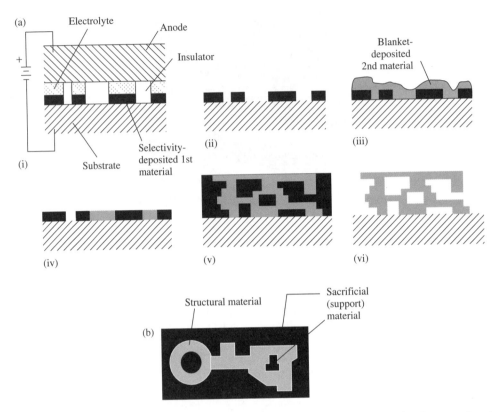

Figure 7.39 (a) Various stages of the EFAB process and (b) structure of a complete layer produced by EFAB

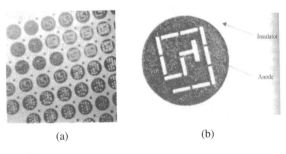

Figure 7.40 Masks for EFAB process: (a) Masks for patterning the cross section of multilayer devices on a common anode and (b) close-up of a mask – the circle has a diameter of 1.3 mm. From Cohen *et al.* (1999)

EFAB is a process still in the early stages of development. The resolution is around 25 µm, and smearing (caused by lapping and misregistration) can also affect the fabrication precision. Moreover, the fabrication speed is still a cause for concern because too many time-consuming electroplating steps are involved. For example, a throughput of two planarised 5 mm layers per hour or about 50 layers per day is predicted.

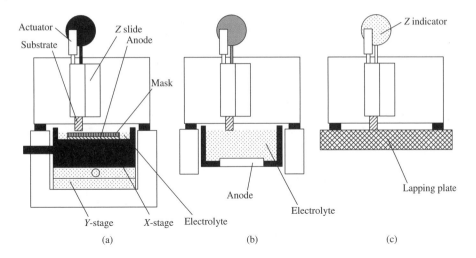

Figure 7.41   Schematic of a manual EFAB machine. From Cohen *et al.* (1999)

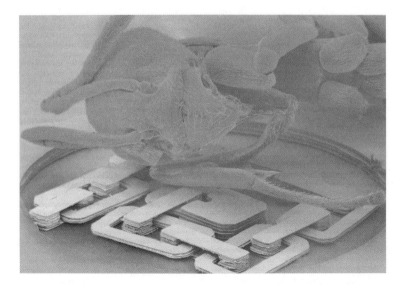

Figure 7.42   3-D metallic microparts fabricated by EFAB. From Cohen *et al.* (1999)

7.7.3 Metal–Polymer Microstructures

Composite metal–polymer microstructures are of some interest in the field of MEMS. A process has been developed by Cabrara *et al.* (1998) that allowed the construction, layer by layer, of a 3-D object with both conducting and nonconducting parts instead of manufacturing the parts separately and assembling them afterwards. For example, to build the cylindrical object described in Figure 7.43 (Cabrera *et al.* 1998), which consists of a metallic element (Part 1) freely rotating inside a polymer housing (Part 2), the major steps involved in this fabrication process include the following:

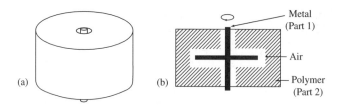

Figure 7.43 Example of a complex metal–polymer part: (a) External view and (b) cross sectional view. From Cabrera *et al.* (1998)

- Electroplating of copper to make Part 1

- Local laser silver plating on polymer to get the conductive base for the electroplating of copper

- MSL with an insoluble resin to make Part 2

- MSL with a soluble resin to make a sacrificial structure between Parts 1 and 2

The details of the layer-by-layer fabrication processes to build the structures shown in Figure 7.44 are described by Cabrera *et al.* (1998).

The electroplating of copper is guided by the soluble polymer that is shaped by MSL. By changing the two-dimensional (2-D) shapes, a 3-D complex metallic structure should be formed in a layer-by-layer fashion. However, electroplating of copper can only occur on a conductive substrate; a local laser silver-plating technique is therefore used to plate silver onto the polymer substrate and connect it to the previously electroplated copper substrate (Figure 7.44, step 10). The insoluble resin is then patterned using MSL. After all layers of plating and MSL, the soluble resin is dissolved off to leave the desired metal–polymer object (Figure 7.45). However, no postfabrication assembly is necessary.

Again, fabrication speed is a concern because slow-plating processes are utilised in every layer.

7.7.4 Localised Electrochemical Deposition

A localised electrochemical deposition apparatus is schematically shown in Figure 7.46 (Madden and Hunter 1996). The tip of a sharp pointed electrode (microelectrode) is placed in a plating solution and brought near the surface where the deposition is required. A potential is applied between the tip and the substrate. The electric field generated for electrodeposition is thus confined to the area beneath the tip as shown in Figure 7.46(a).

In principle, complete 3-D microstructures can be formed using localised electrochemical deposition, provided the object is a conductor and therefore continuously in electrical contact with the substrate electrode. The spatial resolution of this process is determined by the base diameter of the microelectrode.

Another important parameter that needs to be considered in the process is the electrodeposition rate. The vertical deposition rate can be up to 6 μm/s, which is 100 times greater than that of conventional electroplating (Madden and Hunter 1996).

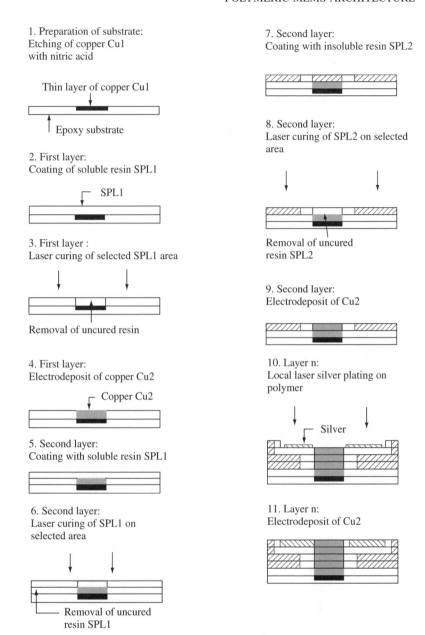

1. Preparation of substrate:
Etching of copper Cu1
with nitric acid

Thin layer of copper Cu1

Epoxy substrate

2. First layer:
Coating of soluble resin SPL1

SPL1

3. First layer :
Laser curing of selected SPL1 area

Removal of uncured resin

4. First layer:
Electrodeposit of copper Cu2

Copper Cu2

5. Second layer:
Coating with soluble resin SPL1

6. Second layer:
Laser curing of SPL1 on
selected area

Removal of uncured
resin SPL1

7. Second layer:
Coating with insoluble resin SPL2

8. Second layer:
Laser curing of SPL2 on selected
area

Removal of uncured
resin SPL2

9. Second layer:
Electrodeposit of Cu2

10. Layer n:
Local laser silver plating on
polymer

Silver

11. Layer n:
Electrodeposit of Cu2

Figure 7.44   Layer-by-layer manufacturing steps of a complex metal–polymer part. From Cabrera
et al. (1998)

The shape geometry of the microelectrode used for localised electrochemical deposition is critical to the deposition profile. As shown in Figure 7.47, two different microelectrode configurations have been modeled and used for experiments by Madden, one in which insulation covers the length of the electrode, leaving a disk-shaped tip exposed, and the second in which a portion of the conical segment above the disk is also left

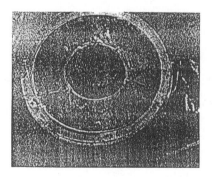

Figure 7.45   Complex metal–polymer part of height 15 mm and diameter 10 mm. From Cabrera *et al.* (1998)

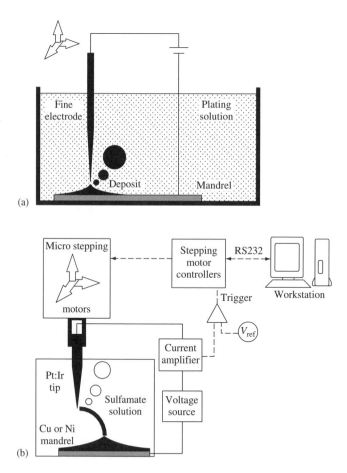

Figure 7.46   (a) The concept of localised electrochemical deposition and (b) apparatus used for position control

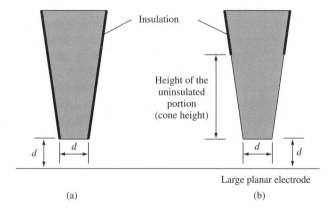

Figure 7.47 Modeled electrode tip geometries: (a) disk and (b) disk and cone

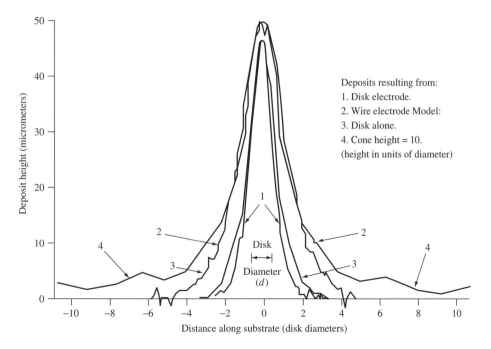

Figure 7.48 Observed deposition profiles resulting from a Pt disk electrode (500 μm diameter) and a Pt wire electrode (500 μm diameter)

exposed. The cone angle is 28°, which is representative of cone angles in the etched tips (Madden and Hunter 1996). The separation between the substrate and the tip is about the size of its base diameter. A comparison of the deposited profiles suggests that the electrode in which the tip alone is exposed has the better lateral resolution (see Figure 7.48).

Through the use of a fine tip (5 to 100 μm diameter), it is possible to make high aspect ratio and 3-D micronickel structures using localised electrochemical deposition as shown in Figure 7.49 (Madden and Hunter 1996).

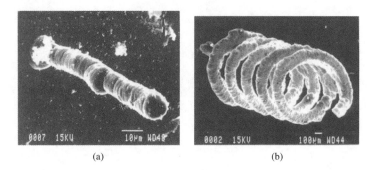

(a)   (b)

Figure 7.49   Nickel microstructures fabricated by localised electrochemical deposition (a) pipe and (b) microspring. From Madden and Hunter (1996)

However, there are two problems to be solved before this process can be widely used in industry, namely, assembly and mass production.

7.8 COMBINED SILICON AND POLYMERIC STRUCTURES

MSL can be readily used to fabricate *polymer* 3-D microstructures; however, silicon microtechnology prevails in the fabrication of integrated circuits (ICs) and microsensors. Therefore, the combination of silicon and polymeric microstructures is an attractive option for smart MEMS devices. Some recent efforts in this direction are described in this section, including 'architecture combination' by the photoforming process, MSL integrated with thick film lithography, and the AMANDA (surface micromachining, moulding, and diaphragm transfer) process.

7.8.1 Architecture Combination by Photoforming Process

Architecture combination is a technology for building complicated structures by mechanically connecting two or more architectures made by different micromachining processes. This technology helps in making a system that consists of, for example, lithography, electroplating, and moulding process (LIGA) linkages driven by a silicon micromotor that are housed in a polymer microstructure (see Figure 7.50; Takagi and Nakajima 1994). The desired housing structure is a complicated one and this function actually requires a coupled mechanism in which one component is made by one process and another component made by another process. Photoforming, as described earlier, has been employed for this purpose not only because of its relatively high resolution but also because of its 3-D fabrication capabilities (Figure 7.51; Takagi and Nakajima 1994).

For example, a microclamping tool has been fabricated using this architecture combination technology (Takagi and Nakajima 1994). The pressure vessel and heating components can be fabricated first using silicon micromachining (Figure 7.51) and then the polymer clamping structures are formed using photoforming on top of the silicon structure. When the pressure vessel is heated up, it induces the flexible diaphragm to deflect and the clamping structure will open; it then closes when the vessel cools. The polymer clamping

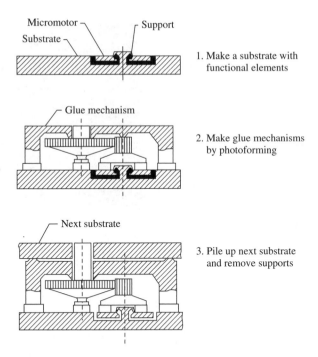

Figure 7.50   Fabrication of combined architecture. From Takagi and Nakajima (1994)

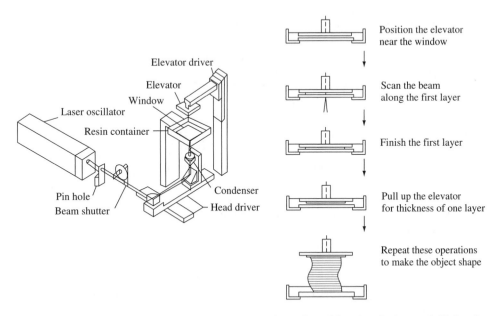

Figure 7.51   Microscale photoforming MSL: (a) configuration of forming devices and (b) forming sequence. From Takagi and Nakajima (1994)

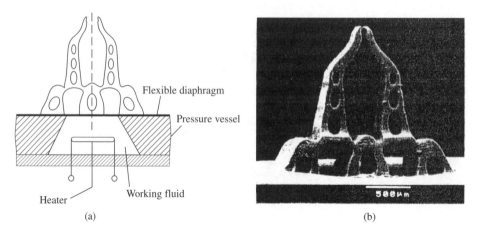

(a)   (b)

Figure 7.52 Design (a) and actual (b) photoformed thermally driven microclamping tool. From Takagi and Nakajima (1994)

structure is shown in Figure 7.52 (Takagi and Nakajima 1994). The clamp is flexible and soft and therefore should not damage the sample that it is holding.

With this architecture combination, the components fabricated with different processes are joined together during the photoforming process – the alignment is then critical to achieve a successful tool.

7.8.2 MSL Integrated with Thick Film Lithography

Many micromechanical components have been fabricated using planar processes such as thin film and bulk-silicon micromachining. High aspect ratio micromachining is available through LIGA, deep reactive ion etching (RIE), and thick resist lithography with high resolutions. But these processes do not allow true 3-D fabrications. On the other hand, MSL can be used to construct complex 3-D microstructures, but with the constraints of a lower resolution and the problems associated with the manipulation and assembly of polymeric microstructures. An approach that seeks to combine MSL and thick resist lithography may provide a technique to build new 3-D microstructures with more functionality (Bertsch *et al.* 1998).

EPON SU-8 resin has been used for thick resin lithography, and structures as thick as 2 mm with an aspect ratio of 20:1 have been obtained, for example, a high-definition monoblock axle-gear master for an injection mould for watch gears. Ideally, the axle of the gears must be conical for the centreing and reduction of the friction torque. One level or multilevel SU-8 structures were built first by lithography (Figure 7.53), and then fixed to an MSL elevator attached to the Z-stage immersed with photopolymerisable resin (Figure 7.54). After careful alignment, the axle grows layer by layer on the SU-8 surface. No assembly step is required.

Figure 7.55 shows a one-level SU-8 structure on which a conical axle has been added. The gear is 400 μm in diameter and 600 μm in height, the conical axle part is 250 μm high, and the diameter of the axle is 80 μm. Figure 7.56 shows a two-level SU-8 structure on which an axle that is 400 μm high and 150 μm in diameter has been added.

Figure 7.53   Master layout in SU-8 before electroplating. From Bertsch *et al.* (1998)

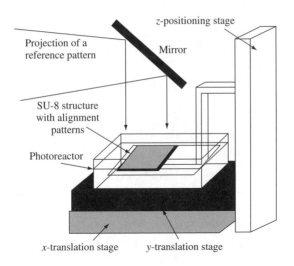

Figure 7.54   Schematic showing the positioning of the SU-8 structure relative to the UV light beam

7.8.3 AMANDA Process

AMANDA is a process that combines surface micromachining, micromoulding, and diaphragm transfer to fabricate microparts from polymers. A flexible diaphragm, made of either functional or structural materials, is deposited and patterned on a silicon substrate using a surface-micromachining process. The moulding process is then used to build the housing for the fabricated diaphragm and the diaphragm is next transferred from the silicon substrate to the polymeric housing. Hence, the AMANDA process allows low-cost production of reliable microdevices by batch fabrication.

Figure 7.55 Combination of MSL with one-level SU-8 lithography: conical axle added on a one-level SU-8 microgear. From Bertsch *et al.* (1998)

Figure 7.56 Two-level SU-8 structure with an added axle. From Bertsch *et al.* (1998)

As an example for the AMANDA process, the fabrication process for a pressure transducer is shown in Figure 7.57 (Schomburg *et al.* 1998). A silicon wafer is covered with 60 nm of gold and 1.5 μm of polyimide by physical vapour deposition (PVD) and spin coating, respectively. The polyimide is then patterned by photolithography and an additional 100 nm of gold is evaporated on top of the polyimide layer. The second layer of gold is patterned, and strain gauges are formed. The polyimide disks, which are 30 μm thick, are built up on the strain gauges by spin coating and photolithography.

The housings of AMANDA devices are usually produced by injection moulding to save time (Schomburg *et al.* 1998). Typically, several housings can be made in a batch. The housing can be moulded from various thermoplastic materials, such as polysulfone, poly (methyl methacrylate) (PMMA), polyacetylene, polycarbonate, polyvinylidene fluoride (PVDF), and polyetherether ketone (PEEK).

The mould inserts for housings are fabricated by milling and drilling using a computer numerically controlled (CNC) machine, LIGA, deep RIE, and so on.

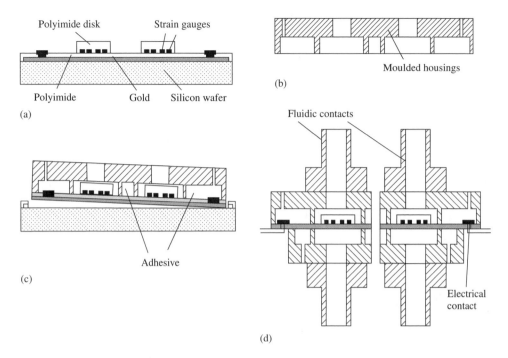

Figure 7.57 Production of pressure sensors with the AMANDA process: (a) diaphragm is surface-micromachined; (b) housings are made by moulding; (c) diaphragm is transferred to the housings; and (d) diced chips with fluidic and electrical contacts. From Schomburg *et al.* (1998)

The diaphragm is transferred to the housings. Adhesive is injected into the cavities, which are part of the microstructure of the housings. In the case of the example shown in Figure 7.57, the housings are adhesively bonded to the polyimide on the wafer. The polyimide around the housings is severed and the housings, together with the polyimide diaphragm, are freed from the wafer. The polyimide can be pulled off the wafer because the adhesion of the first gold layer to the silicon surface is poor. Usually, the diaphragm is encapsulated in a second shell, which is moulded and bonded similar to the first shell. Finally, batches are diced and equipped with fluidic and electric contacts (Figure 7.57).

The dimensional accuracy of the microstructures fabricated by the AMANDA process depends on the lithography, precision of the mould insert and the moulding process, and the alignment and temperature control during the bonding of the moulded part and the diaphragm. The lateral accuracy of the pattern on the diaphragm can be very high because it is fabricated by photolithography. Transfer of the diaphragm to the polymer housing causes overall shrinkage determined by the thermal expansion coefficient of the housing and through the bonding temperature. The precision of the mould insert for housing fabrication can be very high if a LIGA process is used. The precision of a moulding process can be several microns (or even higher) using either injection or hot-embossing moulding. In order to improve the dimensional accuracy of the AMANDA process (Schomburg *et al.* 1998) further, difficulties with the alignment have to be overcome and better shrinkage control is required.

7.9 APPLICATIONS

Various microdevices have been developed to date using the microfabrication processes described earlier. This section describes the fabrication of some microsensors and microactuators using the MSL and the AMANDA processes.

7.9.1 Microactuators Fabricated by MSL

Our first example is a shape memory alloy (SMA) microactuator glued to the 3-D structures fabricated by MSL (Figure 7.58, Ballandras *et al.* 1997). The MSL structure was designed with a clamping area such that the SMA wire can be easily inserted. Once the SMA wire is correctly positioned in the MSL architecture, it is submitted to a mechanical initial stress to store mechanical energy within it, which is performed by using a calibrated weight fixed at one end of the wire, the other end being embedded in a rigid frame. The laser beam is then focused on the clamping area, inducing the polymerisation of the monomer and thus bonding the SMA wire with the polymer structure.

The shape-memory effect has been selected as the basis for actuation here because it offers both high strain and force (Bernard *et al.* 1997). The effect is based on the reversible solid-state phase transformation of the material from the easily deformable martensite at low temperatures to the relatively rigid austenite at high temperatures. For

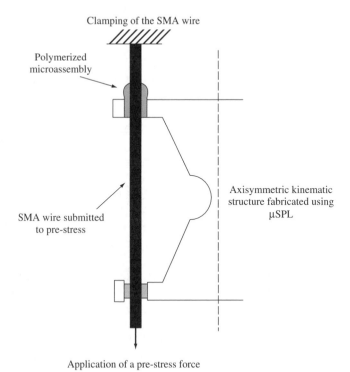

Figure 7.58 Assembly principle of an SMA wire and a μSPL structure. From Ballandras *et al.* (1997)

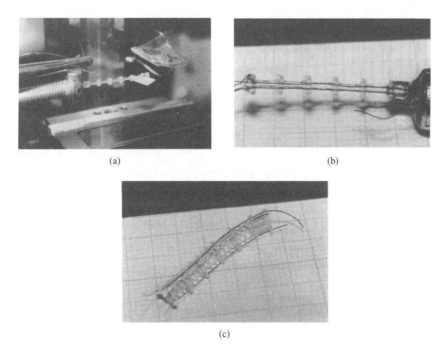

(a)   (b)

(c)

Figure 7.59   Micrographs of SMA actuators made by MSL: (a) spherical elastic joint type placed on experimental bench; (b) single-spring type; and (c) multiple overlapping spring type. From Ballandras *et al.* (1997)

a TiNi alloy, the phase-transformation temperatures are dependent primarily on the ratio of Ti to Ni and are in the range of 10 to 100 °C. The austenite phase has an associated parent shape, so that when deformed martensite is heated, the material exerts a force to re-attain the austenitic shape. This force is the basis of the actuation process. The amount of shape recovery in unconstrained material depends on the level of strain in the deformed martensite.

The assembly technique described in the preceding paragraph permits the fabrication of various single module SMA actuators with dimensions of several millimeters. A device with five modules is shown in Figure 7.59 (Ballandras *et al.* 1997) with a total length of 2 cm. By applying 2 V direct current (DC), a large displacement of 2 mm is obtained, and this corresponds to 20 percent of the total length of the actuator. The maximum current in the SMA wires was found to be 300 mA. The actuator dynamics appears to be very stable with regard to the very large number of actuating cycles, and this means that the MSL structures are stiff enough to ensure the return of the actuators to their initial position. It should be pointed out that the fabricated actuators only weigh a few grams, which is advantageous when using these actuators as manipulators in robots because less energy is needed to move the actuators themselves.

A micropump has also been fabricated by MSL (Corrozza *et al.* 1995). The micropump comprises a pump body and an actuator consists of a disk of piezoelectric ceramic disk glued to a thin brass plate (Figure 7.60). The pump body is fabricated by MSL with a complicated geometry to realise the optimal hydraulic performance and consists of two ball valves, a pumping chamber, and inlet and outlet channels. Each ball valve has a

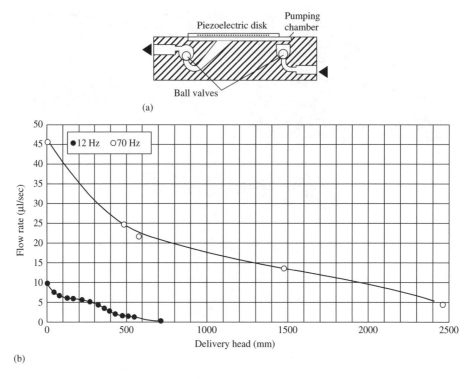

Figure 7.60 Polymer micropump developed using MSL: (a) radial section of the pump and (b) flow rate of pump against delivery head at two drive frequencies

cylindrical chamber connected to a hemispherical chamber that contains a mobile ball. When the actuating diaphragm moves upward, the inlet valve is open and the outlet valve is closed, and the fluid is then sucked into the pump chamber. During the pumping mode, the actuating membrane deflects downward, the inlet valve is closed and the outlet valve opens, letting fluid flow out of the pump. The dimensions of this micropump are listed here:

- Pump body: 18-mm diameter × 5 mm height
- Brass plate: 18-mm diameter × 0.1 mm height
- Piezoelectric ceramic disk: 10-mm diameter × 0.4 mm height
- Diameter of the channel: 0.5 mm

Under an alternating current (AC) input voltage of 300 V and 70 Hz, the back pressure can be 2.45 m H_2O and flow rate about 45 µl/s (Figure 7.60(b), Corrozza *et al.* 1995). This represents a fairly high output when compared with other types of micropumps.

7.9.2 Microconcentrator

A concentration process is necessary for various chemical operations for both detection and purification. With the recent development of various types of biochemical IC chips,

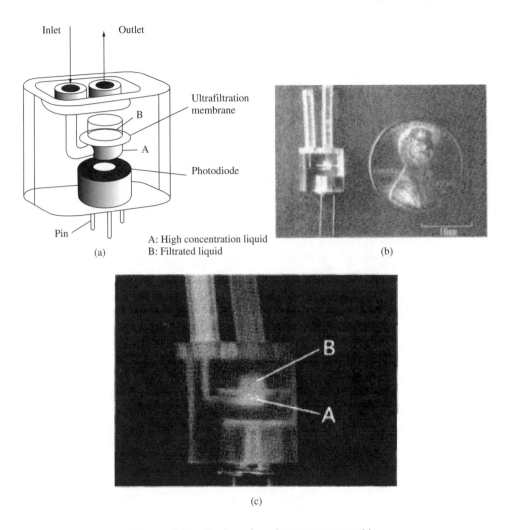

Inlet | Outlet

Ultrafiltration membrane

B

A

Photodiode

Pin

A: High concentration liquid
B: Filtrated liquid

(a)

(b)

(c)

Figure 7.61 Design of a microconcentrator chip

there is a growing demand for a microconcentrator with molecular filtering. The basic design and a prototype of the microconcentrator chip are shown in Figure 7.61 (Ikuta *et al.* 1999). The overall chip size is $8 \times 6 \times 8.5$ mm^3 and it consists of a microreactor chamber divided by an ultrafiltration membrane. The lower part (A) of the microchamber is filled with a high-concentration solution. A photodiode at the bottom can detect changes of colour or intensity for use in biochemical luminescence tests. The transparent nature of the polymer structures is an advantage over silicon. The fabricated device was used to verify the concentration of protein in a biochemical reaction. The firefly protein, luciferase, was used because it emits visible light with a wavelength of 560 nm in a redox reaction with beetle luciferin. Therefore, the concentration of luciferase can be determined from the intensity of the luminescence. In the experiments, the reagents of luciferase, luciferin, and so on were premixed to supply to the concentrator. The luminescence was observed from the chamber (A) at the underside of the ultrafiltration

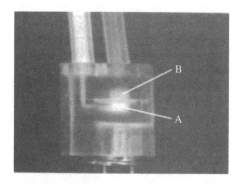

Figure 7.62 Long-exposure photograph to show the luminescence from the concentrator during reagent supply. From Ikuta *et al.* (1999)

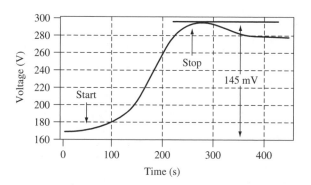

Figure 7.63 Typical output voltage of the photodiode while the reagent is supplied

membrane, indicating that luciferase has been separated by the ultrafiltration membrane successfully (Figure 7.62, Ikuta *et al.* 1999). As expected, the output from the photodiode increased with the supply of reagent and decreased when the supply was stopped (Figure 7.63).

In this example, the MSL process has been used to fabricate the 3-D device, and therefore packaging and leakage problems are eliminated completely. In addition, the transparent polymer was helpful in the testing of the devices. We expect that more advanced microdevices will be developed using this technology in the near future.

7.9.3 Microdevices Fabricated by the AMANDA Process

A micropump has been successfully fabricated by the AMANDA process as shown in Figure 7.64 (Schomburg *et al.* 1998) and comprises two membrane valves and a pump thermopneumatical actuator. The membrane of the two valves and the diaphragm of the actuator are made of flexible polyimide, allowing for large deflections and excellent conformity. The large deflection of the actuation diaphragm provides higher flow rates because the output flow rate (Q) of a diaphragm micropump is proportional to the amplitude (w) of the diaphragm deflection according to a static analysis model

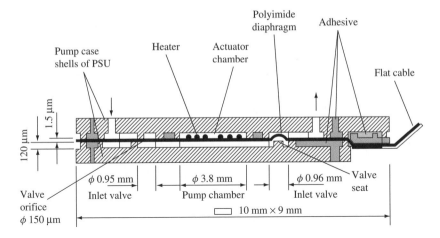

Figure 7.64 Micropump fabricated in a small-scale production line. From Schomburg *et al.* (1998)

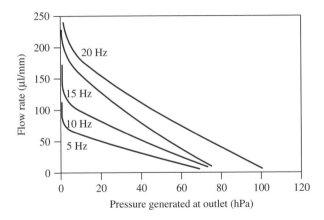

Figure 7.65 Typical flow rate as a function of the micropump pressure at various driving frequencies

(Jiang *et al.* 1998). This states that

$$Q \propto \eta \Delta V f \propto \eta f w A \tag{7.16}$$

where η is a coefficient relating to the valve leakage or efficiency, f is the actuating frequency, ΔV is the difference in volume between the pump mode and the supply mode, and is proportional to wA, where A is the diaphragm area within the pump chamber.

The deflection of the actuation diaphragm in a micropump fabricated by AMANDA process can be one-to-two orders of magnitude higher than that of the silicon micropump. Also, the excellent conformity of the valve membrane leads to a high efficiency η. All these characteristics are associated with a high output flow rate and back pressure (Figure 7.65). However, the working frequency of this micropump is relatively low because thermal pneumatic actuation was adopted, which may not be acceptable in some applications requiring high flow uniformity. The technical data of this type of micropump are listed in Table 7.1. It should be noted that both the housings and the diaphragms of the micropump are made of polymer and can be selected according to the special demands of the particular application, for example, biomedical fluid delivery. Another advantage of AMANDA is that it is a batch-fabrication process and therefore enables the production of a micropump at a relatively low cost.

Several types of microvalves have been developed with AMANDA (Schomburg *et al.* 1998). For example, a bistable microvalve, which consists of a preformed polyimide membrane that is able to occupy two stable positions (Figure 7.66). In the top position, the membrane closes the inlet duct of the fluid chamber, whereas, in the bottom position it clears that opening. In this case, the valve is controlled pneumatically. The technical

Table 7.1 Technical specification of a micropump. From Institute for Mikro

Pump attributes:	Values
Dimensions of the pump (mm)	$9.3 \times 10 \times 1.2$
Maximum height with fluidic ports (mm)	7.9
Outer diameter of fluidic ports (mm)	0.91
Flow rate without back pressure (μl/min)	150–250
Maximum pressure generated (hPa)	70–120
Life under laboratory conditions (load cycles)	>315 million
Drive voltage (V)	10–15
Drive pulse width (ms)	1–2
Drive frequency (Hz)	5–30
Power consumption (mW)	~150
Pump case	Polysulfone
Pump membrane	Polyimide
Adhesive bonds	Epoxy resin
Heater coil	Gold
Fluidic connection tube	Stainless steel

Note: Values for operation at 22 °C.

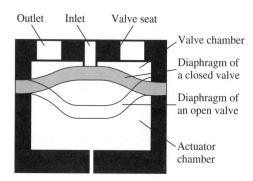

Figure 7.66 Cross section of a bistable micropump (top) and a bistable microvalve (bottom)

Table 7.2 Technical specification of microvalves

Valve part	Value
Valve chamber size (mm)	5 × 5 × 1
Valve chamber material	PMMA
Fluid chamber size (mm)	Diameter 3, height 0.125
Actuator chamber size (mm)	Diameter 3, height 0.125
Inlet opening diameter (µm)	100
Membrane diameter (mm)	3
Membrane thickness (µm)	25
Membrane material	Polyimide
Membrane deflection (µm)	120 max.
Maximum flow rates and inlet pressure	0.49 ml/s at 740 hPa
Lifetime (load cycles)	>285 million

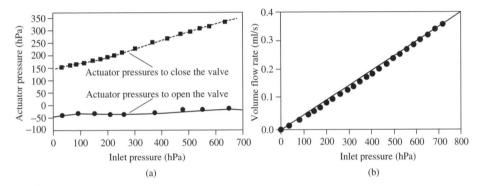

Figure 7.67 Characteristics of a microvalve fabricated by the AMANDA process: (a) actuation pressure and (b) volumetric flow rate

data on these microvalve samples are listed in Table 7.2 and the measured characteristics of the microvalve are presented in Figure 7.67. Applications include integral components of pneumatic and hydraulic systems, systems for chemical analyses of liquids and gases, dosage systems for medical applications, and so on.

AMANDA has also been used to fabricate transducers. For polymer membranes, the low Young's modulus results in large deflections and strains at comparatively low pressure loads. Therefore, polymer pressure transducers are suitable for measuring small differential pressures. A schematic view of a pressure transducer is shown in Figure 7.68 (Martin *et al.* 1998); the outer dimensions of this transducer are 5.5 × 4.3 × 1.2 mm^3. The thin polyimide diaphragm supports strain gauges made of gold, covered by a 30 µm-thick polyimide disk. This disk bends by the pressure dropped across the diaphragm, and the generated strain is measured with a Wheatstone bridge.

A volume flow transducer based on pressure difference measurement is shown in Figure 7.69 (Martin *et al.* 1998). The pressure drop along a capillary is measured and the flow rate is then calculated. These transducers can be easily integrated into the polymer micropump and microvalves developed by the AMANDA process to form a fully integrated microfluidic system.

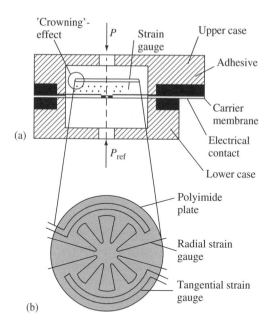

(a)

(b)

Figure 7.68   (a) Schematic cross section of a differential pressure transducer and (b) top view of the polyimide plate and strain gauge pattern

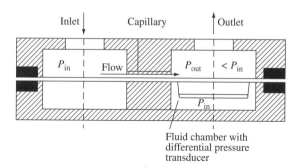

Figure 7.69   Schematic cross section of a volumetric flow rates transducer without electrical contacts. From Martin *et al.* (1998)

7.10 CONCLUDING REMARKS

In this chapter, we have reviewed the emerging field of MSL and its combination with other process technologies. MSL offers the promise of making a variety of microparts and microstructures without the use of vacuum systems and, in the case of polymeric microparts, high temperatures. It is particularly attractive in that it can be used to make in batch process truly 3-D microparts in a wide range of materials, polymers, metals, and ceramics at a modest cost. Because there are many applications in which silicon microstructures are ruled out as a result of, for example, biocompatibility, this technology looks extremely promising, not only for biofluidic but also for other types of MEMS devices. The main disadvantage of MSL is that it takes a long time to write into, and process, a large number of resist layers to fabricate a 3-D component. Although some of the MSL process technologies address this issue, costs must be reduced to compete with simpler methods, such as stamping, making 2-D microstructures.

REFERENCES

Andre, J. C., Le Methanute, A. and de Wittee, O. (1984). French Patent, No. 8411241.

Ballandras, S. *et al.* (1997). "Microstereolithography and shape memory alloy for the fabrication of miniaturised actuators," *Sensors and Actuators A*, **62**, 741–747.

Bau, H. H. *et al.* (1998). "Ceramic tape-based meso systems technology," *ASME MEMS*, **66**, 491–498.

Beluze, L., Bertch, A. and Renaud, P. (1999). "Microstereolithography: a new process to build complex three-dimensional objects," *Proc. SPIE*, **3680**, 808–817.

Bernard, W. L., Kahn, H., Heuer, A. H. and Huff, M. A. (1997). "A titanium-nickel-shape-memory alloy actuated micropump," *IEEE Technical Digest on Transducers '97*, **1**, 361–364.

Bertsch, A., Lorenz, H. and Renaud, P. (1998). "Combined microstereolithography and thick resist UV lithography for three-dimensional microfabrication," *Proc. IEEE MEMS*, 18–23.

Bertsch, A. *et al.* (1997). "Microstereolithography using liquid crystal display as dynamic mask-generator," *Microsyst. Technol.*, **3**, 42–47.

Cabrera, M. *et al.* (1998). "Microphotofabrication of very small objects: pushing the limits of stereolithography," *Mol. Cryst. Liq. Cryst.*, **315**, 223–234.

Cohen, A. *et al.* (1999). "EFAB: rapid, low-cost desktop micromachining of high aspect ratio true three-dimensional MEMS," *Proc. IEEE MEMS*, 244–251.

Corrozza, M. C., Croce, N., Magnani, B. and Dario, P. (1995). "A piezoelectric-driven stereolithography-fabricated micropump," *J. Micromech. Microeng.*, **5**, 177–179.

English, J. M. and Allen, M. G. (1999). "Wireless micromachined ceramic pressure sensors," *Proc. 12th IEEE Int. Conf. Microelectromech. Syst.*, 511–516.

Epstein, A. H. *et al.* (1997). *Proc. Int. Conf. Solid-State Sensors Actuators*, **2**, 753–756.

Griffith, M. L. and Halloran, J. W. (1995). *"Stereolithography of ceramics,"* *Proc. 27th Int. SAMPE Technical Conf.*, 970–979.

Hull, C. (1984). US Patent No. 4,575, August 8, 330.

Ikuta, K. and Hirowatari, K. (1993). "Real three-dimensional microfabrication using stereolithography and metal molding," *Proc. IEEE MEMS*, 42–47.

Ikuta, K., Maruo, S. and Kojima, S. (1998). "New microstereolithography for freely moved three-dimensional microstructure-super IH process with submicron resolution," *Proc. IEEE MEMS*, 290–295.

Ikuta, K., Maruo, S., Fujisawa, T. and Yamada, A. (1999). "Microconcentrator with opto-sense microreactor for biomedical IC chip family," *Proc. IEEE MEMS*, 376–381.

Ikuta, K., Ogata, T., Tsubio, M. and Kojima, S. (1996). "Development of mass productive microstereolithography (mass-IH process)," *Proc. IEEE MEMS*, 301–305.

Jacobs, P. F. (1992). *Rapid prototyping and manufacturing: fundamentals of stereolithography*, Society of Manufacturing Engineers, USA.

Jacobs, P. F. (1996) *Stereolithography and other RP&M technologies: from rapid prototyping to rapid tooling*, Society of Manufacturing Engineers, USA.

Jiang, X. N. *et al.* (1998). "Micronozzle/diffuser flow and its application in micro-valveless pump," *Sensors and Actuators A*, **70**, 81–87.

Jiang, X. N., Sun, C. and Zhang, X. (1999). "Microstereolithography of three-dimensional complex ceramic microstructures and PZT thick films on Si substrate," *ASME MEMS*, **1**, 67–73.

Katagi, T. and Nakajima, N. (1993). "Photoforming applied to fine machining," *Proc. IEEE MEMS*, 173–178.

Kathuria, Y. P. (1996). "Rapid prototyping: an innovative technique for microfabrication of metallic parts," *Proc. 7th Int. Symp. Micromachine Hum. Sci.*, 59–65.

Kodama, H. (1981). "Automatic method for fabricating a three-dimensional plastic model with photo-hardening polymer," *Rev. Sci. Instrum.*, **52**, 1770–1773.

Madden, J. D. and Hunter, J. W. (1996). "Three-dimensional microfabrication by localised electro-chemical deposition," *J. Microelectromech. Syst.*, **5**, 24–32.

Martin, J., Bacher, W., Hagena, O. F. and Schomburg, W. K. (1998). "Strain gauge pressure and volume-flow transducers made by thermoplastic molding and membrane transfer, "*Proc. IEEE MEMS*, 361–366.

Maruo, S. and Kawata, S. (1998). "Two-photon-absorbed near-infrared photopolymerisation for three-dimensional microfabrication," *J. Microelectromech. Syst.*, **7**, 411–415.

Maruo, S. and Kawata, S. (1997). "Two-photon-absorbed photopolymerisation for three-dimensional microfabrication," *Proc. IEEE MEMS*, 169–174.

Monneret, S., Loubere, V. and Corbel, S. (1999). "Microstereolithography using a dynamic mask generator and noncoherent visible light source," *Proc. SPIE*, **3680**, 553–561.

Nakamoto, T. and Yamaguchi, K. (1996). "Consideration on the producing of high aspect ratio microparts using UV sensitive photopolymer," *Proc. 7th Int. Symp. Micromachine Human Sci.*, 53–58.

Polla, D. L. and Francis, L. F. (1996). "Ferroelectric thin films in microelectromechanical systems applications," *MRS Bull.*, 59–65.

Schomburg, W. K. *et al.* (1998). "AMANDA-low-cost production of microfluidic devices," *Sensors and Actuators A*, **70**, 153–158.

Sun, C., Jiang, X. N. and Zhang, X. (1999). "Experimental and numerical study on microstereo-lithography of ceramics," *ASME MEMS*, 339–345.

Suzumori, K., Koga, A. and Haneda, R. (1994). "Microfabrication of integrated FMAs using stereo-lithography," *Proc. IEEE MEMS*, 136–141.

Takagi, T. and Nakajima, N. (1994). "Architecture combination by micro photoforming process," *Proc. IEEE MEMS*, 211–216.

Taylor, C. S. *et al.* (1994). "A spatial forming a three-dimensional printing process," *Proc. IEEE MEMS*, 203–208.

Thornell, G. and Johansson, S. (1998). "Microprocessing at the fingertips," *J. Micromech. Microeng.*, **8**, 251–262.

Varadan, V. K., Varadan, V. V. and Motojima, S. (1996). "Three-dimensional polymeric and ceramic MEMS and their applications," *Proc. SPIE*, **2722**, 156–164.

Wayne, R. P. (1988). *Principles and Applications of Photochemistry*, Oxford University Press, New York.

Zhang, X., Jiang, X. N. and Sun, C. (1999). "Microstereolithography of polymeric and ceramic microstructures," *Sensors and Actuators A*, **77**, 149–156.

Zissi, S. *et al.* (1996). "Stereolithography and microtechnologies," *Microsyst. Technol.*, **2**, 97–102.

8

Microsensors

8.1 INTRODUCTION

A sensor may be simply defined as a device that converts a nonelectrical input quantity $\overline{E}$ into an electrical output signal E; conversely, an actuator may be defined as a device that converts an electrical signal E into a nonelectrical quantity $\overline{E}$ (see Figure 8.1). In contrast, a processor modifies an electrical signal (e.g. amplifies, conditions, and transforms) but does not convert its primary form. A transducer is a device that can be either a sensor or an actuator. Some devices can be operated both as a sensor and an actuator. For example, a pair of interdigitated electrodes lying on the surface of a piezoelectric material can be used to sense surface acoustic waves (SAWs) or to generate them. This device is referred to as an *interdigitated transducer* (IDT). The importance of this device is such that we have dedicated Chapter 13 to describing its applications as a microsensor and Chapter 14 to describing its use in microelectromechanical system (MEMS) devices.

It has been proposed by Middelhoek that a sensor or actuator can be classified according to the energy domain of its primary input–output (I/O). There are six primary energy domains and the associated symbols are as follows:

- Electrical   E
- Thermal   T
- Radiation   R
- Mechanical   Me
- Magnetic   M
- Bio(chemical)   C

For example, Figure 8.2 shows the six energy domains and the vectors that define the conventional types of sensors and actuators, that is, $\overrightarrow{A}$ vector represents a thermal sensor, whereas $\overleftarrow{A}$ represents a thermal actuator. In this way, all the different types of sensors (and actuators) can be classified. In practice, the underlying principles of a sensor may involve several stages; for example, the primary nonelectrical input (radiation) that first transforms into the mechanical domain, then into the thermal domain, and finally into the electrical domain.

Figure 8.3 shows the vectorial representation of this radiation sensor and the three different stages of the conversion.

In theory, a transducer could have a large number of stages, but in practice, this is usually between one and three. For example, an electromagnetic actuator has two: first,

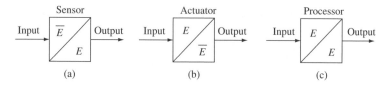

Figure 8.1 Basic input–output representation of (a) a sensor; (b) an actuator; and (c) a processor in terms of their energy domains

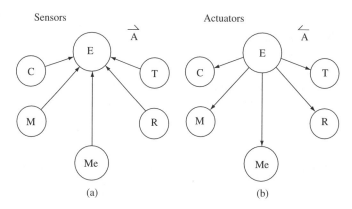

Figure 8.2 Vectorial representation of (a) a sensor and (b) an actuator in energy domain space. A processor would be represented by a vector that maps from E and back onto itself

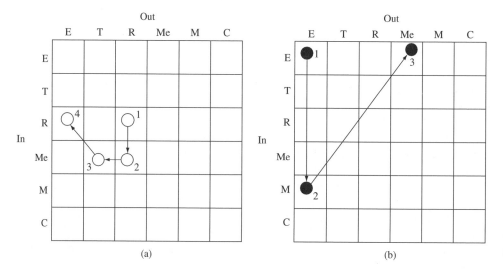

Figure 8.3 Vectorial representation of a multistage transducer in energy domain space: (a) a four-stage radiation sensor and (b) a three-stage magnetic actuator

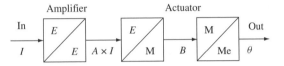

Figure 8.4   Block-diagram representation of the transduction processes within a magnetic actuator (i.e. electromagnetic motor). The front-end power electronic device is also shown

the electrical signal E is converted into the magnetic domain M, and then the magnetic domain is finally converted to a mechanical force that drives the motor and produces motion Me.

This actuator system can also be illustrated in a block diagram (see Figure 8.4) together with a power amplifier on the front end to enhance the small electrical actuating input current signal I. In this case, the current through a coil induces a magnetic field B, which induces a torque on the rotor and hence outputs a rotational motion θ. This block diagram is similar to a control block diagram, and a transfer function can be assigned to each stage of the transduction process to model the system dynamics.

There is another approach that has been adopted here to classify sensors and actuators more precisely in terms of the electrical principle employed. Table 8.1 shows the different names that are derived from the electrical domain and used to describe different types of sensors (and actuators).

The first set of devices is named according to the electrical *property* that is changed, that is, the electrical resistance R, electrical capacitance C, or electrical inductance L. For example, capacitive sensors are widely used because they are voltage-controlled devices[1] (such as metal oxide semiconductor integrated circuits (MOS ICs)) and offer low power consumption – an essential feature for battery-operated devices and instruments.

Table 8.1   Classification of transducers by electrical property or signal type

Property/signal	Descriptor	Example of sensor	Example of actuator
Property:			
Resistance, R	Resistive	Magnetoresistor	Piezoresistor
Capacitance, C	Capacitive	Chemical capacitor	Electrostatic motor
Inductance, L	Inductive	Inductive proximity sensor	Induction motor
Signal:			
Voltage, V	Potentiometric	Thermocouple	Electrical valve
Current, I	Amperometric	Fuel cell	Solenoid valve
Charge, q	Coulombic or electrostatic	Piezoelectric pressure	Electrostatic resonator
Frequency, f	–	Acoustic wave	Stepper motor[a]

[a]Operated with a pulsed rather than alternating current (AC) actuating signal

[1] These voltage-controlled devices normally have high input impedance at low-drive frequencies and so draw low currents.

The second set of devices is named according to the *nature* of the electrical signal. Therefore, a capacitive sensor could be called a *potentiometric sensor* when a change in voltage is recorded or a *coulombic sensor* when a change in electric charge is recorded. In practice, sensors tend to be classified according to both the primary measurand (or actuand) and the basic principle involved, for example, a capacitive pressure sensor. Using this nomenclature, it is possible to describe reasonably clearly the type of device in question.

Many books that have been published on the topic of sensors[2] often focus on one principle, such as thermal, pressure, chemical, and so on. Appendix K lists a number of general books on sensors, but interested readers are referred to two books in particular. First, an introductory text by Hauptmann (1991), which gives an excellent overview of sensors for readers unfamiliar with the field, and second, a more advanced eight-volume book series by Göpel published by Wiley-VCH, which provides the most comprehensive review of sensors to date[3]. There are relatively few books that have been published specifically on the topic of actuators. More commonly, actuators are often described within books on either transducers or, perhaps, instrumentation. Therefore, we recommend the introductory texts on *Transducers* by Norton (1989) and the more advanced instrumentation reference book edited by Noltingk (1995).

In this chapter, we are concerned with miniature sensors, so-called *microsensors*[4], which are fabricated using predominantly the bulk- and surface-micromachining technologies described in Chapters 5 and 6, respectively. Again, there are a number of textbooks already published, which report on the topic of microsensors, but there are very few on microactuators[5]. For example, we recommend the book on *Silicon Sensors* by Middelhoek and Audet (1989) and *Microsensors* by Gardner (1994). The subsequent sections provide an overview of the field of microsensors, and as stated above, the emerging field of IDT microsensors is covered separately in Chapter 13.

Some sensing devices have a part or all of the processing functions integrated onto the same silicon substrate. We refer to these devices as *smart* sensors. We reserve the label of 'intelligent' for devices that have in addition some biomimetic function such as self-diagnostic, self-repair, self-growth, and fuzzy logic. The topic of smart (and intelligent) sensors is dealt with in Chapter 15.

There have been rapid developments in the field of microsensors during the past 10 years, and a sharp increase has taken place in the size of the world market, which has become some billions of euros today (see Chapter 1). Here, we focus upon the main types of microsensors, which have powered this sensing revolution, together with some of the emerging new designs.

8.2 THERMAL SENSORS

Thermal sensors are sensors that measure a primary thermal quantity, such as temperature, heat flow, or thermal conductivity. Other sensors may be based on a thermal

---

[2] This includes books on the topic of transducers (where a sensor is an input transducer).
[3] Wiley-VCH regularly publish books called *Sensors Update* to supplement the original volume series.
[4] Most microsensors are based on silicon technology; however, the term refers to devices with one dimension in the micron range.
[5] Published proceedings of meetings are not regarded here as textbooks.

measurement; for example, a thermal anemometer measures air flow. However, according to our classification of measurand energy domain, this would be regarded as a mechanical sensor and appear under Section 8.2.3. Consequently, the most important thermal sensor is the temperature sensor.

Temperature is probably the single most important device parameter of all. Almost every property of a material has significant temperature dependence. For example, in the case of a mechanical microstructure, its physical dimensions – Young's modulus, shear modulus, heat capacity, thermal conductivity, and so on – vary with operating temperature. The effect of temperature can sometimes be minimised by choosing materials with a low temperature coefficient of operation (TCO). However, when forced to use standard materials (e.g. silicon and silica), the structural design can often be modified (e.g. adding a reference device) to compensate for these undesirable effects.

It is often necessary to use materials that are not based on complementary metal oxide semiconductor (CMOS), such as magnetoresistive, chemoresistive, ferroelectric, pyroelectric; these compounds tend to possess strong temperature-dependencies[6]. In fact, the problem is particularly acute for chemical microsensors, as most chemical reactions are strongly temperature-dependent.

Many nonthermal microsensors (and MEMS devices) have to operate either at a constant temperature – an expensive and power-intensive option when requiring heaters or coolers, – or in a mode in which the temperature is monitored and real-time signal compensation is provided. Clearly, microdevices that possess an integrated temperature microsensor and microcontroller can automatically compensate for temperature and thus offer a superior performance to those without. This is why temperature sensors are a very important kind of sensors and are commonly found embedded in microsensors, microactuators, MEMS, and even in precision microelectronic components, such as analogue-to-digital converters.

8.2.1 Resistive Temperature Microsensors

Conventionally, the temperature of an object can be measured using a platinum resistor, a thermistor, or a thermocouple. Resistive thermal sensors exploit the basic material property that their bulk electrical resistivity ρ, and hence resistance R, varies with absolute temperature T. In the case of metal chemoresistors, the behaviour is usually well described by a second-order polynomial series, that is,

$$\rho(T) \approx \rho_0(1 + \alpha_T T + \beta_T T^2) \text{ and } R(T) \approx R_0(1 + \alpha_T T + \beta_T T^2) \qquad (8.1)$$

where ρ_0/R_0 are the resistivity or resistance at a standard temperature (e.g. $0\,°C$) and α_T and β_T are temperature coefficients. α_T is a sensitivity parameter and is commonly known as *the linear temperature coefficient of resistivity or resistance* (TCR) and is defined by

$$\alpha_T = \frac{1}{\rho_0} \frac{d\rho}{dT} \qquad (8.2)$$

---

[6] The properties of common metals, semiconductors, and other materials are tabulated in Appendices F, G, and H.

Platinum is the most commonly used metal in resistive temperature sensors because it is very stable when cycled over a very wide operating temperature range of approximately -260 to $+1700\,°C$, with a typical reproducibility of better than $\pm0.1\,°C$. In fact, platinum resistors are defined under a British Standard BS1904 (1964), made to a nominal resistance of $100\,\Omega$ at room temperature, and referred to as Pt-100 sensors. Platinum temperature sensors are very nearly linear, and α_T takes a value of $+3.9 \times 10^{-4}/K$ and β_T takes a value that is four orders of magnitude lower at $-5.9 \times 10^{-7}/K^2$. In contrast, thermistors, that is, resistors formed from semiconducting materials, such as sulfides, selenides, or oxides of Ni, Mn, or Cu, and Si have highly nonlinear temperature-dependence. Thermistors are generally described by the following equation:

$$\rho(T) \approx \rho_{\text{ref}} \exp[B(1/T - 1/T_{\text{ref}})] \tag{8.3}$$

where the reference temperature is generally $25\,°C$ rather than $0\,°C$ and the material coefficient β is related to the linear TCR by $-B/T^2$. The high negative TCR means that the resistance of a pellet falls from a few megaohms to a few ohms over a short temperature range, for example, $100\,°C$ or so.

8.2.2 Microthermocouples

Unlike the metal and semiconducting resistors, a thermocouple is a potentiometric temperature sensor in that an open circuit voltage V_T appears when two different metals are joined together with the junction held at a temperature being sensed T_s and the other ends held at a reference temperature T_{ref} (see Figure 8.5).

The basic principle is known as the *Seebeck effect* in which the metals have a different thermoelectric power or Seebeck coefficient P; the thermocouple is conveniently a linear device, with the voltage output (at zero current) being given by

$$V_T = (V_B - V_A) = (P_B - P_A)(T_s - T_{\text{ref}}) = (P_B - P_A)\Delta T \tag{8.4}$$

Thermocouples are also widely used to measure temperature, and their properties are defined in British and US standards for different compositions of metals and alloys, for

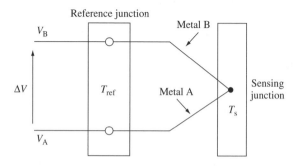

Figure 8.5   Basic configuration of a thermocouple temperature sensor (a type of potentiometric thermal sensor)

example, types B, E, J, K, N, R, S, and T. Typically, they can operate from -100 to $+2000\,°C$ with an accuracy of between 1 and 3 percent for a full-scale operation (FSO).[7]

Here, we are mainly interested in whether a temperature sensor can be integrated in a silicon process to become either a temperature microsensor or part of a silicon-based MEMS device. Table 8.2 summarises the typical properties of conventional temperature sensors and, more importantly, whether they can be integrated into a standard integrated circuit (IC) process.

As is apparent from Table 8.2, it is possible to integrate resistive temperature sensors such as the platinum Pt-100. However, the deposition of platinum or the thermistor oxide is a nonstandard IC process and therefore requires additional pre- or post-IC processing steps. The inclusion of nonstandard materials during, for example, a CMOS process, which is 'intermediate' CMOS, is generally regarded as highly undesirable and should be avoided if possible.

It is possible to fabricate silicon resistors in standard silicon IC process, as described in Chapter 4. For example, five or more resistors can be made of doped silicon in a standard bipolar process, such as a base resistor, emitter resistor, or an epi-resistor, and two or three resistors can be made in a CMOS process (see Figure 4.15). The resistivity of a single crystal of silicon varies with temperature and doping level, as illustrated in Figure 8.6, and the lightly doped silicon provided the highest TCR. In practice, it is difficult to make single-crystal silicon with an impurity level below $\sim 10^{12}$ cm^{-3}; therefore, it will not behave as an intrinsic semiconductor with a well-defined Arrhenius temperature-dependence because the intrinsic carrier concentration is about 10^{10} cm^{-3} at room temperature. In highly doped silicon resistors ($\sim 10^{18}$ cm^{-3}), the temperature-dependence approximates reasonably well to the second-order polynomial given in Equation (8.1). Nevertheless, the temperature-dependence of a silicon resistor is nonlinear and depends upon the exact doping level, making it less suitable for use as a temperature sensor than other

Table 8.2 Properties of common temperature sensors and their suitability for integration. Modified from Meijer and van Herwaarden (1994)

Property	Pt resistor	Thermistor	Thermocouple	Transistor
Form of output	Resistance	Resistance	Voltage	Voltage
Operating range (°C)	Large -260 to $+1000$	Medium -80 to $+180$	Very large -270 to $+3500$	Medium -50 to $+180$
Sensitivity	Medium 0.4%/K	High 5%/K	Low 0.05 to 1 mV/K	High ~ 2 mV/K
Linearity	Very good $< \pm 0.1$ K	Very nonlinear	Good ± 1 K	Good ± 0.5 K
Accuracy:				
–absolute	High over wide range	High over small range	Not possible	Medium
–differential	Medium	Medium	High	Medium
Cost to make	Medium	Low	Medium	Very low
Suitability for IC integration	Not a standard process	Not a standard process	Yes	Yes–very easily

[7] The sensitivity diminishes significantly below $-100\,°C$.

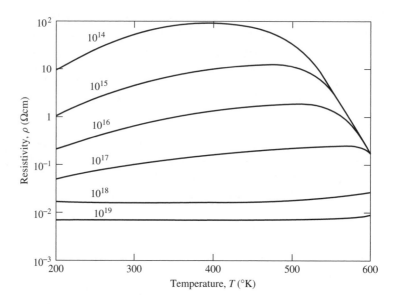

Figure 8.6 Temperature-dependence of single-crystal silicon doped at various levels (n-type). From Wolf (1969)

types of device. Therefore, the preferred approaches are to make a microthermocouple out of silicon or, better still, use the inherent temperature sensitivity of a silicon diode or transistor.

The Seebeck coefficient of single-crystal silicon varies with both temperature and doping concentration (p-type) as shown in Figure 8.7. Doping has the effect of reducing the temperature variation of the coefficient itself; hence, the response of a silicon-based thermocouple becomes more linear. As a variety of doping levels are possible in a planar IC process, a Seebeck coefficient ranging from $+0.5$ to $+5$ mV/°C is achievable.

In theory, the Seebeck coefficient of a doped semiconductor is given by

$$n\text{-type: } P_{n-\text{Si}} = -\frac{k_B}{q}\{[\ln(N_c/n) + 2.5] + (1 + s_n) + \phi_n\}$$

$$p\text{-type: } P_{p-\text{Si}} = -\frac{k_B}{q}\{[\ln(N_v/p) + 2.5] + (1 + s_p) + \phi_p\}$$

(8.5)

where k_B is the Boltzmann's constant, q is the carrier charge, N_c and N_v are the density of states at the bottom of the conductance band and top of the valence band, n and p are the donor and acceptor concentrations, s is a parameter related to the mean free time between collisions and the charge carrier energy and its value varies between -1 and $+2$ depending on whether the carriers can move freely or are trapped, and finally ϕ is a phonon drag term for the carrier. In practice, the Seebeck coefficient can be readily estimated from the silicon resistivity rather than the carrier concentrations and is simply given by

$$P_{\text{Si}} \approx \frac{m k_B}{q} \ln \frac{\rho}{\rho_0}$$

(8.6)

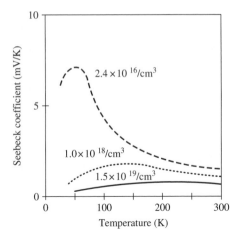

Figure 8.7 Variation of Seebeck coefficient for single-crystal silicon doped with temperature at different concentrations of boron (i.e. p-type). Adapted from Geballe and Hull (1955)

where m is a dimensionless constant (negative for n-type and positive for p-type) and is typically around 2.6 and ρ_0 is a resistivity constant of 5×10^{-6} Ωm.

Therefore, a silicon thermocouple can be made in an IC process with doped silicon and a standard metal contact, for example, aluminum. Figure 8.8 shows such a thermal microsensor and consists of a series of N identical p-Si/Al thermocouples.

The theoretical voltage output V_{out} of this thermopile is given subsequently (from Equation (8.4)) and agrees well with experimental values.

$$V_T = N(V_{p-\text{Si}} - V_{\text{Al}}) = N(P_{p-\text{Si}} - P_{\text{Al}})\Delta T \qquad (8.7)$$

As the absolute Seebeck coefficient of p-type silicon is positive (e.g. $+1$ mV/K for a sheet resistance of 200 Ω/sq at 300 K) and that for aluminum is negative (i.e. -1.7 μV/K

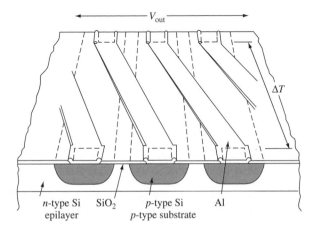

Figure 8.8 Example of a temperature microsensor: a p-Si/Al thermopile integrated in an n-type epilayer employing a standard bipolar process. From Meijer and van Herwaarden (1994)

at 300 K), an output on the order of n millivolts per degree can be achieved from a thermopile. Polysilicon/gold thermocouples have also been made with an output of about $+0.4$ mV/K in which the n-type (phosphorous) polysilicon has a lower Seebeck coefficient of -176 μV/K (for a sheet resistance[8] of 100 Ω/sq at 300 K) and the gold has a standard value of $+194$ μV/K. However, these are not standard IC process materials and so polysilicon-based thermocouples are not the preferred fabrication route for low-cost temperature microsensors.

8.2.3 Thermodiodes and Thermotransistors

The simplest and easiest way to make an integrated temperature sensor is to use a diode or transistor in a standard IC process. There are five ways in a bipolar process and three ways in a CMOS process to make a p-n diode (see Table 4.2). The I-V characteristic of a p-n diode is nonlinear (Figure 4.19) and follows Equation (4.14), which is repeated here for the sake of convenience:

$$I = I_s[\exp(\lambda q V / k_B T) - 1] \tag{8.8}$$

where I_s is the saturation current, typically 1 nA and λ is an empirical scaling factor that takes a value of 0.5 for an ideal diode. Rearranging Equation (8.8) in terms of the diode voltage gives

$$V = \frac{k_B T}{q} \ln\left(\frac{I}{I_s} + 1\right) \tag{8.9}$$

Therefore, when the diode is operated in a constant current I_0 circuit (see Figure 8.9(a)), the forward diode voltage V_{out} is directly proportional to the absolute temperature[9] and

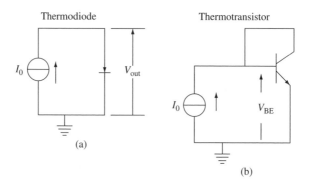

Figure 8.9 Basic temperature microsensors: (a) a forward-biased p-n diode and (b) an n-p-n transistor in a common emitter configuration with V_{CE} set to zero

---

[8] The resistance of a square piece of material is independent of its size.
[9] Sometimes called a proportional to absolute temperature (PTAT) device.

the voltage sensitivity S_T is a constant depending on the drive current:

$$V_{out} = \frac{k_B T}{q} \ln\left(\frac{I_0}{I_s} + 1\right) \text{ and } S_T = \frac{dV_{out}}{dT} = \frac{k_B}{q} \ln\left(\frac{I_0}{I_s} + 1\right) \tag{8.10}$$

The overall temperature sensitivity of the diode depends on the relative size of the drive current and saturation. When the drive current is set to a value well above the saturation current, Equation (8.10) becomes

$$V_{out} \approx \frac{k_B T}{q} \ln \frac{I_0}{I_s} \text{ and } S_T \approx \frac{k_B}{q} \ln \frac{I_0}{I_s} \text{ when } I_0 \gg I_s \tag{8.11}$$

Let us suppose that the forward current is 0.1 μA and about 100 times the diode saturation current of approximately 1 nA; then, the expected temperature sensitivity is +0.2 mV/K. However, in practice, the temperature-dependence of a diode depends on the strong temperature-dependence of the saturation current itself. The actual value can be obtained experimentally from the temperature-dependence of the forward junction voltage of a silicon diode[10], that is, -2 mV/°C, and therefore $V_{f0} \propto T$.

In a similar way, a bipolar transistor can be used as a temperature sensor. For example, Figure 8.9(b) shows an n-p-n transistor in a common-emitter configuration and constant current circuit. From our basic theory, the base-emitter voltage V_{BE} is proportional to the absolute temperature and simply related to the collector current I_C by

$$V_{BE} = \frac{k_B T}{q} \ln\left(\frac{I_C}{I_{C0}}\right) \text{ where } I_{C0} = A_E J_s \tag{8.12}$$

where A_E is the area of the emitter, J_s is the saturation current density, and I_{C0} is the reverse saturation current. More accurate models can be developed from, for example, those discussed previously for a bipolar transistor (Equation (4.20)), but the base-emitter current approximates well in practice to

$$V_{BE} \approx V_{BE0} + \lambda T \tag{8.13}$$

where λ is an empirical constant that depends on the current density and process parameters and the offset voltage V_{BE0} has a typical value of 1.3 V when the base-collector voltage V_{BC} is set to zero.

To make a truly PTAT sensor, it is necessary to fabricate two transistors – one with an emitter area A_{E1} and the other with A_{E2}. Then the difference in their base-emitter voltages is directly proportional to the absolute temperature and is given by

$$\Delta V_{BE} = (V_{BE1} - V_{BE2}) = \frac{k_B T}{q} \ln\left(\frac{I_{C1} J_{s2} A_{E2}}{I_{C2} J_{s1} A_{E1}}\right) \approx \frac{k_B T}{q} \ln\left(\frac{A_{E2}}{A_{E1}}\right) \tag{8.14}$$

When the two transistors are identical, the collector currents and saturation current densities are equal, and the ratio of the emitter areas only determines the sensor's response.

---

[10] Typically 0.7 V at 25 °C for silicon (and 0.25 V for germanium).

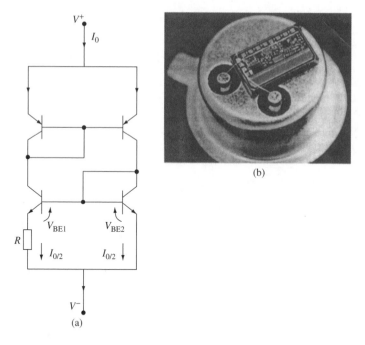

Figure 8.10 (a) Basic transistor circuit for a PTAT sensor and (b) photograph of a commercial integrated silicon temperature IC. From Wolffenbuttel (1996)

Transistors are the most attractive elements for measuring temperature either in a discrete device or in a part of a standard IC. For example, Figure 8.10(a) shows a simple PTAT circuit that uses two identical p-n-p transistors to divide the current equally into two[11] n-p-n transistors with different emitter areas. The voltage dropped across the resistor R is simply the difference in base-emitter voltages for the n-p-n transistors; therefore, the current I_o flowing out is

$$I_o \approx \frac{2k_B T}{qR} \ln\left(\frac{A_{E2}}{A_{E1}}\right) \tag{8.15}$$

Figure 8.10(b) shows a commercially available integrated temperature IC based on the PTAT circuit. In the temperature IC (RS 590kH), the output current I_o has been set by laser-trimming of the resistor to 298.2 ± 2.5 µA for a temperature of 298.2 K, and the temperature sensitivity S_T is 1.0 µA/°C over the range of -55 to $+150$ °C.

There are a number of variations in this type of PTAT circuit, such as using a set of eight identical n-p-n transistors of equal emitter area and adding a reference offset voltage to have an output closer to zero at room temperature. However, this type of temperature sensor is simple to make in a standard IC process and has a good sensitivity and low dependence upon process variation because of the ratiometric principle employed. Therefore, it is an attractive option in many cases.

---

[11] Commonly referred to as a current mirror.

8.2.4 SAW Temperature Sensor

In certain circumstances, it may be cost-effective to use other technologies. For example, the remote wireless sensing of temperature – perhaps on a rotating part – requires a temperature IC, radio frequency (RF) transmitter, and a battery power supply. An alternative approach would be to use a wireless SAW temperature sensor. Figure 8.11 shows a schematic drawing of such a device. The SAW sensor consists of a thin lithium niobate piezoelectric layer on top of a ceramic, glass, or silicon substrate (Bao *et al.* 1994). A thin aluminum film is patterned using optical lithography (1-mask process) to form a pair of interdigital electrodes connected to a small microwave antenna and a pair of reflectors. The basic principle is that a frequency modulated (FM) electromagnetic signal is transmitted remotely and is picked up by the small antenna, which then drives the SAW via the interdigital electrodes down to the reflectors. The reflectors return the wave that then drives the FM antenna and sends back two signals to the remote location. The time-delays of the two signals are measured using suitable electronic circuitry, for example, a mixer to obtain the phase differences from the reference signal and a microcomputer to interpret the output.

The difference in phase angle $\Delta\varphi$ between the two signals is linearly related to the temperature by the two time delays τ_1 and τ_2 (Bao *et al.* 1994),

$$\Delta\varphi \approx \omega_0\alpha(\tau_2 - \tau_1)\Delta T \tag{8.16}$$

where the original FM signal has a frequency $(2\pi\omega_0)$ equal to 905 MHz, the time-delays of the reflected signals are 1 μs and 1.1 μs at room temperature, and the temperature coefficient of lithium niobate is $9.4 \times 10^{-5}/°C$. The temperature sensitivity of the SAW-IDT microsensor is calculated to be 3.1 degrees/°C. Experimental results are shown in Figure 8.12 and it can be seen that they agree well with the theory. The resolution of the sensor is about 1 degree of angle or 0.33 °C.

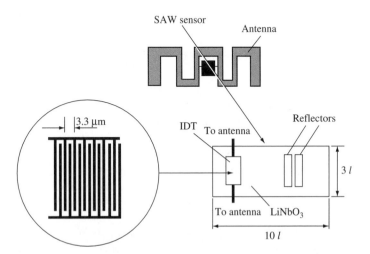

Figure 8.11   Integrated passive SAW-IDT microsensor for wireless temperature sensing of remote components. After Bao *et al.* (1994)

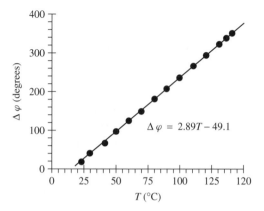

Figure 8.12 Effect of temperature on the phase shift from an integrated wireless SAW sensor. After Bao *et al.* (1994)

This temperature microtransducer has great practical value in that the sensing part is passive, that is, it requires no power supply; second, it is wireless and therefore can be embedded in rotating or moving parts, such as car tyres, turbine blades, helicopter rotors, and so on; and finally, it is very inexpensive to make. SAW-IDT microsensors have many more important applications than simply acting as a temperature sensor. Consequently, the microsensors are subsequently covered in further detail; Chapter 12 provides fabrication details and Chapter 13 summarises the possible applications of IDT microsensors.

8.3 RADIATION SENSORS

Radiation sensors can be classified according to the type and energy of the measurand, as illustrated in Figure 8.13, in which the energy E_R in electron volts (eV) of the electromagnetic radiation is simply related to its frequency f in Hertz, and wavelength λ in meters by

$$E_R \text{ (in eV)} = \frac{hf}{e} = \frac{hc}{\lambda e} \qquad (8.17)$$

where h is Planck's constant, c is the velocity of light in a vacuum (i.e. 3.0×10^8 m/s), and e is the charge on an electron (i.e. 1.60×10^{-19} C).

Radiation can come in the form of particles, such as neutrons, protons, and alpha and beta particles, emitted from the decay of nuclear material. However, the detection of nuclear particles and high-energy electromagnetic radiation (i.e. gamma rays and X rays) generally requires sophisticated instrumentation that cannot be readily integrated into a miniature device. The exception to this general observation is perhaps the detection of low-energy X rays or electrons through a solid-state photoelectric detector, the principle for which is covered in Section 8.3.2.

The most common types of radiation microsensor detect electromagnetic radiation with energies or wavelengths from the ultraviolet-to-near-infrared (UV-NIR) region, which includes visible, through the NIR and thermal-infrared region and into the microwave and radio regions. The most important regions are the visible light region and the NIR

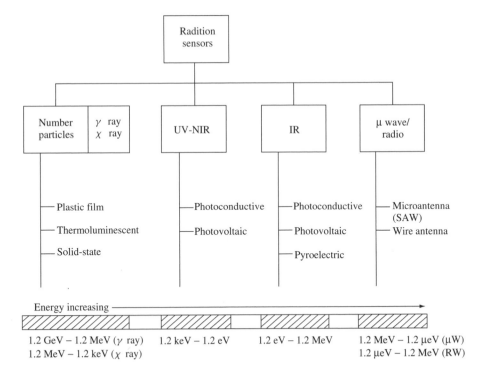

Figure 8.13 Classification of radiation sensors according to their electromagnetic energy

region because these are the wavelengths at which signals are transmitted down fibre-optic cables in modern telecommunication systems. Readers interested in the general field of fibre-optic sensors are referred to Udd (1991) and those interested in the field of biosensing and chemical sensing are referred to Boside and Harmer (1996). The real interest to us here is whether the optical components and any optical interconnects can be integrated into a microtransducer or MEMS device (see Tabid-Azar 1995). In this book, we have concentrated on devices that have electrical outputs (sensors), electrical inputs (actuators), and on electrical interconnects, as described earlier on in Sections 4.5 and 4.6 (e.g. printed circuit boards (PCBs) and multichip modules (MCMs)). This was done as it reflects the bulk of past developments, but it is now becoming increasingly evident that the optical signal domain will become the more significant for both land-based signal transmission and for the operation of sensors within hazardous environments.

Radiation microsensors can be distinguished by their underlying operating principle, namely, photoconductive, photovoltaic (or Photoelectric), pyroelectric, and microantenna. We will employ this distinction to discuss these four types of radiation sensor in turn.

8.3.1 Photoconductive Devices

The basic principle of a photoconductive cell (or conductive radiation device) is shown in Figure 8.14. The radiation excites a number of electrons from the valence band of a semiconductor material into its conduction band and thus creates both electrons and

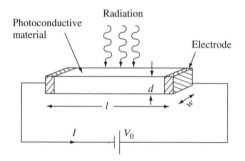

Figure 8.14 Basic layout of a photoconductive cell with a constant voltage V_o drive circuit

holes that can contribute to the conduction process. This Photoconductive effect is the dominant process when the energy of the radiation is above the band gap energy of the semiconducting material. At higher energies, of 100 keV and above, other effects, such as Compton scattering, occur but these only apply for detecting hard X rays and gamma rays.

If the radiation produces N_t carriers per second in a slab of material of length l, width w, and depth d, its change in electrical conductivity $\Delta\sigma$ and change in electrical conductance G is given by

$$\Delta\sigma = eN_t(\mu_n\tau_n + \mu_p\tau_p) \quad \text{and so} \quad \Delta G = eN_t(\mu_n\tau_n + \mu_p\tau_p)\frac{wd}{l} \tag{8.18}$$

where μ_n and μ_p are the mobilities of the electrons and holes and τ_n and τ_p are their lifetimes. The conductance I/V_o can be measured in a constant voltage V_o circuit with the cell resistance typically falling almost linearly with illuminance from megaohms in the dark (band gap exceeds the $1/40^{\text{th}}$ eV of thermal energy at room temperature) to a few ohms.

The response of a photoconductive cell to radiation is determined by the choice of semiconducting material. Figure 8.15 shows the various materials used to cover the UV to IR range. Cadmium sulfide is commonly used to make a photoconductive cell for the visible region (0.4 to 0.7 μm) because it is inexpensive and easy to process, although other materials are used for IR photoconductors, such as PbS with a peak response at 2.2 μm, PbSe with a peak response at 2.2 μm, and HgCdTe (MCT) with a response tailored within the range of 12 to 16 μm.

Photoconductive cells are commercially available at low cost and are commonly employed in a wide range of applications; examples include light-activated switches for night lights, dimmers, and children's toys.

8.3.2 Photovoltaic Devices

Photovoltaic, or photoelectric, sensors form the second class of radiation microsensors and are potentiometric radiation sensors. The basic principle is that the radiation (i.e. photons) induces a voltage across a semiconductor junction; this effect is known as the *photovoltaic effect*. The materials most commonly used to make photovoltaic sensors are Si for the visible/NIR region, and Ge, InGaAs, InAs, or InSb for the NIR-to-IR region.

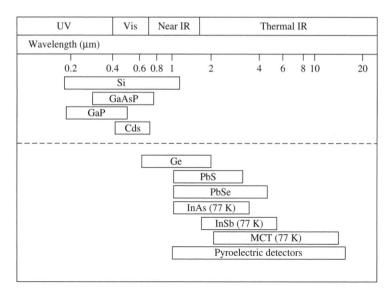

Figure 8.15   Some common semiconducting materials used in radiation microsensors and their dynamic range within the UV-to-IR spectrum

The most obvious advantage of a photovoltaic cell over a photoconductive cell is that it is compatible with a bipolar process (e.g. Si or GaAs). However, the reduced size and integrated electronics lead to a higher sensitivity, faster response time (µs instead of ms), and better stability.

Photosensitive diodes, known as photodiodes, can be made in a standard vertical bipolar process (see Section 4.3.4) such as a *p-n* diode, or variations on this process, such as a *p*-Si/insulator/*n*-Si PIN diode, Schottky-type diode, and silicon avalanche diode (see Figure 8.16).

The basic principle of all these photodiodes is that the photon creates an electron-hole pair in the space-charge region of the junction. These charges are then separated by the local field to the different doped regions, and they modify the diode voltage V_d. The diode voltage is the open circuit voltage V_{oc} and can be measured by reverse-biasing the diode and finding the voltage dropped across a high external load resistor R_L. Then the output voltage is given by

$$V_{oc} \approx \frac{k_B T}{e} \ln\left(1 + \frac{I_R}{I_s}\right) \text{ and so } V_{oc} \propto \ln(I_R) \tag{8.19}$$

where I_R is the photocurrent and is proportional to the intensity of incident radiation and, as usual, I_s is the reverse saturation current. Indeed, this equation is the same one that applied to a thermodiode (see Section 8.2.1 and Equation (8.9)), and so the device will normally be temperature-sensitive. Thus, the performance of a photodiode can be improved by either running it at a constant voltage – as low as possible to reduce junction noise – or it must be temperature-compensated.

An alternative approach is to fabricate a phototransistor rather than a diode. Like a PTAT device, an IC can be built using two identical transistors; the first produces a photoinduced collector current I_{CR} and the other produces a reference collector current I_{C0}.

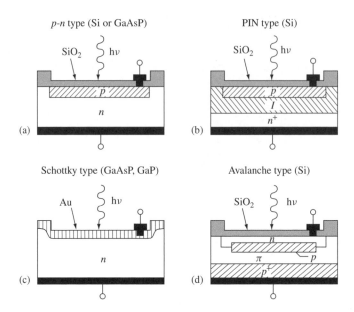

Figure 8.16 Some different types of radiation microsensors: (a) *p-n* diode; (b) PIN diode; (c) Schottky diode; and (d) Avalanche diode

Then the difference in base-emitter voltages is simply related to I_{CR}, which is proportional to the illuminance I_R, by

$$\Delta V_{BE} = (V_{BE1} - V_{BE2}) = \frac{k_B T}{e} \ln \left(\frac{I_{CR}}{I_{C0}} \right) \propto \ln(I_{CR}) \qquad (8.20)$$

Thus, photodiodes and phototransistors not only have fast response times but can also have integrated circuitry with, for example, a linearised output voltage or internal temperature compensation.

8.3.3 Pyroelectric Devices

A pyroelectric material can also be used to make a coulombic radiation sensor and it works on the principle that the radiation heats up the surface of a pyroelectric crystal (usually $LiTaO_3$) and reduces the normal polarisation state of the crystal, thereby inducing the charge to flow off its surface and creating a voltage. Pyroelectric sensors are not very sensitive to the wavelength of the radiation and cover a wide dynamic range in the IR spectrum, as illustrated in Figure 8.15. Their dynamic response depends on the thermal time constant of the device and the electrical time constant of the associated circuit. The read-out circuit needs to have a very high input impedance and so is usually CMOS technology.

Pyroelectric sensors are best-suited for detecting short-term changes in the IR signal, and the most popular application is to detect the slow movement of a human body in

Table 8.3 Some commercial radiation microsensors and their typical characteristics

Device	Peak wavelength	Supplier/ Part	Size/ Package	Typical output	Typical rise/fall time	Price (euro[a])
Photoconductive:						
CdS	Visible (0.53 μm)	NORP12	15 mm plastic	1 MΩ to 100 Ω for 0.1 to 10 000 Lux	18 ms/ 120 ms	2.4
Photovoltaic:						
p-n diode	Visible (0.75 μm)	IPL1002 0BW	TO-18 metal can	0.01 to 100 mW/cm^2	250 ns	5.4
PIN diode	Visible (0.85 μm)	Siemens SFH206	TO-92 plastic	0.5 mW/cm^2	20 ns	1.8
Diode & amp.	Vis–NIR (0.90 μm)	OSI5K	TO-5 metal can	30 mV/μW/cm^2	40 μs/ 40 μs	30.2
n-p-n phototransistor	NIR (0.86 μm)	Siemens SFH 309	3 mm plastic	50 μW/cm2	8 μs/ 8 μs	1.0
Pyroelectric:						
LiTaO3	IR (7 to 15 μm)	Sentel DP-2101-101	5 mm metal case	1800 V/W	100 ms	8.0

[a]Note that 1 euro is worth approximately 1.1 US dollars here

a burglar alarm system. The use of a band-pass filter removes sensor drift caused by changes in the ambient conditions.

Photoelectric and pyroelectric sensors are made using a relatively mature technology and so there is a very wide variety of commercially available devices based on different semiconductor materials, processes, and packages. Table 8.3 gives our choice of the discrete devices that are commercially available together with their typical characteristics.

The prices shown are based on a one-off price for 1999 and depend on the choice of package (e.g. metal can versus plastic package). The rise times are often a function of radiation intensity, with higher levels producing faster responses.

Figure 8.17 shows a photograph of three different radiation microsensors: a CdS photoconductive sensor (NORP12) that has a relatively large active area of 12 mm^2 and a slow response time; a p-n photodiode (OSI5K) with an active area of 5 mm^2 and an integrated operational amplifier, and a pyroelectric sensor.

Radiation microsensors, together with temperature ICs, are a relatively mature technology and, therefore, can be readily employed within an integrated optoelectronic system. In fact, the technology has advanced to such an extent that arrays of photovoltaic sensors can be fabricated with associated electronics to make digital line-scan and array cameras. These devices may be regarded here as smart microsensors because they usually have integrated digital read-out electronics.

8.3.4 Microantenna

Finally, microwave or short-wavelength radio waves can be detected using a small metal strip patterned using UV lithography onto a planar surface. The miniature antenna can

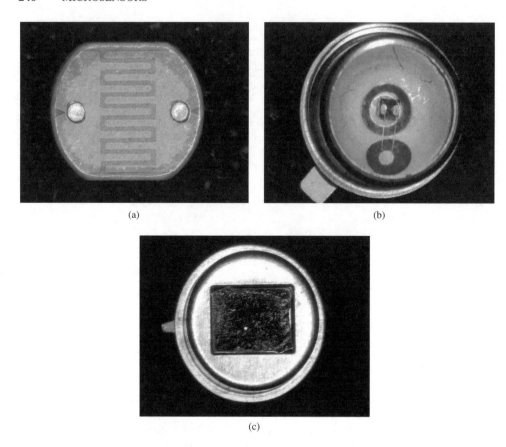

(a)   (b)

(c)

Figure 8.17 Examples of radiation microsensors: (a) CdS photoconductive cell; (b) *p-n* silicon photodiode; and (c) LiNbO$_3$ pyroelectric IR sensor

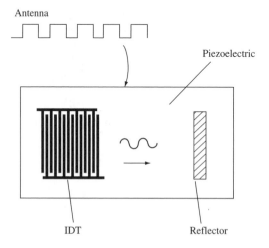

Antenna

Piezoelectric

IDT

Reflector

Figure 8.18 Wireless communication and powering of a SAW-IDT microsensor with an aluminum microantenna

detect low-energy microwave signals with a suitable design of the loop. The signal can then be used, for example, to generate a SAW in a piezoelectric material for a wireless mechanical sensor or simply to sense the electrical signal and pass it onto a decoder (Figure 8.18).

As the microstrip can be made of aluminum, it is compatible with standard microtechnology and can be deposited along with the aluminum interconnects. However, the microantenna can also be used as a transmitter, in which case it is acting as a radiation microactuator.

The use of microantennae in SAW-IDT microsensors is described in Chapter 13 and again in the concept of a smart electronic tongue in Chapter 15.

The way in which a machine interfaces with a person is important and is likely to be a key issue in the future when microsensors and MEMS devices become smaller and more autonomous. Therefore, the integrated microwave antenna may prove to be a very useful tool in which a human operator can communicate with and remotely control a small MEMS structure implanted in some inaccessible environment, such as inside the human body!

8.4 MECHANICAL SENSORS

8.4.1 Overview

Mechanical microsensors are, perhaps, the most important class of microsensor because of both the large variety of different mechanical measurands and their successful application in mass markets, such as the automotive industry. Table 8.4 lists some 50 or so of the numerous possible mechanical measurands and covers not only static and kinematic parameters, such as displacement, velocity, and acceleration, but also physical properties of materials, such as density, hardness, and viscosity.

Figure 8.19 shows a classification scheme for mechanical microsensors together with an example of a device type.

Table 8.4 List of mechanical measurands. Adapted from Gardner (1994)

Acceleration	Flow rate (mass)	Momentum	Sound level
Acoustic energy	Flow rate (volumetric)	Orientation	Stiffness
Altitude	Force (simple)	Path length	Tension
Angle	Force (complex)	Pitch	Thickness
Angular velocity	Frequency	Position	Torque
Angular acceleration	Friction	Pressure	Touch
Compliance	Hardness	Proximity	Velocity
Deflection	Impulse	Reynimity number	Vibration
Deformation	Inclination	Roll	Viscosity
Density	Kinetic energy	Rotation	Volume
Diameter	Length	Roughness	Wavelength
Displacement	Level	Shape	Yaw
Elasticity	Mass	Shock	Young's modulus

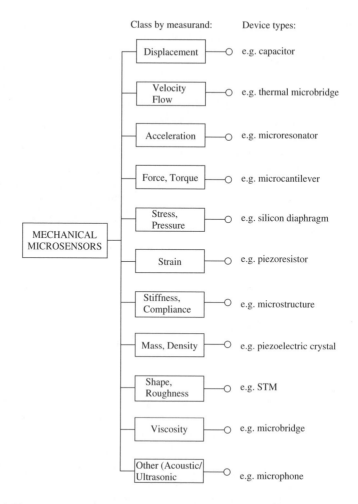

Figure 8.19 Classification scheme for mechanical microsensors. From Gardner (1994)

The most important classes of mechanical microsensors to date is a subset of only six or so and these constitute the majority of the existing market for micromechanical sensors.[12] Thus, the main measurands of mechanical microsensors are as follows in alphabetical order:

- Acceleration/deceleration
- Displacement
- Flow rate
- Force/torque
- Position/angle
- Pressure/stress

[12] See Chapter 1 for details of the main markets for mechanical and other types of microsensor.

Therefore, we describe in detail here four of the most important types of mechanical microsensors, namely,

- Pressure microsensors   (Section 8.4.5)
- Microaccelerometers   (Section 8.4.6)
- Microgyroscopes   (Section 8.4.7)
- Flow microsensors   (Section 8.4.8)

8.4.2 Micromechanical Components and Statics

There are a number of micromechanical structures that are particularly important because they are used as the basic building blocks for a whole host of different microsensors (and for microactuators and MEMS). In their simplest form, these micromechanical structures are simply

- A cantilever beam
- A bridge
- A diaphragm or a membrane[13]

Figure 8.20 shows a schematic diagram of each of these three types of microstructures in both plane view and in cross section. The physical dimensions of the structures are defined in the figures, such as length l, width w, thickness d, and breadth b. In this example, the bridge is shown on the top of the supporting structure, which as we shall see subsequently, is a common feature of microbridges.

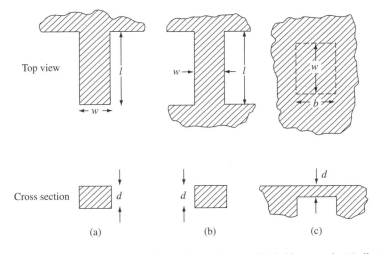

Figure 8.20 Basic microstructures: (a) cantilever beam; (b) bridge; and (c) diaphragm or membrane

[13] A membrane is formed by tension and a diaphragm is formed by stiffness; therefore, if a structure exhibits elasticity, it is a diaphragm.

If we assume that these microstructures are made of a uniform, homogeneous, and elastic material, we can apply a simple theory to describe the way in which they deform when a mechanical load is applied, such as a force, torque, stress, or pressure.

For example, the free end of a cantilever beam[14] will deflect by a distance Δx when a point load F_x is applied to it (Figure 8.21(a)). If we ignore gravitational forces and assume that there is no residual stress in the beam, then the deflection is simply given by

$$\Delta x = \frac{l^3}{3E_m I_m} F_x \tag{8.21}$$

where E_m is the Young's modulus of the material and I_m is the second moment of area of the beam and related to its width and thickness by

$$I_m = \frac{wd^3}{12} \tag{8.22}$$

Equation (8.21) may be rewritten as,

$$F_x = k_m \Delta x \text{ where } k_m = \frac{3E_m I_m}{l^3} \tag{8.23}$$

where the constant of proportionality k_m is called the *stiffness* or *spring constant*.

The simple cantilever beam can thus be used to convert a mechanical force into a displacement. In a similar way, a cantilever beam, bridge, and diaphragm can be used to measure not only a point force but also a distributed force, such as stress. In addition, a diaphragm can be used to measure a hydrostatic or barometric pressure. However, in all these cases, the deflection of the structure has a more complex analytical form, which depends on the precise nature of the built-in supports.

Basic theory assumes that there is no inherent stress in the microstructure itself because that would cause movement in the absence of an applied load. However, the structure could itself be regarded as a sensor for material stress, and so cantilever beams are often used as test structures on silicon wafers to show that the film deposited has no residual stress. In fact, bridges can be used to measure the axial compressive load F_y rather than the

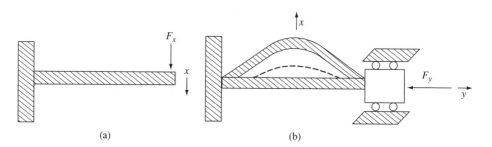

(a)   (b)

Figure 8.21   (a) Deflection of a cantilever beam by a vertical point force F_x and (b) buckling of a beam by an axial compressive load F_y

[14] We assume that the built-in end is clamped on both sides.

transverse load shown earlier (Figure 8.21(b)). In this case, the deflection is not linearly related to the applied load but a solution can be obtained from the buckling equation of Euler where

$$\frac{d(\Delta x)}{dy^2} = -\frac{F_y \Delta x}{E_m I_m} \tag{8.24}$$

The solution for a bridge of length l fixed at both ends is a sinusoidal one where

$$\Delta x = A \sin(\sqrt{F_y/E_m I_m} y) \tag{8.25}$$

where A is a constant. The beam will buckle when the load reaches a critical value of

$$F_c = \pi^2 E_m I_m / l^2 \tag{8.26}$$

In a similar manner, analytical equations that relate the static deflection of a microstructure of different geometry to point and distributed loads and indeed rotations caused by an applied moment or torque can be found.

Therefore, the important mechanical parameters and properties in the design of static microstructures can be summarised as shown in Table 8.5.

8.4.3 Microshuttles and Dynamics

The dynamics of microstructures are also very important for several reasons. First, the dynamic response of a cantilever beam, microbridge, or diaphragm determines the bandwidth of the microsensor, that is, the time taken for the structure to respond to the applied static load or follow a dynamic load. Second, the kinetics of the structure can be used in *inertial* sensors to measure linear and angular accelerations. Finally, there is a class of mechanical sensors based on a microstructure or a microshuttle that is forced to resonate at some characteristic frequency.

In an inertial or resonant sensor, the cantilever beam (or bridge) is redesigned so that the mass is more localised and the supports more convoluted. Figure 8.22 shows the basic design of an inertial or so-called microflexural type of structure together with its equivalent lumped-system model.

Table 8.5 Some important mechanical parameters and material properties in defining the static deflection of microstructures

Parameters/properties	Nature
Point/distributed force, torque, stress, pressure	Load applied/measurand
Width, breadth, thickness, length	Size of structure
Young's modulus, yield strength, buckling strength, Poisson's ratio, density	Material properties
Spring constant, strain, mass, moment of inertia	Calculable parameters
Lateral/vertical deflection, angular deflection	Response

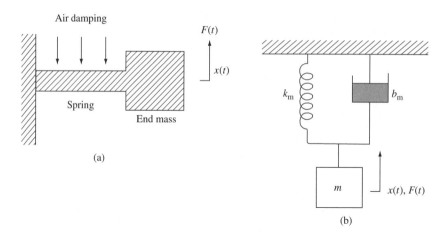

Figure 8.22   (a) A simple microflexure comprising a heavy end or load mass and a thin springlike support and (b) lumped system equivalent model of the microflexure

The dynamic equation of motion of a simple microflexural structure can be approximated to a one-dimensional lumped-system model and is given by

$$m\ddot{x} + b_{\mathrm{m}}\dot{x} + k_{\mathrm{m}}x = F_x(t) \tag{8.27}$$

where b_{m} is the damping coefficient, k_{m} is the stiffness constant (see Equation (8.23)), and m is the end mass (ignoring the thin support). Equation (8.27) can readily be solved using the method of Laplace transforms, provided the three coefficients are constant and hence independent of both displacement x and time t. The Laplace transform of the displacement $X(s)$ is related to the Laplace transform of the applied force $F(s)$ by the characteristic transfer function $H(s)$ of the structure. Hence, the dynamical response of the structure is described using the complex Laplace parameter s by

$$X(s) = H(s)F(s) = \frac{1}{ms^2 + b_{\mathrm{m}}s + k_{\mathrm{m}}}F(s) \tag{8.28}$$

The application of a sinusoidal force at a drive frequency (angular) ω causes the mass to oscillate; its characteristic curve, obtained from Equation (8.28), is shown in Figure 8.23. The resonant frequency ω_0 and damping factor ξ of the simple microflexural system are given by

$$\omega_0 = \sqrt{k_{\mathrm{m}}/m} \text{ and } \xi = b_{\mathrm{m}}/2\sqrt{k_{\mathrm{m}}m} \tag{8.29}$$

Thus, the response or gain[15] of the structure varies with the stiffness and damping of the structure as well as the nature of the applied load, including a gravitational or inertial force.

There are a number of different designs of microflexural structures and three of the basic types – the hammock, folded, and crab-leg – are shown in Figure 8.24. The folding of the supports changes its effective stiffness and so extends the linear range of deflection.

[15] Here the gain is the ratio of the amplitudes of the output to the input signals.

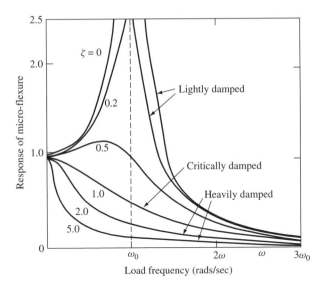

Figure 8.23 Dynamic response of an ideal microflexural structure to a sinusoidal driving load

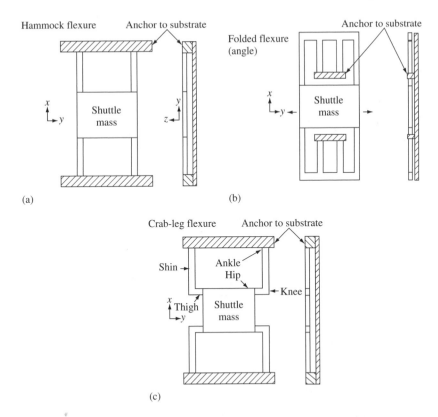

Figure 8.24 Three common microflexural designs: (a) hammock; (b) folded; and (c) crab-leg. Adapted from Gardner (1994)

Table 8.6 Some mechanical characteristics of three different microflexures

Parameter	Hammock	Folded	Crab-leg
Bending deflection y	Axial and bending stress	Bending stress	Bending stress
Spring constant k_y	Nonlinear in y	Constant and independent of y	Constant and independent of y
Axial deflection x	Axial stress only	Axial and bending stress	Axial and bending stress
Spring constant k_x	Stiff: $4E_m A/l$	Quite stiff	Stiff

Table 8.7 Some important mechanical parameters and material properties that define the dynamic deflection of microflexures

Parameters/properties	Nature
Point/distributed force, torque, stress, pressure	Load applied/measurand
Width, breadth, thickness, length	Size of structure
Young's modulus, yield strength, buckling strength, Poisson's ratio, density, viscosity, friction	Material properties
Spring constant, strain, mass, moment of inertia, natural frequency, damping coefficient	Calculable parameters
Lateral/vertical deflection, angular deflection, resonance, bandwidth	Response

Table 8.6 provides the characteristic properties of these flexures, in which their dynamical response is much more complicated than that of a simple end mass and is often determined using computational methods, for example, a finite-element or finite-difference analysis.

When designing dynamic structures, we need to consider some additional parameters, which are listed in Table 8.7.

8.4.4 Mechanical Microstructures

The two most important questions that now need to be asked by the designer are as follows. First, if these mechanical structures can be made on the micron scale and second, if they still follow classical theory, for example, the linear theory of elasticity.

We know that microbeams, microbridges, and microdiaphragms can all be made in silicon using the bulk- and surface-micromicromaching techniques, which were described in Chapters 5 and 6. In fact, a number of worked examples of the process flow to fabricate the following microstructures were given:

- A cantilever beam made of undoped silicon   (WE 5.1)
- A thin cantilever beam   (WE 5.4)
- A free-standing polysilicon beam   (WE 6.1)
- An array of thin diaphragms/membranes   (WE 5.3)
- A comb resonant structure   (WE 6.7)

Table 8.8 Some mechanical properties of bulk materials used to make micromechanical sensors

Material Property	Si (SC)	Si (poly)	SiO$_2$	Si$_3$N$_4$	SiC	Diamond	Al	PMMA
Young's modulus (GPa)	190[a]	160	73	385	440	1035	70	–
Yield strength (GPa)	6.9	6	8.4	14	10	53	0.05	0.11
Poisson's ratio	0.23	0.23	0.20	0.27	–	–	0.35	–
Fracture toughness (MN/m^2)	0.74	–	–	4–5	3	30	30	0.9–1
Knoop hardness (10^{9}kg/m^2)	0.8	–	0.8	3.5	–	7.0	–	–

[a]For [111] Miller index (168 GPa for [110], 130 GPa for [100]). Shear modulus 58 GPa for [111], 62 GPa for [110], and 79 GPa for [100]

Clearly, the microstructures can then be fabricated from single-crystal silicon, polycrystalline silicon, and also from metals and other types of material. The processes shown also demonstrate that the residual strain is negligible because the cantilevers and diaphragms shown are neither curling nor buckling when free from any external load.

Table 8.8 summarises the mechanical properties of some of the materials that have been used to make micromechanical structures and are important in their practical design and usage. Other important physical properties of these materials, such as density, thermal conductivity, and heat capacity, may be found in the Appendices F (metals), G (semiconductors), and H (ceramics and polymers).

As stated in Table 8.8, the question that must be asked is whether a material behaves on the micron scale in the same way as it does on the macro scale? The answer to this important question is 'yes' for pure single-crystal silicon. In this case, there are very few defects and so structures on the micron scale have the same fundamental properties as on the large scale. In fact, the same rule also applies for polycrystalline materials *provided* the average grain size is much smaller than the smallest dimension of the microstructure. As the typical grain size in low-pressure chemical vapour deposition (LPCVD) polysilicon is 50 to 80 nm, the material will behave elastically down to about the micron level. The same rule can be applied to other polycrystalline materials, such as metals.

Accordingly, we can apply classical geometric scaling rules to structures down to a few microns in size without a breakdown in the laws. For example, a reduction in the size of a cantilever structure will increase its resonant frequency by a factor K but reduce its mass by K^3, deflection by K, spring constant by K, and so on.

Finally, we must consider the types of transducer for a microstructure that convert its deflection into an electrical quantity. There are a number of different ways in which the movement could be detected such as

- Capacitive (electrostatic) pickup

- Resistive (conductive) pickup

- Inductive (amperometric) pickup

The two most commonly used forms of transduction are capacitive and resistive. Figure 8.25 shows a microflexure in which its end is capacitively coupled to a stationary sense electrode.

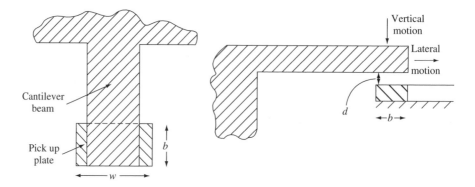

Figure 8.25   Capacitive measurement of the deflection of a simple cantilever beam

The capacitance C and change in capacitance δC are given by

$$C = \frac{\varepsilon A}{d} \text{ and hence } \frac{\delta C}{C} = \frac{\delta \varepsilon}{\varepsilon} + \frac{\delta A}{A} - \frac{\delta d}{d} \tag{8.30}$$

Therefore, a change in capacitance is related to changes in the plate separation d, area of overlap A, and dielectric permittivity ε. The capacitance of a structure with a 200 μm square area and a separation of 4 μm is about 0.1 picofarads. Therefore, it is necessary to measure changes in capacitance to a resolution on the order of 10 fF or less!

Many silicon mechanical microsensors use this principle to measure a *vertical* deflection (with A and ε constant) because the area can be made relatively large and the gap size small, that is, a few microns. This means that the change in capacitance can be measured using integrated electronics with an acceptable sensitivity. Another advantage of a capacitive pickup is that the input impedance is high and so little current is consumed; hence, the method is suitable for use in battery-operated devices with integrated CMOS circuitry. However, it is difficult to sense *lateral* deflections of silicon structures fabricated by standard surface-micromachining techniques because the resulting structures are only a few microns high. Comb structures are often used to increase the area of overlap, and the change in area of overlap is used to measure the deflection. Even so, very large structures are needed to achieve useful values of the capacitance. That is why lithography, electroplating, and moulding process (LIGA) and other techniques, such as deep reactive ion etching (RIE), are required to make much thicker structures and therefore measure lateral deflections in a more practical way. However, this basic problem applies whether one tries to sense the deflection of a microflexural structure or drive it electrostatically in a microactuator.

The other important type of pickup is through a piezoresistor (see Figure 8.26). Piezoresistors can be made easily either as a region of doped single-crystal silicon (SCS) in a bulk-micromachined structure or as a doped polysilicon region in a surface-micromachined structure. The gauge factor K_{gf} of a strain gauge defines its sensitivity and simply relates the change in fractional electrical resistance ΔR to the mechanical strain ε_m

$$\frac{\Delta R}{R} = K_{gf}\varepsilon_m \tag{8.31}$$

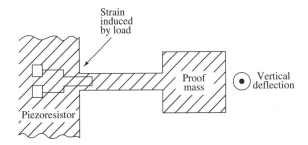

Figure 8.26 Piezoresistive measurement of the deflection of a cantilever beam

Doped silicon resistors (piezoresistors) can be made at a very low cost and have a strain gauge factor that is much higher ($\sim$50 to 100) than that for metals ($\sim$2). However, it is harder to control the exact resistance of the silicon piezoresistor and, more importantly, its actual gauge factor is strongly dependent on both the doping level and the ambient temperature. Consequently, an embedded temperature sensor is essential for a precise measurement of the strain and hence any static displacement by this method. This problem is not so critical in a dynamic structure where it is only necessary to measure the frequency of oscillation; however, care is still needed because the deposition of the piezoresistor may itself induce stress in the microstructure and cause a shift in its natural resonant frequency!

8.4.5 Pressure Microsensors

Pressure microsensors were the first type of silicon micromachined sensors to be developed in the late 1950s and early 1960s. Consequently, the pressure microsensors represent probably the most mature silicon micromechanical device with widespread commercial availability today. The largest market is undoubtedly the automotive, and Table 8.9 shows the enormous growth in the world market for automotive silicon micromachined sensors from 1989 to 1999. The two most important silicon sensors are the pressure and microaccelerometer (Section 8.4.6) sensors, with substantial growth expected for gyrometers (Section 8.4.7), which will be used for navigation.

Table 8.9 Worldwide growth for automotive silicon micromachined sensors. From Sullivan (1993)

Year	Revenue[a] (MEuro)	Growth-rate (%)	Year	Revenue (MEuro)	Growth-rate (%)
1989	175	–	1995	376	21
1990	283	62	1996	463	23
1991	323	14	1997	564	22
1992	321	−1	1998	679	20
1993	285	−11	1999	804.2	18
1994	312	10			

[a] 1 euro $\equiv$ \$1.1 for September 2000

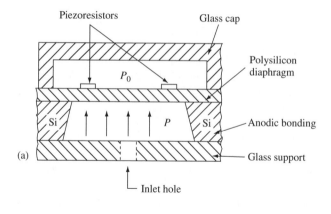

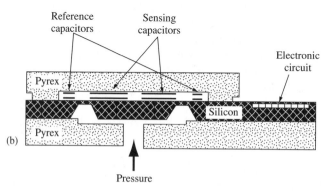

Figure 8.27   Basic types of silicon pressure sensors based on a vertical deflection: (a) piezo-resistive (polysilicon) and (b) capacitive (single-crystal silicon)

The two most common methods to fabricate pressure microsensors are bulk and surface micromachining of polysilicon. Silicon diaphragms can be made using either technique as described earlier. Figure 8.27 illustrates the basic principles of a piezoresistive sensor and a capacitive pressure sensor.

The deflection in the diaphragm can be measured using piezoresistive strain gauges located in the appropriate region of maximum strain, as shown in Figure 8.27(a). The strain gauges are usually made from doped silicon and are designed in pairs with a read-out circuit such as a Wheatstone bridge. The change in strain can be related to the applied pressure $(P - P_0)$ and stored in a lookup table. The precise relationship depends on the relevant piezoresistive coefficient Π of the diaphragm material.

$$V_{\text{out}} \propto \Delta R \propto \Pi(P - P_0) \qquad (8.32)$$

A single crystal of silicon is a desirable material to use for the diaphragm because neither creep nor hysteresis occurs. The piezoresistive constant (Π_{44}) is typically $+138.1$ pC/N and that makes measuring pressure in the range of 0 to 1 MPa relatively straightforward.

Figure 8.27(b) shows the general arrangement of a single-crystal silicon pressure sensor with capacitive pickup. In this case, a capacitive bridge can be formed with two reference capacitors and the output voltage is related to the deflection of the membrane Δx and hence the applied pressure $(P - P_0)$.

$$V_{\text{out}} \propto \Delta C \propto \Delta x \propto (P - P_0) \tag{8.33}$$

In this case, the accurate positioning of the pickup electrodes is crucial.

By controlling the background pressure P_0, it is possible to fabricate the following basic types of pressure sensors:

- An *absolute* pressure sensor that is referenced to a vacuum ($P_0 = 0$)
- A *gauge*-type pressure sensor that is referenced to atmospheric pressure ($P_0 = 1$ atm)
- A *differential* or relative type (P_0 is constant).

There are advantages and disadvantages of capacitive against piezoresistive pressure sensors and these are summarised in Table 8.10.

The main advantage of using bulk micromachining is that the electronic circuit can be more readily integrated. There are many examples of capacitive pressure sensors with digital readout. Readers are directed toward Worked Example 6.8 for the process flow of an air gap capacitive pressure sensor with digital readout. An example of a capacitive pressure sensor is shown in Figure 8.28 with a 100 μm polysilicon diaphragm and integrated capacitance circuit (Kung and Lee 1992). The output voltage from the integrated n-type metal oxide semiconductor (nMOS) circuit is also shown against air pressure in non-SI units of PSI. This design achieves a high resolution by using integrated electronics.

An alternative approach to enhance the sensitivity of silicon pressure sensors was proposed by Greenwood in 1988 and comprised the use of a resonant microstructure. Figure 8.29 shows the micromechanical structure bulk-micromachined out of single-crystal silicon (Greenwood 1988).

The basic principle is the change of resonant frequency of oscillation of this structure when the pressure on the diaphragm causes it to curve. In turn, this curvature creates tension in the shuttle mass supports and this shifts its resonant frequency. The dynamical equation that governs the behavior is a modified version of Equation (8.27) to include a tension term, which affects the effective spring constant k_{m}. The resonant (torsional)

Table 8.10 Relative merits of capacitive and piezoresistive static deflection pressure sensors

	Advantages	Disadvantages
Capacitive	More sensitive (polysilicon)	Large piece of silicon for bulk micromachining
	Less temperature-sensitive	Electronically more complicated
	More robust	Needs integrated electronics
Piezoresistive	Smaller structure than bulk capacitance	Strong temperature-dependence
	Simple transducer circuit	Piezocoefficient depends on the doping level
	No need for integration	

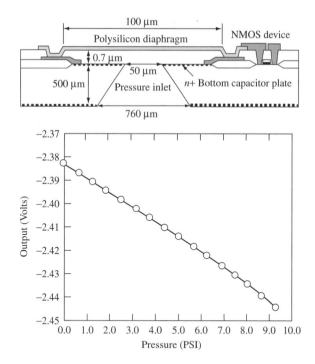

Figure 8.28 Polysilicon capacitive pressure sensor: (a) cross section with integrated electronics, (b) voltage response from a 100 μm square diaphragm of thickness 1 μm. From Kung and Lee (1992)

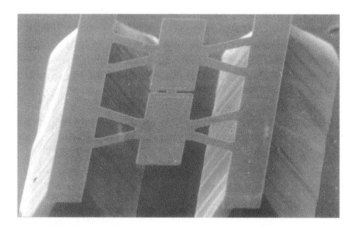

Figure 8.29 A vertical resonant capacitive pressure sensor based on the torsional oscillation of a strained bulk-micromachined structure. From Greenwood (1988)

pressure sensor proved to have excellent resolution (a few centimeters in air) and stability (parts per million (ppm) per year) through the running of the resonator in a partial vacuum. Accordingly, it is possible to achieve a high mechanical Q factor, here about 18 000 at a pressure of approximately 1 Pa, and hence achieve very high pressure sensitivities.

Further efforts have been made to fabricate a lateral resonant capacitive sensor employing thin film polysilicon technology. Figure 8.30(a) shows a resonant capacitive sensor fabricated in polysilicon along with its response (Figure 8.30(b)). The nonlinear response is fitted using a high-order polynomial and temperature effects are compensated for.

Here, the microstructure behaves as a nonlinear resonator and Equation (8.27) is extended to describe a hard spring (Duffin's equation) so that

$$m\ddot{x} + b_{\mathrm{m}}\dot{x} + k_{\mathrm{m}}^0 x + k_{\mathrm{m}}^1 x^3 = F_x(t) \tag{8.34}$$

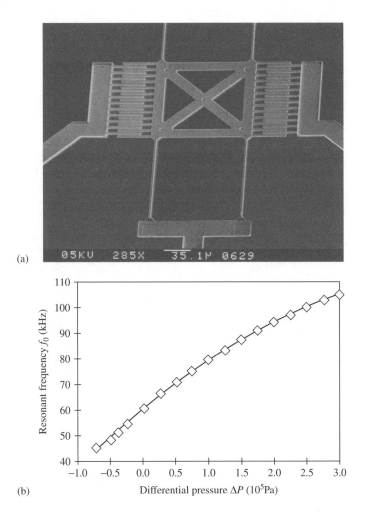

(a)

(b)

Figure 8.30 (a) Lateral resonant capacitive pressure sensor based on the linear oscillation of a strained surface micromachined structure; (b) its response to barometric pressure (from Welham and coworkers (1996)); (c) current silicon process; and (d) latest device with piezoresistive pickup (Welham *et al.* 2000)

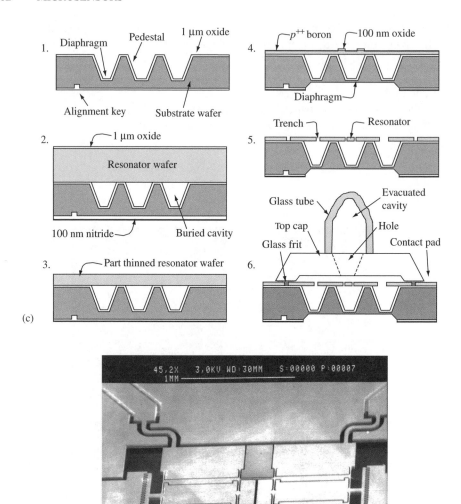

Figure 8.30 (*Continued*)

The solution to Equation (8.34) is interesting because it has two possible deflections at certain frequencies. However, running the oscillator at low deflections using closed-loop feedback avoids this stability issue. The problem with this structure is that the capacitances for drive and sensing are too low because the microshuttle is only 1 to 5 μm thick. However, recent developments of LIGA and deep RIE now make resonant lateral structures a practical device. The resonator has now been redesigned by Welham *et al.* (2000) to overcome these problems together with a piezoresistive pickup. Figure 8.30(c) shows the new silicon process and the fabricated device is shown in Figure 8.30(d). These

Table 8.11 Applications of silicon pressure sensors in 1997. Adapted from Madou (1997)

Application	Cost of device (Euro)	Market (MEuro)	Pressure range (kPa)	Year introduced
Manifold pressure	9	30	0–105	Current
Barometric pressure	9	100	50–105	Current
Exhaust gas recirculation	9	3.3	0–105	1989
Fuel pressure	9	97	0–105	1994
Tyre pressure	n/a	455	500	1994–1995
Active suspension hydraulics	7	14	20 000	1994–1995
Climate control	9	19	50–105	Current

devices have an accuracy of 0.01 percent root-mean-square (rms) or better, which, so far, exceeds that for static pressure sensors. The product is being commercialised by Druck Ltd (UK) as a precision pressure sensor because it is relatively expensive to make.

Nevertheless, the preferred technology today is bulk silicon-micromachined piezo-resistive pressure sensors because of low cost, robustness and ease of circuit integration. Table 8.11 summarises the current automotive pressure sensor applications (Madou 1997).

Clearly, the automotive market for pressure sensors is enormous and commercial devices are available today from Motorola, NovaSensor, SSI Technologies, and other manufacturers. As costs are driven down, the move toward piezoresistive polysilicon is desirable but creates some stability and precision issues. Therefore, we may see the appearance of alternative technologies to make diaphragms such as silicon on insulator (SOI).

8.4.6 Microaccelerometers

The second most important type of mechanical microsensors is inertial and measures, for example, linear acceleration and angular velocity. Inertial sensors are again a mass market in the automotive industry, second only to pressure sensors.

Microaccelerometers are based on the cantilever principle in which an end mass (or shuttle) displaces under an inertial force. Thus, the dynamics can be described in simple terms by the second-order system of a mass-spring damper described earlier.

Figure 8.31 shows the basic principle of the two most important types: capacitive pickup of the seismic mass movement and piezoresistive pickup.

The capacitive polysilicon surface-micromachined and single-crystal-micromachined devices are probably the most prevalent and generally come with high g and low g variations. Microaccelerometers are now produced in their millions with sophisticated damping and overload protection. For example, Lucas Novasensor make a bulk microactuator for self-testing. Analog Devices introduced a capacitive polysilicon surface-micromachined device in 1991 (AXDL-50).

The main markets of microaccelerometers are in automatic braking system (ABS) and suspension systems (0 to 2 g) and air bag systems (up to 50 g). Table 8.12 gives the data about the US market in automotive accelerometers over the past decade. The market today is worth some €200 M in the United States alone. It should be noted that the

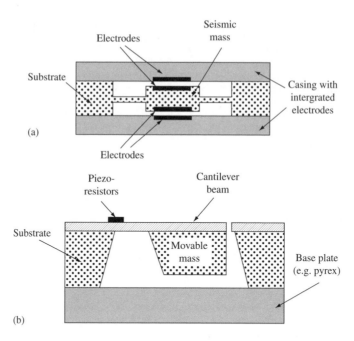

Figure 8.31 Basic types of microaccelerometers: (a) capacitive and (b) piezoresistive. Adapted from Fatikow and Rembold (1997)

Table 8.12 US market for automotive microaccelerometers in million euros. Adapted from MIRC (1990)

Application	1991	1992	1993	1994	1995	1996	1997	1998	1999	2000
Air bag-cars and vans	21	55	89	88	151	127	129	131	133	135
ABS[a]	0	8	8	16	24	31	50	52	54	56
Suspension	0	6	13	18	26	19	19	20	21	22
Total	21	69	110	122	201	177	198	203	208	213

[a] Automatic braking system

unit price has fallen from €100 to €9 during this period, so unit sales have dramatically risen.

The more recent ADXL 250 (Analog Devices) employs a folded flexure structure for improved linearity and provides two-axis measurement. In contrast, Figure 8.32 shows the CSEM MS6100 precision low-power capacitive accelerometer (170 μA at 3 V) with high dynamic stability (2 mg for 2 g sensor) but poor low-temperature stability (typical offset is 200 μg/°C).

Table 8.13 provides a comparison of the specification of some commercially available microaccelerometers.

Through increasing the damping and stiffness of the microresonators, it is possible to increase the dynamic range further; therefore, microaccelerometers are also used in military applications, such as missile control.

Figure 8.32   The MS 6100 capacitive microaccelerometer with associated electronics on a hybrid package

Table 8.13   Comparison of the specification of some commercially available microaccelerometers

Manufacturer	Motorola	Bosch	Bosch	Analog devices	Analog devices
Part no.	XMMAS4 01G1D	SMB050	SMB060	ADXL50JH	ADXL250JQC
Number of axes	1	1	2	1	2
Full-scale range	$\pm40\ g$	$\pm35\ g$	$\pm35\ g$	$\pm50\ g$	$\pm1\ g$
Zero g offset drift	20 g	–	–	1.5 g	0.3 g
Temp. range ($^\circ$C)	−40 to 105	–	–	–	−40 to 85
DC current (mA)	5	7	12	10	3.5
3 dB bandwidth (Hz)	400	400	400	1300	1000
Nonlinearity (% FS)	0.5	0.5	0.5	0.2	0.2
Axis alignment ($^\circ$)	N/A	N/A	1	N/A	0.1
Noise density (mg/Hz)	7.8	–	–	6.6	2
Package[a]	16-pin DIP, SIP	28-pin PLCC	28-pin PLCC	10-pin TO-100	14-pin Cerpak
Price(€)[b]	11	N/A	N/A	13.6	18.1

[a]See Chapter 4 on packeges
[b]Unit price for 100 pcs

8.4.7 Microgyrometers

The second type of inertial sensor is the gyroscope that measures the change in orientation of an object. Silicon-micromachined gyroscopes have been fabricated on the basis of coupled resonators. The basic principle is that there is a transfer of energy from one resonator to another because of the Coriolis force. Thus, a simple mass m supported by springs in the x- and y-axes and rotated around the z-axis at an angular velocity Ω has the following equations of motion.

$$m\ddot{x} + b\dot{x} + k_x x - 2m\Omega\dot{y} = F_x$$
$$m\ddot{y} + b\dot{y} + k_y y + 2m\Omega\dot{x} = F_y \tag{8.35}$$

where the terms $2m\Omega\dot{x}$ and $2m\Omega\dot{y}$ describe the Coriolis forces and the resonant frequencies are

$$\omega_{0x} = \sqrt{k_x/m} \quad \text{and} \quad \omega_{0y} = \sqrt{k_y/m} \tag{8.36}$$

Now assume that the resonators are excited and behave harmonically with the amplitudes $a(t)$ and $b(t)$. By fixing the amplitude of one oscillator (a_0) by feedback and then for synchronous oscillators ($\omega_{0x} = \omega_{0y}$), the equations simply reduce to

$$\frac{db}{dt} + \left(\frac{c}{2m}\right)b + \Omega a_0 = 0 \tag{8.37}$$

Under a constant rotation, the steady-state solution to Equation (8.37) is a constant amplitude b_0 where

$$b_0 = -\left(\frac{2m}{c}\right)a_0\Omega \tag{8.38}$$

Therefore, the amplitude of the undriven oscillator is linearly proportional to the rotation or precession rate Ω.

The first silicon coupled resonator gyrometer was developed by Draper Laboratory in the early 1990s and its arrangement is shown in Figure 8.33. The device is bulk-micromachined and supported by torsional beams with micromass made from doped (p^{++}) single-crystal silicon (SCS). The outer gimbal was driven electrostatically at a constant amplitude and the inner gimbal motion was sensed. The rate resolution was only 4 deg s^{-1} and bandwidth was just 1 Hz.

More advanced gyroscopes have been fabricated using surface micromachining of polysilicon. There are a number of examples of coupled resonator gyroscopes such as the MARS–RR1 gyroscope reported by Geiger et al. (1998). The performance of this device is provided in Table 8.14.

There are reports of a number of other types of device to measure precision rates; the IDT MEMS device described in Chapter 14 is one such example. Another is the ring gyroscope that again works by the Coriolis force transferring energy from one mode into another at 45° (Ayazi and Najifi 1998). The basic approach is attractive but does require a deep etch to produce viable devices. Figure 8.34 shows two ring gyroscopes. The first was made at the University of Michigan (Ayazi and Najifi 1998), whereas the second is a prototype made by DERA Malvern, (UK).

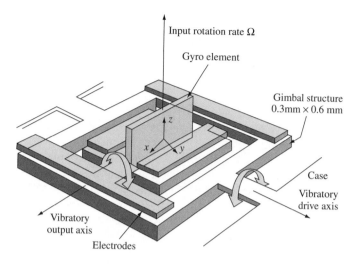

Figure 8.33 Early example of a silicon-micromachined coupled resonant gyrometer. From Grieff *et al.* (1991)

Figure 8.34 Some examples of ring microgyroscopes: (a) University of Michigan, USA and (b) DERA (UK). From McNie *et al.* (1998)

Table 8.14 Performance of a polysilicon-coupled resonant gyroscope; the MARS–RR1 is taken from Geiger *et al.* (1998)

Parameter	Specification
Bias stability	0.018 deg s^{-1}
Noise	0.27 deg/h
Sensitivity	10 mV/(deg s^{-1})
Linearity	<0.2%
Supply voltage	15 V
Current (discrete electronics)	20 mA
Shock survival	1000 *g*

There is considerable interest in silicon gyroscopes in the defence industries for controlling missiles, but low-cost commercial devices for nonmilitary applications (e.g. automotive) are now appearing.

8.4.8 Flow Microsensors

The measurement of the flow rate of a gas (or liquid) is important in a number of different fields from automotive and aerospace to the chemical industries. For example, it is important to know the amount of fuel flowing into an engine or domestic gas supplied to boilers in homes. Indeed, there are a number of traditional ways to measure flow directly (e.g. a rotating vane) and indirectly through the differential pressure (orifice plate, Venturi tube, and Pitot tube). Here, we are interested in the measurement of flow using novel micromechanical sensors. One possible method is to use an ultrasonic technique such as a SAW device but, as these are covered elsewhere, we will focus here on other types of micromechanical devices.

The most commonly used principle to detect flow in gases and liquids using microsensors is based on the concept of a thermal flow sensor that was first postulated by Thomas in 1911. The basic principle is shown in Figure 8.35 in which the heat transferred per unit time (P_h) from a resistive wire heater to a moving liquid is monitored at two points via thermocouple temperature sensors.

When a steady state has been achieved, the mass flow rate Q_m is related to the difference in the temperatures ($T_2 - T_1$) and is given by

$$Q_m = \frac{dm}{dt} = \frac{P_h}{c_m}(T_2 - T_1) \tag{8.39}$$

where c_m is the specific heat capacity of the fluid, assuming that there is no heat loss from the wall of the tube. The mass flow rate of the liquid can be converted to the volumetric flow rate Q_V via its density ρ_m:

$$Q_V = \frac{dV}{dt} = \frac{Q_m}{\rho_m} \tag{8.40}$$

The placing of the heating coil and temperature sensors within the walling of a pipe (Figure 8.35(b)) makes more practical sense and the embodiment is the so-called

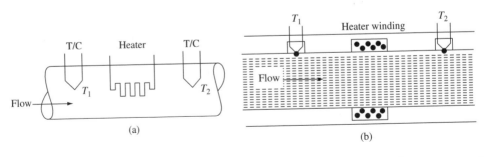

(a)

(b)

Figure 8.35 Principle of a thermal flow sensor: (a) original Thomas flow meter and (b) boundary-layer version

boundary-layer flow meter. Unfortunately, the relationship between the flow rate and temperature difference is more complex and the precise relationship should be determined experimentally for different fluids. For more details on boundary layer thermal sensors, readers are referred to Meijer and van Herwaarden (1994).

A number of different silicon-micromachined flow sensors have been reported since the late 1980s. For example, Johnson and Higashi reported in 1987 on the use of bulk micromachining to make a pair of silicon microbridges. There is a thermistor on top and each microbridge and a resistive heater is split between the two as shown in Figure 8.36(a). The silicon microbridge structure is advantageous because it is thermally isolated and, therefore, the power loss of the heater is minimised. In this case, 1 mW of power produces a 15 °C rise in temperature in a gas. The temperature difference is measured in a Wheatstone bridge and the voltage output is shown in Figure 8.36(b) for air, carbon monoxide, and methane.

The device can be used to measure flow velocities of up to 30 m/s in a 5 μm by 250 μm channel. The mass flow rate is related to the velocity v by

$$Q_m = \rho_m A v \tag{8.41}$$

In a slightly different approach, Stemme (1986) replaced the microbridge by a cantilever beam and incorporated a layer of polyimide to help thermally isolate the heater from the substrate. In this case, the thermistors have been replaced by thermodiodes and the sensor has integrated CMOS circuitry. Flow velocities of up to 30 m/s can again be measured but this time with about half the power consumption.

Since the late 1980s, commercial silicon flow-rate sensors have been manufactured in large numbers. For example, Honeywell make several different air flow-rate sensors using the thermal flow silicon microbridge approach. Table 8.15 summarises the specification of two current products that can be used to measure the volumetric flow rate of gases. The

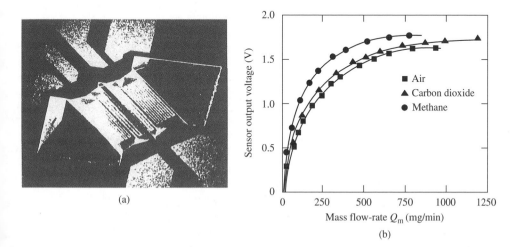

Figure 8.36 (a) A bulk-micromachined silicon microbridge thermal flow-rate sensor and (b) its characteristic response. From Johnson and Higashi (1987)

Table 8.15   Specification of two commercial silicon microbridge flow microsensors (Honeywell)

Specification	AWM 3100V	AWM 3300V
Operating range (ml/s)	0 to 200	0 to 1000
Power consumption (mW)	30	30
Voltage output at FSO (V)	5.0 DC	5.0 DC
Null voltage (V DC)	1.00 ± 0.05	1.00 ± 0.10
Repeatability (% FSO)	± 0.5	± 1.0
Response time, max. (ms)	3	3
Operating temperature range (°C)	-25 to $+85$	-25 to $+85$
Weight (g)	10.8	10.8
Approximate cost for 1 off (euro)	130^a	130^a

[a]Price in October 2000 in the United Kingdom

commercial flow-rate microsensor has onboard thermal compensation but the unit must be calibrated for different gases (or gas mixtures) because of their different densities and thermal capacities.

There are now a number of other types of flow sensors that have been reported. They range from relatively simple devices, for example, a low-cost CMOS-based wind meter, which was first described by Oudheisen and Huijsing in 1990 to the more complex devices, for example, a recently improved version, which has a thermal sigma–delta modulation system (Makinwa and Huijsing 2000). However, the application of silicon flow sensors is likely to increase significantly over the next few years because of the growing importance of the field of microfluidics in MEMS. There is a need to monitor the flow-rate of different compounds – when they flow inside microfluidic systems that are now referred to as micro total analysis systems (μ-TAS). These systems permit entire chemical reactions to take place on the silicon wafer and are pioneering by a number of research laboratories such as the Institute of Microtechnology in Neuchatel, Switzerland. The advantages of making reactions take place on silicon are clear in that the volumes of expensive reagents are very low and the process can be fully automated on chip – an approach that is commonly referred to these days as 'laboratory-on-a-chip.'

8.5 MAGNETIC SENSORS

In this section, we focus on the basic principles of these different kinds of magnetic microsensors together with some examples of commercial products and research devices. Figure 8.37 shows the various kinds of magnetic microsensors that can be used to measure the magnetic flux density $\boldsymbol{B}$. The devices are classified here according to the form of the *output* signal rather than the energy domain of the input signal used earlier.

Clearly, some of these magnetic microsensors are also employed within other types of sensors; for example, a Hall effect device can be used to measure the proximity of a magnet and it then becomes a mechanical (magnetic) sensor, and so on.

The typical characteristics of magnetic microsensors are summarised in Table 8.16 and the details are discussed in the subsequent sections.

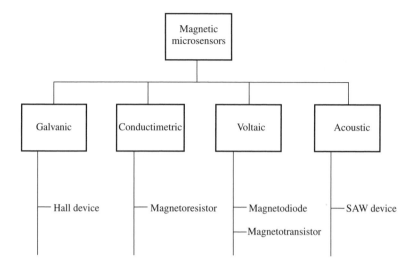

Figure 8.37 Classification of magnetic microsensors by principle

Table 8.16 Typical characteristics of some magnetic microsensors

Sensor type	B range (T)	Sensitivity S_I	Cost	Comments
Hall plate (general)	μ to m	5–500 V/T	Low	Good linearity with InSb more sensitive than Si, fabricate in bipolar or nMOS process, offsets at low fields
Hall effect IC (304367, RS Components)	±50 m	7.5–8.6 V/T	Low	100 kHz bandwidth
Hall effect IC (KSY10, Siemens)	0 to 1	200 V/T	Low	Good linearity and sensitivity (±0.7%)
Magnetoresistor (general)	n to d	~500%/T	Low	High sensitivity but less linear than Hall IC and not standard IC process
Magnetoresistor (F830L, Siemens)	0 to 1	~700%/T	Low	Nonlinear (parabolic) output, lower bandwidth than IC devices.
Magnetodiode (general)	>m	25 V/T	Low	Reasonable sensitivity with CMOS device, SOS device is lower at 5 V/T and more expensive to make
Magnetotransistors (general)	>m	0.5–7%	Medium	Various types made using lateral bipolar technology, tend to be temperature-dependent.
SAW delay-line	μ–m	1 Hz/μT	High	Needs a magnetoelastic material and measures induced strain, high sensitivity but needs microwave technology
SQUID	10f–μ	Quanta	High	Extremely sensitive device that can be made in IC process but operates at 77 K

8.5.1 Magnetogalvanic Microsensors

The magnetogalvanic effect was first discovered by Hall in 1879 and is the fundamental principle that governs the operation of a Hall plate magnetic sensor. Figure 8.38 shows the schematic drawing of a Hall device. When a current I_x is passed down a slab of material of length l and thickness d and a perpendicular magnetic flux density B_z is applied, a voltage V_H appears across the slab perpendicular to I_x and B_z and can be measured.

The theoretical Hall voltage V_H can be obtained from the simple transport theory and is given by

$$V_H = \left(-\frac{1}{ne}\right)\frac{I_x B_z}{d} = R_H \frac{I_x B_z}{d} \tag{8.42}$$

where n is the carrier density and R_H is called the Hall coefficient. The Hall coefficient of metals is rather small and has a typical value of approximately -1×10^{-4} cm^3/C that is close to the theoretical value. The Hall coefficient is much larger in a semiconducting material, and Figure 8.39 shows the variation of the Hall voltage with carrier concentration for n- and p-type materials.

The Hall voltage of a semiconducting slab is generally defined as

$$V_H = R_H k_g \frac{I_x B_z}{d} \approx -\frac{r_n}{ne} k_g \frac{I_x B_z}{d} (n\text{-Si}) \text{ or } +\frac{r_p}{ne} k_g \frac{I_x B_z}{d} (p\text{-Si}) \tag{8.43}$$

$r_{n/p}$ is the first correction factor that depends on the carrier scattering process and energy-band structure and varies from about 1.15 for n-Si to about 0.7 for p-Si, whereas k_g is a second correction factor that corrects for the finite geometry of the slab and takes a value of 0.7 or so. Equation (8.43) then shows the approximate Hall voltage for an extrinsic semiconductor, from which the sensitivity S_I can be found, that is,

$$S_I = \frac{1}{I_x}\frac{dV_H}{dB_z} = -\frac{r_n k_g}{ned}(n\text{-Si}) \text{ and } +\frac{r_p k_g}{ped}(p\text{-Si}) \tag{8.44}$$

Thus, a low carrier concentration (and hence high resistivity) generally produces a higher Hall coefficient.

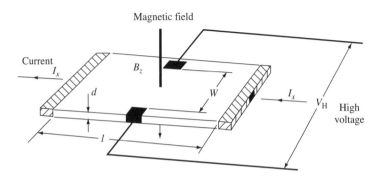

Figure 8.38   Schematic diagram of a Hall plate sensor in which the Hall voltage V_H is related to the magnetic flux density B_z

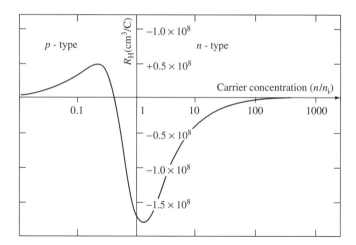

Figure 8.39 Variation of the Hall coefficient R_H of silicon with its doping type and concentration. From Middelhoek and Audet (1989)

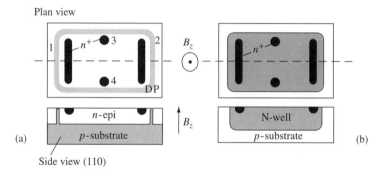

Figure 8.40 Schematic layout of a Hall effect magnetic microsensor fabricated in (a) a bipolar IC process and (b) an nMOS IC process

A Hall plate sensor can be fabricated using standard IC processes and generally employs an n-type material that has higher carrier mobility. For example, Figure 8.40 shows the layout of a substrate Hall plate sensor made from a bipolar process and a CMOS process. The lateral bipolar Hall effect IC has an n-type epilayer plate isolated by a deep p-type diffusion to be about 200 μm square with an electron density n of 10^{15} cm^{-3} and thickness of 5 μm or so. The Hall contacts are then defined by a shallow n^+ diffusion region and the metal interconnect (not shown) is patterned on top. The design of the nMOS Hall effect IC device is similar except that the plate is defined by an N-well. Both ICs measure the horizontal magnetic flux density (i.e. in the plane of the IC).

Therefore, Hall devices can be made using either a standard bipolar IC process or a standard N-well CMOS process (see Chapter 4). Bipolar devices tend to have a sensitivity S_I of about 50 V/AT, whereas nMOS devices can be as high as 1000 V/AT. The higher value is achieved by using an ultrathin Hall layer d of around 10 nm, which more than compensates for lower carrier mobility.

A large variety of Hall effect devices are commercially available (e.g. from Siemens) with different designs, packages, and output signals. For example, a general-purpose linear Hall effect IC (RS 304-267) is available in a 4-pin dual-in-line (DIL) plastic package to detect magnetic flux densities of ± 40 mT. The supply voltage is 4 to 10 V direct current (DC) with a current output in the milliampere range and a sensitivity of about 10 V/T. Linear Hall effect devices are also commercially available and these have an integrated transconductance amplifier to provide a linear output voltage. For example, the RS 650-532 (RS Components Ltd) comes in a 3-pin in-line surface-mounted package with its thick film resistors laser-trimmed in a hybrid[16] (ceramic) circuit to give a voltage sensitivity of 75 ± 2 V/T.

One problem associated with a Hall plate device is that the offset voltage becomes significant at low magnetic flux densities and, therefore, an alternative is to use a magnetoresistive device (see next section) that has a higher sensitivity. This offset problem can be reduced by spinning the Hall plate and averaging the signal (Bellekom 1998), but a more practical solution is to employ a magnetoresistor or, when needing a standard IC process, to employ a magnetodiode or magnetotransistor instead (see Section 8.5.3).

8.5.2 Magnetoresistive Devices

The resistance of a semiconducting material is influenced by the application of an external magnetic flux density B_z. In this effect, known as *magnetoresistivity*, the resistance R of a slab of the material (see Figure 8.38 for basic layout) depends on the Hall angle θ_H and is given by

$$R = R_0(1 + \tan^2 \theta_H) \text{ and } \theta_H \approx k_g \mu^2 B_z^2 \quad (8.45)$$

R_0 is the resistance of the slab at zero flux density. The Hall angle is the angle by which the direction of the current I_x is rotated as a result of the Lorentzian force that acts on the charge carriers and is related to the mobility μ of the carrier. The ideal, theoretical value is corrected by a geometrical factor k_g that depends on the actual aspect ratio of the slab. Ideally, the slab should be much wider than longer and the Hall voltage must be shorted out.

The low carrier mobility in silicon ($\mu_n \sim 1600$ cm^2/Vs) makes the effect rather small and so other materials are used that exhibit a giant magnetoresistive effect. For example, InSb has a very high electron mobility of 70 000 cm^2/Vs and when doped with NiSb can produce a high sensitivity of about 700%/T as illustrated in Figure 8.41.

The strong temperature-dependence of the response of a giant magnetoresistor needs compensation. Magnetoresistors are relatively inexpensive to make but are not compatible with an IC process like a Hall effect IC or magnetodiode/transistor. Therefore, the integration of a giant magnetoresistor into a silicon MEMS structure would be difficult. More recently, research has been directed toward perovskite materials that exhibit a so-called 'colossal' magnetoresistance and may also be used as the dielectric material in submicron ICs because of their high dielectric constant and high breakdown field strength.

---

[16] Hybrid technologies are described in Section 4.6.

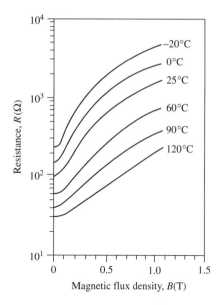

Figure 8.41   Response of a InSb/InSb magnetoresistive device to an applied magnetic flux density

8.5.3 Magnetodiodes and Magnetotransistors

Figure 8.42 shows the schematic layout of two types of diode. The first is made using a silicon-on-sapphire (SOS) IC process in which carriers are injected from both the n^+ and p^+ junctions and drift under the action of the electric field. The magnetic flux density is applied perpendicular to the carrier flow, and a difference in the recombination rates at the two junctions occurs (Suhl effect). The difference in recombination rates modifies the I-V characteristics of the diode, and the forward voltage sensitivity is given by,

$$S_V = \frac{\mathrm{d}V_\mathrm{f}}{\mathrm{d}B_z} \approx \frac{e(\mu_n + \mu_p)\tau_\mathrm{eff}(v_2 - v_1)}{8kTl} V_\mathrm{f}^2 \qquad (8.46)$$

where μ_n and μ_p are the electron and hole mobilities, τ_eff is the effective lifetime of the carrier, v_1 and v_2 are the recombination rates at the two junctions, and the thickness d of the n^- diffusion region is less than the ambipolar diffusion length. The SOS structure provides high recombination on one side (Si–Al$_2$O$_3$) and low on the other (Si–SiO$_2$) to give a sensitivity S_V of about 5 V/T (Lutes *et al.* 1980). However, this device is highly nonlinear and has an output that depends strongly on temperature or the roughness of the two surfaces. The second structure can be made from a standard IC process, for example, CMOS. The device looks like a transistor but is operated as a diode, and the reversed biased *p-n* (collector) junction becomes the high recombination surface. When the magnetic field is applied, the collector current I_C changes and so does the base resistivity and hence the base-emitter voltage V_BE. This can produce a forward voltage sensitivity of about 25 V/T (Popovic *et al.* 1984) that is five times greater than that of the SOS device.

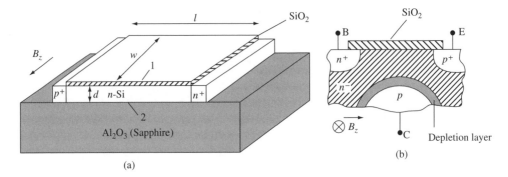

Figure 8.42 Schematic layout of two types of magnetodiode: (a) fabricated by an SOI (sapphire) process and (b) fabricated by a standard CMOS process. After Popovic *et al.* (1984)

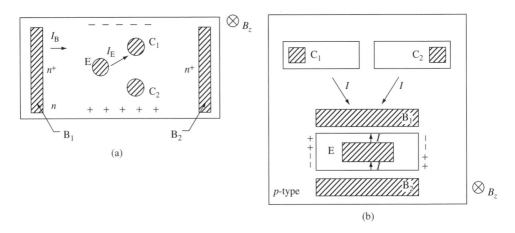

Figure 8.43 Schematic layout of two types of magnetotransistors fabricated in a bipolar process: (a) the drift-aided lateral and (b) the injection-modulation lateral *p-n-p* magnetotransistor. From Gardner (1994)

The poor control of the recombination rates in a magnetodiode makes them problematic. A better approach is to fabricate a magnetotransistor. Two effects can take place in a magnetotransistor as a result of the effect of the Lorentz force on the current carriers; first, generation of a Hall voltage in the active region, such as a Hall plate device, causes a modulation of the emitter injection current; and second, the deflection of the injected minority carrier current can be measured using a split collector. The domination of one process or another depends on the design of the device. For example, Figure 8.43 shows the structures of two different devices. The first device is a drift-aided lateral *p-n-p* transistor fabricated using a standard bipolar process with two collectors, C_1 and C_2. A voltage applied between the base contacts b_1 and b_2 causes it to behave like a Hall plate; the emitter current I_E deflected by the Lorentz force causes a difference in the collector currents according to

$$\Delta I_C = (I_{C1} - I_{C2}) = k_g(\mu_n + \mu_p)B_z I_E \tag{8.47}$$

In the second device, the collectors are now outside the base pads and the Lorentz force creates opposing Hall fields in the emitter and base areas. Therefore, the injected emitter current is modulated differently and so is the current flowing through the collectors. The second device has a higher sensitivity (7%/T) than the first device (0.5 to 5%/T).

A bipolar device can also be designed in which both processes occur. Figure 8.44(a) shows a photograph of such a device together with a plot of the difference in the collector current against magnetic flux density (Avram *et al.* 1998). The plot shows that, when in a common-base configuration, the collector current difference of the magnetotransistor is a linear function of the applied flux density, as predicted by Equation (8.46), and has a sensitivity of about 250 μA/T.

Magnetotransistors, like magnetodiodes, can also be made from a standard lateral or vertical CMOS or diffusion-channel MOS (DMOS) process, which has the advantage of higher sensitivities. Interested readers are referred to Middelhoek and Audet (1989) for further details. Despite the promise of these latter devices, the most successful commercial device is the Hall effect IC, which is simple and inexpensive to process.

8.5.4 Acoustic Devices and SQUIDs

Although a Hall effect IC is of practical use in many situations, there is a general problem of low sensitivity. One possible solution, that is rather attractive, is to use magnetic microsensors based on a delay-line SAW device. Figure 8.45 shows the basic structure of a SAW magnetic sensor (Hanna 1987).

> *Worked Example E8.1:* **Magnetostrictive Strain Gauge**
>
> A thin film IDT magnetostrictive strain gauge can be fabricated using acoustic materials and a wet metal etch-based on the propagation of Love waves[17].
>
> **Process Flow:**
>
> 1. The substrate consists of a single-crystal nonmagnetic (111) gadolinium gallium garnet wafer on top of which a thin epitaxial layer of garnet film ($Y_{1.5}Lu_{0.3}Sm_{0.3}Ca_{0.9}Ge_{0.9}Fe_{4.1}O_{12}$) is grown.
>
> 2. This step is followed by the thermal evaporation of a layer of aluminum of about 100 nm thick.
>
> 3. A thin layer of ZnO is then sputtered down to improve the electromechanical coupling of the Love waves to the magnetic garnet film.
>
> 4. Finally, an IDT metallisation layer is deposited by thermal evaporation[18] and the electrode structure is patterned using UV photolithography and a wet etch. The distance between the IDT transducers is set to be 6000 μm.

The propagation of the SAW is modified by the magnetoelastic coupling between the magnetic spin and the strain fields. In other words, the SAW device is operating simply as a strain sensor in which the strain is induced in the garnet film by magnetostriction.

---

[17] Love waves, and other surface acoustic waves, are explained in Chapter 10.
[18] An alternate lift-off process is described in Chapter 12.

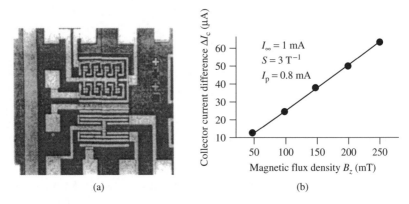

Figure 8.44 Photograph (a) and characteristic (b) of a bipolar magnetotransistor in which both current difference and injection-modulation effects occur. From Avram *et al.* (1998)

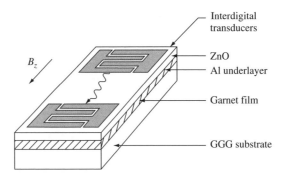

Figure 8.45 Schematic layout of a SAW delay-line magnetic sensor. After Hanna (1987)

The change in the acoustic velocity v of the wave results in a change in the resonant frequency v_0 of the SAW oscillator and hence,

$$\frac{\Delta f}{f_0} = \frac{\Delta v}{v_0} = F(B_z) \tag{8.48}$$

The shift in oscillator frequency is a nonlinear function $F(\cdot)$ of the magnetic flux density B_z. Figure 8.46 shows the variation of the frequency of a SAW oscillator with the centre frequency of 105 MHz. The sensitivity of the SAW device is high and is enhanced by a DC magnetic field with a value of about 250 Hz/mT, and the resolution is therefore on the order of microtesla. The high sensitivity of a SAW magnetic sensor is a significant advantage over a Hall effect IC or magnetodiode, and the process could be made compatible with SOI. Furthermore, the addition of a microantenna and reflectors would enable the creation of a wireless magnetic sensor similar to the wireless SAW temperature sensor (Bao *et al.* 1994).

The most sensitive of all magnetic sensors is known as a *superconducting quantum interference device* (SQUID). A SQUID comprises a ring structure made from a superconducting material (e.g. a high-temperature superconductor) that is interrupted by either one

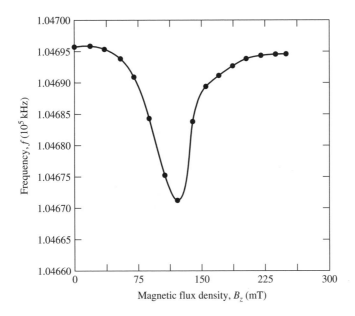

Figure 8.46   Response of a SAW delay-line magnetic sensor. After Hanna (1987)

(RF SQUID) or two (DC SQUID) Josephson junctions. In a DC SQUID, a bias current I_{bias} is applied, with currents passing through each Josephson junction. The magnetic flux inside the loop ϕ_{int} must be quantised in units of the quantum flux ϕ_0, which takes a value of $h/2e$ or 2.07 fWb. Thus, the internal flux is a measure of the external flux ϕ_{ext} and the difference in phase δ at the two junctions,

$$\phi_{int} = \phi_{ext} + \frac{L I_{bias}}{2}(\sin \delta_2 - \sin \delta_1) \qquad (8.49)$$

where L is the self-inductance of the ring. The corresponding oscillation in junction voltage can be averaged and the typical V–I characteristic of a DC SQUID is shown in Figure 8.47(a), where I_0 is the peak current flowing through the junctions and R is the junction resistance.

An integrated DC SQUID has been reported (Koch 1989) and the basic layout is shown in Figure 8.47(b). The SQUID ring is made of a thin superconducting film that is shaped like a square washer, a large pickup coil and a smaller input coil. The washer acts as a ground plane to the input coil and hence improves the coupling between the spiral input coil and the SQUID ring. The input coil has an inductance of 100 nH, whereas the SQUID loop is about 1000 times less at only 100 pH. The integrated SQUID can then be run in an AC bridge with the deflection as a measure of the external magnetic flux and hence the flux density. Flux densities as low as 10^{-14} T can be measured, which is comparable with those emitted from the human heart or brain. Moreover, as the SQUID can be run at radio frequencies, it is possible to power and measure the flux density in a wireless fashion.

To sum up, magnetic microsensors are currently employed in a range of other mechanical sensors, such as proximity sensors, speed sensors, flow sensors, and so on. Therefore,

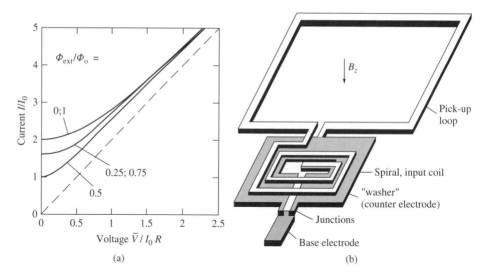

Figure 8.47 (a) Typical I-V characteristics of a DC SQUID magnetometer and (b) schematic arrangement of an integrated SQUID. After Koch (1989)

their main use is when they can be integrated into another microcomponent, such as a MEMS device. For example, a magnetic microsensor could be used in a micromotor to either monitor or control the movement of the rotor. Ideally, the magnetic microsensor should be placed within the feedback arm of the closed-loop control of a microactuator or MEMS device.

8.6 BIO(CHEMICAL) SENSORS

The general topic of chemical sensors is well covered in the standard literature on sensors (Madou and Morrison 1989; Göpel *et al.* 1989–1998; Taylor *et al.* 1996).

The basic components of a bio(chemical)[19] sensor are illustrated in Figure 8.48 and comprise a chemically sensitive layer interfaced to a sensing transducer. The analyte molecules interact with the chemically sensitive layer and produce a physical change that is detected by the transducer and are converted into an electrical output signal.

The nature of this interaction is determined by the type of material used and can be either a reversible process or an irreversible reaction (see Figure 8.49). In a reversible binding reaction, the analyte is typically bound to specific sites within the sensitive layer, and when the external concentration is removed, the analyte molecules dissociate and there is no net change. An example of this would be the adsorption and desorption of an organic vapour in a polymeric material. In an irreversible reaction, the analyte undergoes a chemical reaction catalysed by the sensitive layer and therefore is consumed in the process. In this case, removal of the external analyte concentration still reverses the process but the associated time-constant may be considerably longer. The irreversible

---

[19] A bio(chemical) sensor can be either a biosensor or a chemical sensor. In general, a biosensor either detects biological material or uses it in the active layer.

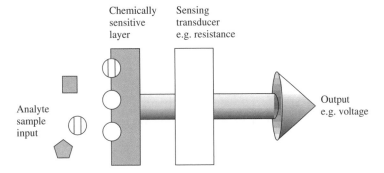

Figure 8.48 Basic components of a bio(chemical) sensor: analyte molecules, chemically sensitive layer, and transducer. After Gardner and Bartlett (1999)

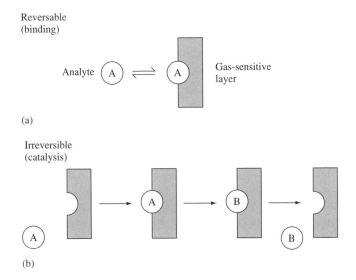

Figure 8.49 Basic mechanisms of a biochemical sensor: (a) reversible binding of the analyte A to a site at the chemically sensitive layer and (b) irreversible reaction of the analyte A at a site to produce molecule B at the chemically sensitive layer. After Gardner and Bartlett (1999)

reaction is in fact more common and its sensitivity and selectivity varies with the shape and charge distributions of the analyte molecule and sensitive layer. The most selective reactions tend to be those like the key-lock mechanism that operates in a biological sensor (biosensor), such as a glucose sensor. However, the poor stability of biological materials makes them unsuitable for use in a real sensor that operates many thousands or millions of times with a lifetime of a year or more but more suited to a single measurement, that is, a disposable sensor. For this reason, we concentrate predominantly here on the field of chemical sensors; anyone interested specifically in biosensors is referred to books by Cass (1990) and Taylor and coworkers (1996).

Figure 8.50 shows the main types of bio(chemical) sensors classified according to the operating principle of the sensitive layer, that is, the signal transduced. The signals that are

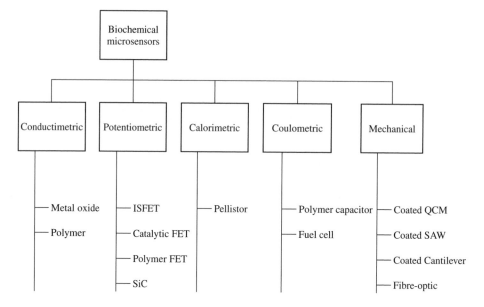

Figure 8.50 Classification of the main types of bio(chemical) sensors. Devices listed are those that can be regarded as microsensors

measured can be the change in electrical resistance (i.e. conductimetric), change in work function (i.e. potentiometric), change in the heat of reaction (i.e. calorimetric), and so on. Here, we are going to discuss the two most important types of chemical microsensor, their commercial availability, and the developments associated with current research devices. The topic of *mechanical* bio(chemical) microsensors is already covered elsewhere in this book with the principles of acoustic sensors described in Chapters 9 to 12 and that of chemical acoustic microsensors in Chapter 13. Moreover, we expand upon and introduce new concepts of chemical sensors through a description of electronic noses and electronic tongues in Chapter 15 on smart sensors.

8.6.1 Conductimetric Devices

Conductimetric gas sensors are based on the principle of measuring a change in the electrical resistance of a material upon the introduction of the target gas. The most common type of gas sensor employs a solid-state material as the gas-sensitive element (Moseley and Tofield 1987). The principal class of material used today is semiconducting metal oxides, with tin oxide (SnO_2) being the most popular. For example, Figure 8.51 shows the structure of a Taguchi-type tin oxide gas sensor; millions of these are sold by Figaro Engineering Inc. (Japan). Complete details of Taguchi devices can be found in a book on stannic oxide gas sensors by Ihokura and Watson (1994).

The device consists of a wire-wound platinum heater coil inside a ceramic former onto which a thick layer of porous tin oxide is painted manually. The film is then sintered at a high temperature so that the appropriate nanocrystalline structure is formed. The electrical resistance of the sintered film is then measured by a pair of gold electrodes and basic potential divider circuit.

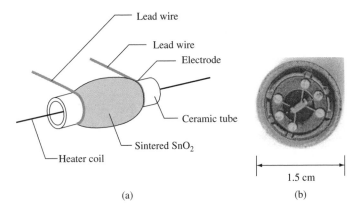

Lead wire

Lead wire

Electrode

Ceramic tube

Sintered SnO$_2$

Heater coil

1.5 cm

(a)   (b)

Figure 8.51 (a) Basic structure of a Taguchi-type tin oxide gas sensors and (b) photograph of a series-8 commercial gas sensor (Courtesy of Figaro Engineering, Japan)

Tin oxide devices are operated at various high temperatures and doped with different materials to enhance their specificity. The response of a tin oxide sensor, in terms of its relative conductance G_s/G_0, where G_s is the conductance of a gas of fixed concentration and G_0 is the conductance in air, is shown in Figure 8.52 (Yamazoe *et al.* 1983). The devices are operated at high temperatures (typically between 300 and 400 °C) for several reasons. First, and most important, the chemical reaction is more specific at higher temperatures, and, second, the reaction kinetics are much faster, that is, the device responds in just a few seconds. Finally, operating the device well above a temperature of 100 °C ameliorates the effect of humidity upon its response – a critical factor for many chemical sensors.

The basic reactions that occur within the porous sintered film can be represented by the following reactions. First, vacant sites within the nonstoichiometric tin oxide lattice react with atmospheric oxygen to abstract electrons out of the conduction band of the tin oxide creating chemisorbed oxygen sites such as O^-, O_2^-, and so on.

$$\tfrac{1}{2}mO_2 + \{vacancy\} + e^- \overset{k_1}{\longleftrightarrow} \{O_m^-\} \tag{8.50}$$

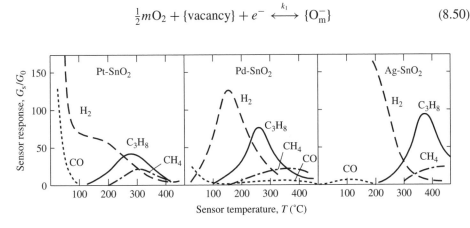

Figure 8.52 Variation of the response of three doped tin oxide gas sensors with temperature for four different gases. Adapted from Yamazoe *et al.* (1983)

Next, this reversible reaction is disturbed when the analyte molecule X reacts with the chemisorbed oxygen species to release electrons and promulgate further reactions:

$$X + \{O_m^-\} \xrightarrow{k_2} \{XO_m\} + e^- \tag{8.51}$$

In a simple physical description, the tin oxide behaves like an n-type semiconductor and, therefore, there is an increase in the electron carrier density n, and hence in the electrical conductivity σ, of the material with increased gas concentration where

$$\Delta\sigma = \mu_n e \Delta n \tag{8.52}$$

where μ_n is the electron mobility.

In fact, changes in the order of magnitude in device conductance that is observed (see Figure 8.53) cannot be explained by the very small change in carrier concentration and, therefore, the common model is one in which the electrons modulate a space charge region (depletion region) that surrounds nanometer-sized grains within the sintered material. It is then a reduction in the height of the intergranular barriers V_s that increases the electron hopping mobility and hence the conductance of the tin oxide film (Williams 1987).

This change in device conductance can be approximately related to the gas concentration C from the chemical rate constants k_1 and k_2 defined in Equations (8.50) and (8.51).

$$\Delta G \propto \frac{k_1}{k_2} C^r \tag{8.53}$$

where the exponent r has a value that lies between 0.5 and 0.9 and depends on the kinetics of the reaction (Gardner 1989; Ihokura and Watson 1994).

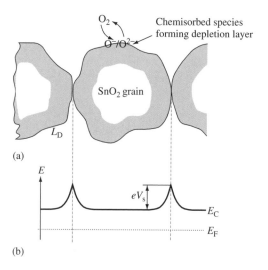

Figure 8.53 (a) Schematic diagram showing a series of nanometre-sized grains in a sintered tin oxide film and (b) band diagram showing the effect of the oxygen-induced depletion regions. From Pike (1996)

Table 8.17 Some commercial gas sensors based on semiconducting metal oxide

Manufacturer	Model	Material	Measurand	Range (PPM)	(Power mW)	Cost[a] (euro)
Figaro Inc. (Japan)	TGS842	Doped SnO$_2$	Methane	500–10 000	835	13
Figaro Inc. (Japan)	TGS825	Doped SnO$_2$	Hydrogen sulfide	5–100	660	50
Figaro Inc. (Japan)	TGS800	Doped SnO$_2$	Air quality (smoke)	<10	660	13
FiS (Japan)	SB5000	Doped SnO$_2$	Toxic gas - CO	10–1000	120	13
FiS (Japan)	SP1100	Doped SnO$_2$	Hydrocarbons	10–1000	400	15
Capteur[b] (UK)	LGS09	Undoped oxide	Chlorine	0–5	650	25
Capteur (UK)	LGS21	Undoped oxide	Ozone	0–0.3	800	25

[a] Price for 1 to 9 units 1 euro is $1.1 here
[b] Now part of First Technology plc (UK)

Table 8.17 lists some tin oxide gas sensors that are commercially available together with their properties.

The requirement to run this type of gas sensor at a high temperature causes the power consumption of about 0.8 W of a Taguchi-type device to be a problem for handheld units. Consequently, there has been considerable effort since the late 1980s toward the use of silicon planar technology to make micropower gas sensors in volume at low cost (less than €5). Designs of silicon planar microhotplates started to appear around the late 1980s when Demarne and Grisel (1988) and later Corcoran *et al.* (1993) reported on the first silicon-based tin oxide gas microsensors. There are two basic configurations of a microhotplate; these are illustrated in Figure 8.54. The first comprises a resistive heater (e.g. platinum) embedded between layers that make up a solid diaphragm (Gardner *et al.* 1995) or a resistive heater (e.g. doped polysilicon) embedded between layers in a suspended microbridge configuration.

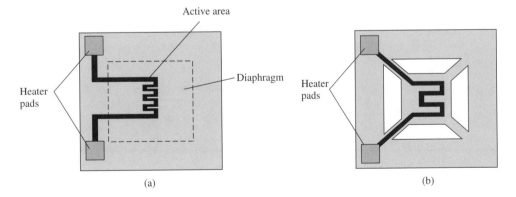

Figure 8.54 Two basic designs of silicon gas sensors: (a) a solid diaphragm and (b) a suspended bridge that contains a meandering resistive heater

We now provide, via a worked example, the process sequence for a resistive gas microsensor.

> ### *Worked Example E8.2:* Silicon-Resistive Gas Sensor Based on a Microhotplate
>
> **Objective:**
>
> To fabricate a resistive planar gas microsensor from bulk silicon micromachining techniques based on a solid diaphragm microhotplate. The small thin diaphragm (less than 1 μm thick) should result in low power consumption.
>
> **Process Flow:**
>
> A five-mask process has been used to fabricate a resistive gas microsensor with the main steps shown in Figure 8.55 (Pike 1996). The initial substrate was a 3″, 280 μm thick, single-sided polished, ⟨100⟩ oriented, SCS wafer. Before processing, the wafers were given an identity mark with a diamond scribe and then subjected to a standard multistage cleaning process, which removed any organic contaminants, adsorbed layers, and particulates. A standard cleaning process was used before each thin film deposition to ensure adequate adhesion[20].
>
> 1. An 80 nm dry SiO_2 film was thermally grown on the wafer at 1100 °C. Note that the intrinsic stress in the thermally grown SiO_2 films must be negligible at 1100 °C – an essential requirement for mechanical stability of the membrane structure.
>
> 2. A 250 nm thick layer of low-stress SiN_x[21] was then deposited by LPCVD.
>
> 3. The microheaters were defined by patterning a thin platinum film with Mask 1 using a lift-off technique. Specifically, a photoresist was first spin-coated onto the wafer and exposed to UV light using Mask 1. Before the photoresist was developed, it was exposed to chlorobenzene to harden the photoresist surface. Hence, during developing, the photoresist was undercut slightly because of the surface modification. This profile ensures that no side coverage occurs during the metallisation deposition, so that when the photoresist was removed, it did not interfere with the metallisation that had bonded to the substrate. The photoresist layer acts as a sacrificial layer, which was removed later with acetone, revealing the mask image patterned into the metallisation[22]. To improve metal adhesion to the substrate, it is common to use a thin adhesion layer of a more reactive metal. Therefore, before sputtering down 200 nm of Pt, a 10 nm tantalum (Ta) adhesion layer was first deposited.
>
> 4. A standard cleaning process prepared the substrate for a second 250-nm layer of LPCVD low-stress SiN_x, which insulates the microheater electrically from the electrodes deposited in a later stage. Mask 2 was used to open up the contact windows in the SiN_x. This required the plasma etching of the SiN_x to reveal the Pt heater contact pads. The patterned photoresist layer used was then stripped off before another cleaning stage.

---

[20] See Chapter 4 for details of wafer cleaning.
[21] The nitride is slightly silicon-rich and so nonstoichiometric.
[22] This is a lift-off process such as that used to make SAW IDT microsensors (Chapter 12).

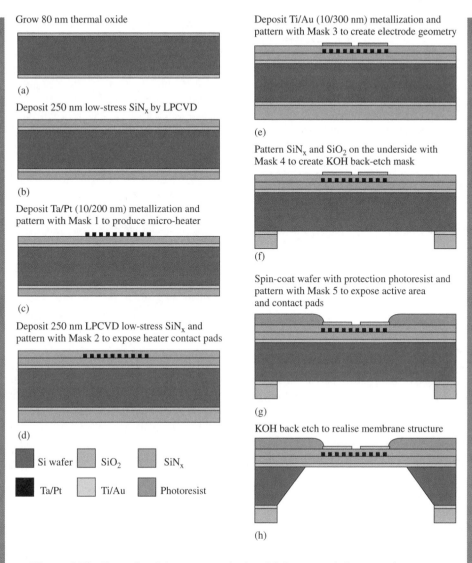

Grow 80 nm thermal oxide

(a)

Deposit 250 nm low-stress SiN$_x$ by LPCVD

(b)

Deposit Ta/Pt (10/200 nm) metallization and
pattern with Mask 1 to produce micro-heater

(c)

Deposit 250 nm LPCVD low-stress SiN$_x$ and
pattern with Mask 2 to expose heater contact pads

(d)

■ Si wafer □ SiO$_2$ □ SiN$_x$

■ Ta/Pt □ Ti/Au ■ Photoresist

Deposit Ti/Au (10/300 nm) metallization and
pattern with Mask 3 to create electrode geometry

(e)

Pattern SiN$_x$ and SiO$_2$ on the underside with
Mask 4 to create KOH back-etch mask

(f)

Spin-coat wafer with protection photoresist and
pattern with Mask 5 to expose active area
and contact pads

(g)

KOH back etch to realise membrane structure

(h)

Figure 8.55 Example of the steps required to fabricate a resistive gas microsensor

5. The next step involved the deposition of the gold (Au) thin film and patterning with
   Mask 3 to define the sensing electrode structure. A lift-off technique was again
   used; therefore, a 10 nm titanium (Ti) adhesion layer followed by 300 nm Au film
   was sputtered over the patterned photoresist. The photoresist was then removed with
   acetone to leave the electrodes on the surface.

6. To create the ultrathin membrane structure required an anisotropic KOH back-etch
   through the SCS. Most photoresists are inappropriate for defining features any deeper
   than 20 μm in KOH etch conditions; therefore, the SiN$_x$ and SiO$_2$ on the wafer
   underside were patterned with Mask 4 using plasma etching to form a suitable
   KOH mask.

7. Before KOH anisotropic etching, the topside protection resist had to be processed. The wafers were held onto a spinner by a vacuum and a layer of Shipley Microposit 1813[23] was spin-coated over the wafers. This protecting layer was then photolithographically patterned with Mask 5 to expose the active diaphragm area and the four contact pads. The photoresist was hard-baked for 1 hour at 180 °C after developing, which made it more resistant to chemical attack. Clearly, the resist will not stand up to attack by organic solvents or high temperatures. This layer has been replaced recently by plasma-enhanced chemical vapour deposition (PECVD) nitride, which permits the definition of a precise gas-sensitive area above the sensing electrodes. Moreover, the nitride passivation layer can withstand the high operating temperatures created by the heater.

8. The final processing stage was a KOH anisotropic bulk back-etch that creates the diaphragm (membrane) structure and a thermal SiO_2 as an etch stop on the topside. To prevent the wafer topside from being exposed to the etchants, the wafer was mounted in a suitable holder during etching.

9. The back-etch also opened up V-grooves (not shown) in the wafer that allows the wafer to be easily snapped up into individual silicon dies. This method is a much more gentle a method than dicing up with a diamond saw.

10. Finally, the gas-sensitive layer is drop-deposited across the electrodes and sintered[24].

Figure 8.56 shows two silicon micromachined resistive gas sensors with embedded platinum resistive microheaters. The first design comprises an array of three microhotplates, each with two sets of resistive gold-sensing electrodes (referred to here as device no. SRL108, Gardner *et al.* (1995)). The second design (IDC 50) comprises a single cell with one microhotplate and one set of resistive electrodes. A small drop of doped tin oxide has been carefully deposited on the surface at Tübingen University (process details are in Al Khalifa (2000)). Both devices were fabricated at the Institute of Microtechnology (Switzerland).

The platinum microheater has a resistance R_{Pt} that depends linearly on its absolute temperature T, namely,

$$R_{Pt}(T) = R_0[1 + \alpha_T(T - T_0)] \tag{8.54}$$

where R_0 is the resistance of the heater at room temperature T_0 and α_T is the linear temperature coefficient of resistance, the values of 190 Ω and $1.7 \times 10^{-3}/°C$[25], respectively, were measured for the device SRL108, which is shown in Figure 8.56(a). The platinum heater not only supplies the power to heat up the diaphragm but also acts as an accurate linear temperature sensor.

Figure 8.57(a) shows the total electrical power required to heat up the microhotplates of a microdevice (SRL108) to temperatures of up to 350 °C above ambient ($T_0 = 22$ °C). A simple quadratic fit to the data is shown.

Heat losses are caused in general by thermal conduction through the membrane, convection/conduction to air, and radiation. The power loss of a microhotplate P_H based on these

---

[23] Shipley 1816 has now replaced 1813.
[24] Other methods include sputtering of thin oxide films and sol-gel.
[25] The bulk value for platinum is higher at $3.8 \times 10^{-3}/°C$.

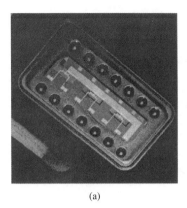
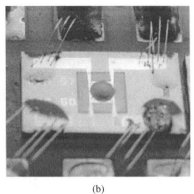

(a)   (b)

Figure 8.56 Photographs of two examples of silicon resistive gas sensors: (a) array of three micro-diaphragms, each 1100 μm × 3500 μm and about 0.6 μm thick with two sets of sensing electrodes per cell and (b) single microdiaphragm of 1500 μm square with a drop of doped tin oxide located on top of a single set of sensing electrodes and a single 750 μm square microheater. Both devices are mounted on a DIL header with 0.1″ spacing

three mechanisms is given by

$$P_H = a_{cond}\,(T - T_0) + b_{conv}\,(T - T_0)^2 + c_{rad}\,\left(T^4 - T_0^4\right) \tag{8.55}$$

with a, b and c being constants. The actual contributions from each of these three mechanisms has been determined by running a device (SRL108) in a vacuum, and Figure 8.57(b) shows that the results are a good fit to the terms in Equation (8.51) (Pike and Gardner 1997).

It can be seen that devices operated at about 350 °C lose most of their heat through convection to air and a negligible amount in radiation. In this case, the DC power consumption of the microhotplate is typically 120 mW at 300 °C or 60 mW per resistive sensor. The thermal response time of the microhotplate was measured to be 2.8 ms for a 300 °C change in operating temperature (Pike and Gardner 1997). Both the power consumption of the device and its thermal time constant will scale down with the size of the diaphragm; hence, power consumptions and time constants of less than 10 mW and 1 ms, respectively, are quite realizable.

Figure 8.58 shows the characteristic response of an undoped and a doped tin oxide resistive gas microsensor operated at a constant temperature of 367 °C to ppm pulses of NO_2 in air at 38% relative humidity (RH). The doped devices clearly show a higher response to NO_2 and it should be noted that the resistance here increases in the presence of the oxidising gas. The resistance falls in the presence of reducing gases such as CO or hydrogen. The rise time of a tin oxide sensor tends to be faster than its decay time; this becomes more apparent when detecting larger molecules such as ethanol. The response is also not well approximated by a first-order process; therefore, an accurate model of the dynamic response requires a multiexponential model (Llobet 1998).

However, the fast thermal response time of the microhotplate permits the rapid modulation of its operating temperature – this can be used to reduce the average power consumption of the device by a factor of approximately 10 when powering up for only 100 ms

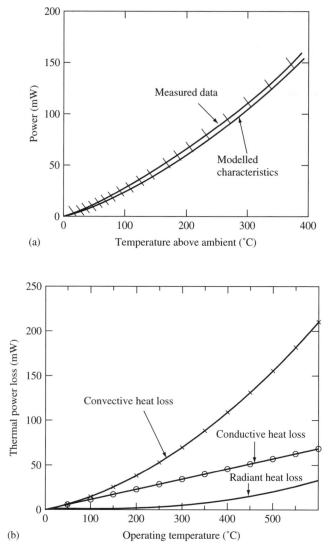

Figure 8.57 Power consumption of a microhotplate-based resistive gas microsensor (SRL108) (a) observed against a simple analytical model and (b) relative contributions of conductive, convective, and radiative heat losses. From Pike (1996)

in every second and thus achieve an average power consumption of below 12 mW for SRL108 or below 1 mW for smaller hot plates. An interesting and alternative approach is to modulate the heater temperature with a sinusoidal AC drive voltage[26] and then relate the harmonic frequency content of the AC tin dioxide resistance signal to the gas present. This approach has been successfully demonstrated by researchers (Heilig *et al.* 1997; Al-Khalifa *et al.* 1997); in this approach, the coefficients of a Fourier analysis are learnt in

---

[26] Strictly speaking, the temperature rise is not a sine wave but is periodic.

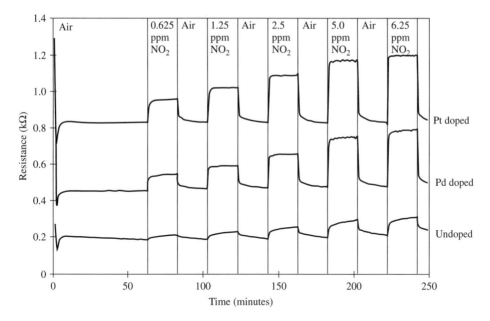

Figure 8.58 Typical response of doped and undoped resistive tin oxide gas microsensors to pulses of NO_2 in air. From Pike (1996)

a simple back-propagation neural network. It is particularly exciting to note that, using this dynamical approach, a *single* microsensor can predict the concentration of a *binary* mixture of gases from the different rate kinetics.

In the past few years, there has been an enormous increase in the number of research groups from Germany, Korea, and China reporting on the fabrication of microhotplate-based resistive gas sensors. These show some general improvements in the device performance, such as a lower power consumption, greater robustness, and so on. Much of this recent interest has stemmed from the fact that Motorola (USA) set up a fabrication facility to make a low-cost CO gas sensor with Microsens (Switzerland) in the mid-1990s that was based on a suspended polysilicon microhotplate design (Figure 8.54(b)). The device was aimed at the automotive market with a nominal price of €1. Since then, the company has been relocated to Switzerland and become independent. The main competition to such silicon gas sensors is from the commercial screen-printed thick-film–based planar devices, such as those sold in medium volume by Capteur Ltd (UK).

A variety of different materials have been studied for use in solid-state resistive gas sensors. These materials are not only semiconducting oxides (e.g. SnO_2, ZnO, GaO, and TiO_2) that tend to operate at high temperatures but also organometallic materials such as phthalocyanines that operate around 200 °C and organic polymers that operate near room temperature (Moseley and Tofield 1987; Gardner 1994). However, the successful application of these other materials in gas sensors has not yet been realised. Instead, some of these materials – conducting polymers, in particular – are being used as nonspecific elements within an array to detect vapours and even smells (Gardner and Bartlett 1999). Details of these devices, or so-called electronic noses, are given in Chapter 15 on Smart Sensors.

8.6.2 Potentiometric Devices

There is a class of field-effect gas sensors based on metal-insulator semiconductor structures in which the gate is made from a gas-sensitive catalytic metal (Lundström 1981).
There are two basic devices, as illustrated in Figure 8.59, in which the structure is configured as either field-effect transistor or gas-sensitive capacitor.

The most common device is an n-channel metal oxide semiconductor field-effect
transistor (MOSFET) device configured in a common source mode, as shown in
Figure 8.59(b). When the device is in saturation, the drain current i_D is simply related to
the gate voltage V_{GS} by

$$i_D = \mu_n C_{ox} \frac{w}{l} (V_{GS} - V_T)^2 \qquad (8.56)$$

where μ_n is the electron mobility, C_{ox} is the capacitance per unit area of the oxide, w and
l are the channel width and length, respectively, and V_T is the threshold voltage (about
0.7 V for silicon). Lundström discovered that when the gate was made of a thin layer of
palladium, the atmospheric hydrogen would dissociate and diffuse through to the interface,
creating a dipole layer and causing a shift in the threshold voltage. Using a circuit to drive
a constant current through the device with common gate and drain terminals leads to a
characteristic voltage response (equal to the shift in threshold voltage) of this type of
device to hydrogen

$$\Delta V_{GDS} = \Delta V_T = \Delta V_{max} \frac{k\sqrt{C_H}}{1 + k\sqrt{C_H}} \qquad (8.57)$$

where k is a constant and C_H is the partial pressure of the hydrogen in air.

The solid palladium gate has subsequently been replaced by an ultrathin discontinuous
metal film so that larger, less diffuse, molecules can reach the oxide surface and be sensed.

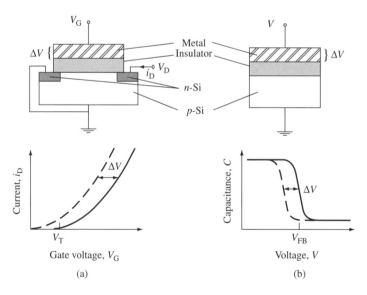

Figure 8.59 Two types of potentiometric gas microsensors (a) n-channel MISFET and
(b) MISCAP. From Lundström *et al.* (1992)

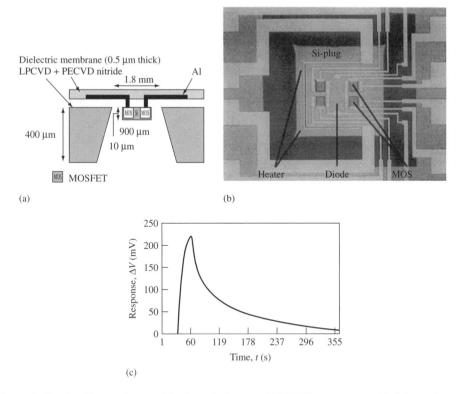

Figure 8.60 A silicon micromachined catalytic gate MOSFET gas sensor: (a) Schematic cross section of a device with a silicon plug, (b) photograph of an array of four polymer-coated n-channel MOSFET devices on a 1800 μm square diaphragm with a 900 μm (10 μm thick) silicon plug to equilibrate temperature, and (c) response of iridium (8 nm) FET at 140 °C. From Briand *et al.* (2000)

This allowed catalytic gate materials (e.g. platinum, palladium, and iridium) to be used to sense gases such as ammonia, ethanol, hydrogen sulfide, and so on. The devices are typically operated at temperatures around 180 °C to increase the activity of the catalyst and rate kinetics. The most recent MOSFET devices use silicon micromachining techniques to define a silicon microplatform and thermal plug (Briand *et al.* 2000). Figure 8.60 shows a schematic cross section of the device and a photograph of the FET device showing a set of four FET sensors with an integrated heater and temperature sensor. A shift in threshold voltage of about 220 mW is observed for a 20-ppm pulse of ammonia in air. The power consumption of the device is greatly reduced to about 100 mW at 200 °C, and again commercial arrays of these devices are being produced by a Swedish company (Nordic Sensors) for gas and odour detection.

There has also been some research effort toward the use of polymeric materials as the gate material to detect organic vapours at room temperature. A similar principle has been developed by Janata (1992) using suspended gate structures to detect the shift in work function of nonconducting polymers. But more recently, the metal gate is no longer suspended, like the catalytic FET device, and is made of a thin porous conducting polymer film (Hatfield *et al.* 2000). Figure 8.61 shows the structure (a) and layout (b) of

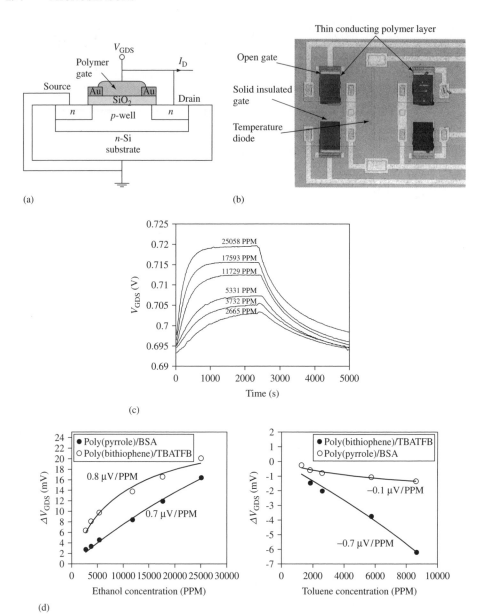

Figure 8.61 PolFET vapour sensor: (a) basic structure, (b) array of four polymer-coated n-channel MOSFETs, (c) typical dynamic response to pulses of ethanol vapour in air, and (d) isotherms of two polymers for ethanol and toluene. From Covington *et al.* (2000)

a conducting polymer FET (polFET) gas sensor and its response (c) to pulses of ethanol vapour in air (Covington *et al.* 2000).

The poly(pyrrole)/butane sulfonic acid (BSA) PolFET has a voltage sensitivity of 0.8 μV/PPM to ethanol and the poly(bithiophene)/TBATFB PolFET has a sensitivity of −0.7 μV/PPM to toluene – its behavior is well described by the Langmuir isotherm.

The data has been fitted to Equation (8.58) and appears approximately linear when kC is much smaller than 1.

$$\Delta V_{GDS} = \Delta V_T = \Delta V_{max} \frac{kC}{1 + kC} \tag{8.58}$$

The main advantages of PolFETs are that they can be operated at ambient temperatures (and therefore require little power) and that they are compatible with CMOS technology. However, the polymers do exhibit a significant humidity dependence and so their future success, as gas (vapour) sensors, will depend on either employing hydrophobic films or on-chip compensation for the humidity variation.

To date, the greatest commercial success of a polymeric potentiometric device has been in the field of biosensors (Scheller *et al.* 1991). The device consists of an enzyme, such as glucose oxidase, which is then attached to an electrode and senses potentiometrically, amperometrically, or impedimetrically. Perhaps the most successful device has been a glucose detector. The device is able to detect the level of glucose in blood by using a

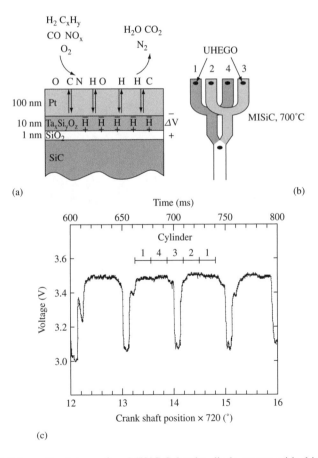

(a)   (b)

(c)

Figure 8.62 (a) Schematic picture of an MISiC Schottky diode sensor with thin platinum as the catalytic layer, (b) positioning of an MISiC sensor in the exhaust system of an automobile engine, and (c) response of an MISiC gas sensor to exhaust gases when one cylinder is injected with excess fuel. From Savage *et al.* (2000)

disposable single-shot sensing element – the enzyme is coupled to a conducting polymer in an electrochemical cell. The sensing strips are fabricated using screen printing, rather than silicon microtechnology, and sold in their millions by companies such as Medisense (UK). Other coatings can be used to detect lipids, peptides, and so on, but to date, their commercial success has been somewhat limited by either the long-term stability of the electrode or the selectivity of the biological coatings.

Finally, there is considerable need in the automotive industry for gas sensors that can monitor the engine combustion process either in-line or at-line – this is extremely demanding and rules out conventional CMOS devices that operate only up to a temperature of about $+125\,°C$. However, there is a field-effect diode made from SiC that can be operated at more than $700\,°C$ and responds in milliseconds (Svenningstorp *et al*. 2000, 2001). Figure 8.62 shows the response of a MISiC Schottky diode to the exhaust gases from a car engine (Savage *et al*. 2000). This device enables the real-time monitoring of the combustion process in each cylinder as it fires in turn and could well be used as a diagnostic sensor.

8.6.3 Others

There are a number of other principles of transduction that can be used to make chemical microsensors. For example, the most obvious type of sensor to make is a capacitive one because the device requires little power and fits in well with CMOS technology. Early work by Gopel on polymeric capacitors had a limited success because of the relatively

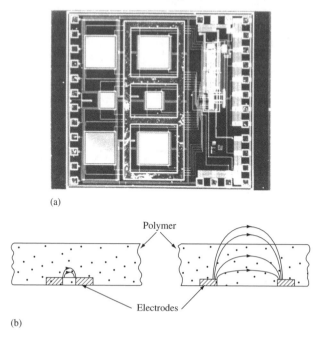

(a)

(b)

Figure 8.63 (a) Array of gas-sensitive polymeric capacitors on a CMOS chip (from Baltes and Brand 2000) and (b) two electrode geometries to discriminate between polymer dielectric constant and swelling changes

poor sensitivity and high noise levels. However, more recent research by Baltes (Baltes and Brand 2000) at ETH (Zurich) has been reported in which polymeric capacitors are integrated on to a CMOS chip to enhance the signal gain, and, with different electrode geometries (Figure 8.63(a)), improve the selectivity.

Interestingly, they report on the use of a pair of electrodes, as shown in Figure 8.63(b), to enhance the selectivity – an approach similar to that proposed by Gardner (1995) with resistive microsensors, and which is based on the fact that there are two mechanisms by which the polymers can work. When the analyte dissolves into the nonconducting polymer, it changes the dielectric constant of the material and hence the capacitance changes. However, the polymer also swells as it absorbs the analyte, which provides a second competing mechanism. Now, the capacitor with the narrow electrodes will not measure the swelling effect as the electric field is contained within the film, whereas the

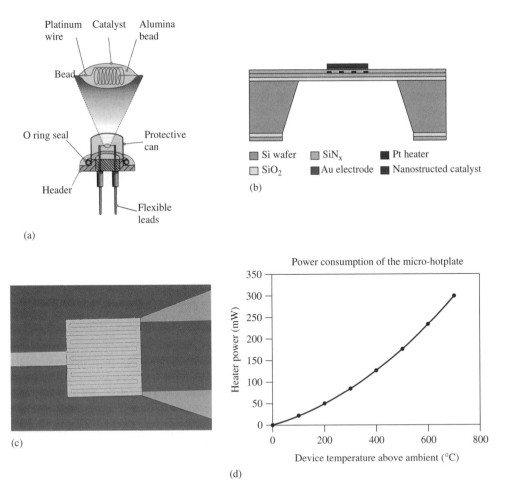

Figure 8.64 (a) A conventional commercial pellistor (City Technology Ltd, UK), (b) cross section of a silicon planar pellistor (SRL162g), (c) photograph of a platinum microheater and gold electrode area, and (d) measured power consumption up to 700 °C. From Lee *et al.* (2000)

capacitor with the wide electrodes will measure the entire swelling that is taking place. Therefore, the selectivity of the capacitive gas microsensor is enhanced.

There is also potential in the development of microcalorimetric gas sensors. For example, conventional pellistors are made in large quantities to detect the presence of combustible gases, such as methane in air (Figure 8.64(a)). They work by sensing a change in temperature of a small catalyst-coated bead held at around 450 °C via a platinum heater or temperature sensing coil. There have been efforts to make a silicon planar pellistor (Gall 1991) but the demands on the silicon technology are considerable because of their very high operating temperatures of around 500 °C. However, recent improvements in microhotplate technology have led to stable operating temperatures of 600 °C, and combined with novel nanoporous gas-sensitive membranes pioneered by Southampton University (Lee *et al.* 2000), Figure 8.64 shows both (b) a schematic cross section of a silicon micromachined microcalorimeter (micropellistor) and (c) a photograph of the actual device. The heater can be seen faintly below the solid gold electrode with a single pad (left) used to electrochemically coat with a nanoporous film of palladium.

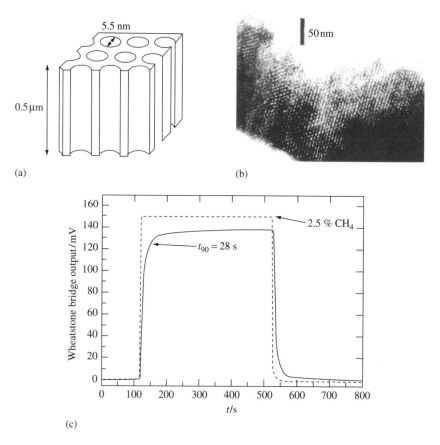

(a)

(b)

(c)

Figure 8.65 (a) Cartoon of nanoporous palladium formed form the hexagonal close-packed phase of a lyotropic film, (b) SEM picture of structure, and (c) typical response of a nanoporous palladium micropellistor to methane

Table 8.18 Relative characteristics of the gas microsensors

Sensing material	Transducer	Sensitivity	Power/operating temp. (°C)	Unit cost	CMOS-compatible or not?
Metal oxide (best for combustible gases)	Resistive	High	High/300	High	SOI only with on-chip anneal
Conducting polymers (best for polar VOCs)	QCM	High	All very low/ambient	Medium	No
	Resistive	Medium		Low	Yes
	MOSFET	Medium		Low	Yes
	Calorimeter	Low		High	Yes, with silicon etch
Catalytic metals (best for reactive gases)	MOSFET	Low	Medium/200	Low	Yes, with etch
	Pellistor	Low	High/500	High	No
Solvating polymer coatings (best for nonpolar VOCs)	SAW	High	All very low/ambient	Medium	Nonstandard
	QCM	High		Medium	No
	Capacitive	Medium		Low	Yes
	Resist	Medium		Low	Yes
	Calorimetric	Low		Medium	Yes, with etch
	Cantilever	Low		High	Yes, with etch

The power consumption of the device, shown in Figure 8.64(d), is about 150 mW at an operating temperature of 450 °C. The microheater has an active area of 570 μm square.

The catalyst-coated bead was formed electrochemically using a lyotropic[27] self-assembling material as a template (Attard *et al.* 1997). The resulting structure is made up of a regular array of 5.5-nm pores in palladium and has a very high surface area, approximately 1000 m^2/g (Figure 8.65 (a,b)). Preliminary results of the micropellistor are shown in Figure 8.65(c). Its response to 2.5 percent methane in air (Lee *et al.* 2000) suggests that commercial silicon micropellistors may become available in the near future.

In conclusion, there are a wide variety of different materials and different types of microsensors reported, which respond to gases and vapours – including those based on acoustic principles and described in other chapters. Table 8.18 summarises some of the key characteristics of the different types of gas microsensors and their potential for integration into a standard process.

---

[27] Made from two phases, literally soaplike.

8.7 CONCLUDING REMARKS

In this chapter, we have presented an overview of some of the different types of microsensors in the literature, such as mechanical, magnetic, and chemical. Most of these devices are fabricated using silicon technology with the addition of either bulk or surface micromachining techniques. However, there are three other important classes of microsensors, and these are discussed elsewhere in the book. First, IDT microsensors, based on acoustic materials, are described in Chapter 13 and offer the attractive possibility of remote wireless communication. Second, IDT-based MEMS sensors can be found in Chapter 14, which describes the use of non-silicon-based micromechanical structures with IDTs. Finally, Chapter 15 explores the expanding field of smart and intelligent devices that will become commonplace in the next few years.

REFERENCES

Al-Khalifa, S. (2000). Identification of a binary gas mixture from a single resistive microsensor, Ph.D. Thesis (Advisor J.W. Gardner), University of Warwick, Coventry, UK.

Al-Khalifa, S., Gardner, J. W. and Craine, J. F. (1997). "Characterisation of a thermal wave microsensor for the intelligent analysis of atmospheric gases," in A. T. Augousti and N. M. White, eds., *Sensors and their Applications*, 8th ed., Iop Publishing, Bristol, pp. 89–94.

Attard, G. S. *et al.*, (1997). "Mesoporous platinum films from lyotropic liquid crystalline phases," *Science*, **278**, 838–840.

Avram, M., Neagoe, O. and Simion, M. (1998). "An optimised integrated bipolar magnetotransistor," *Eurosensors XII*, Vol. **2**, IOP Publishing, Bristol, pp. 991–994.

Ayazi, F. and Najifi, K. (1998). Design and fabrication of high-performance polysilicon vibrating ring gyroscope, *Proc. of the 11th Int. Workshop on MEMS*, Heidelberg, January, pp. 621–626.

Bao, X. Q., Varadan, V. V. and Varadan, V. K. (1994). "Wireless surface acoustic wave temperature sensor," *J. Wave-Material Interact.*, 19–27.

Baltes, H. and Brand, O. (2000). CMOS-based microsensors, *Proceedings of Eurosensors XIV*, Copenhagen, Denmark, August 27–30, pp. 1–8.

Bellekom, S. (1998). "CMOS versus bipolar Hall plates regarding offset correction," *Eurosensors XII*, Vol. 2, IOP Publishing, Bristol, pp. 999–1002.

Boside, G. and Harmer, A. (1996). *Chemical and Biochemical Sensing with Optical Fibers and Waveguides*, Artech House, Boston.

Briand, D. *et al.*, (2000). New response patterns for MOSFET gas sensors using low power devices, *Proceedings of Eurosensors XIV*, Copenhagen, Denmark, August, 27–30, pp. 737–740.

Cass, A. E. G., ed. (1990). *Biosensors: A Practical Approach*, Oxford University Press, Oxford.

Corcoran, P., Shurmer, H. V. and Gardner, J. W. (1993). "Integrated tin oxide sensors of low power consumption for use in gas and odour sensing," *Sensors and Actuators B*, **15–16**, 32–37.

Covington, J. A. *et al.* (2000). "Array of MOSFET devices with electrodeposited conducting polymer gates for vapour and odour sensing," *Proc. of the 7th Int. Symp on Olfaction and Electronic Noses*, Brighton, July, pp. 20–24.

Demarne, V. and Grisel, A. (1988). "An integrated low power thin film CO gas sensor on silicon," *Sensors and Actuators*, **13**, 301–313.

Fatikow, S. and Rembold, U. (1997). *Microsystem Technology and Microrobots*, Springer, Berlin, p. 408.

Gall, M. (1991). "The Si planar pellistor: a low power pellistor sensor in Si think film technology," *Sensors and Actuators B*, **4**, 533–538.

Gardner, J. W. (1989). "A diffusion-reaction model of electrical conduction in tin oxide gas sensors," *Semicond. Sci Technol.*, **4**, 345–350.

Gardner, J. W. (1994). *Microsensors: Principles and Application*, John Wiley & Sons, Chichester.

Gardner, J. W. (1995). "Intelligent gas sensing using an integrated sensor pair," *Sensors and Actuators B*, **27**, 261–266.

Gardner, J. W. *et al.* (1995). "Integrated chemical sensor array for detecting organic solvents," *Sensors and Actuators B*, **26**, 135–139.

Gardner, J. W. and Bartlett, P. N. (1999). *Electronic Noses: Principles and Applications*, Oxford University Press, Oxford.

Geballe, T. H. and Hull, G. W. (1955). "Seebeck effect in silicon," *Phys. Rev*, **98**, 940–947.

Geiger, W. *et al.* (1998). "New design of micromachined vibrating rate gyroscope with decoupled oscillation modes," *Sensors and Actuators A*, **66**, 118–124.

Gopel, W., Hesse, J. and Zemel, J. N. (series eds.). *Sensors: A Comprehensive Review*, in eight volumes 1989–98, Wiley-VCH, Weinheim.

Greenwood, J. C. (1988). "Silicon in mechanical sensors," *J. Phys E.: Sci. Instrum*, **21**, 1114–1128.

Grieff, P., Boxenhorn, B., King, T. and Niles, L. (1991). Silicon monolithic micromechanical gyroscope, *Proceedings of Transducers '91*, July, San Fransisco, pp. 966–969.

Hanna, S. M. (1987). "Magnetic field sensors based on SAW propagation in magnetic films," *IEEE Trans. UFFC*, **34**, 191–194.

Hatfield, J. V., Covington, J. A. and Gardner, J. W. (2000). "GasFETs incorporating conducting polymers as gate materials," *Sensors and Actuators B*, **65**, 253–256.

Hauptmann, P. (1991). *Sensors: Principles and Applications*, Prentice Hall, New York.

Heilig, A. *et al.* (1997). "Gas identification by modulating temperatures of tin dioxide based thick-film gas sensors," *Sensors and Actuators B*, **43**, 45–51.

Ihokura, K. and Watson, J. (1994). *The Stannic Oxide Gas Sensor*, CRC Press, Florida, p. 245.

Janata, J. (1992). "Microsensors based on modulation of work function," in J. W. Gardner and P. N. Bartlett, eds., *Sensors and Sensory Systems for an Electronic Nose*, NATO ASI Series, 212, Kluwer Academic Publishers, Dordrecht, pp. 103–116.

Johnson, R. G. and Higashi, R. E. (1987). "A highly sensitive chip microtransducer for air flow and differential pressure sensing application," *Sensors and Actuators*, **11**, 63–72.

Koch, H. (1989). In W. Göpel, J. Hesse and J. N. Zemel, eds. *SQUID Sensors in Sensors: A Comprehensive Survey, Vol. 5: Magnetic Sensors*, pp. 381–445.

Kung, J. T. and Lee, H. S. (1992). "An integrated air cap capacitor pressure sensor and digital read-out with sub 100 attofarad resolution," *J. Microelectromech Syst.*, **1**, 121–129.

Lee, S. M. *et al.* (2000). Silicon planar pellistors based on nanoporous films and microhotplates, *Proc. of Euspen Technology Workshop*, University of Warwick, UK, September, pp. 18–21.

Llobet, E. (1998). "Selectivity enhancement of metal oxide semiconductor chemical sensors through the study of their transient response to a step-change in gas concentration," Ph.D. Thesis, University of Barcelona, Spain.

Lundström, I. (1981). "Hydrogen sensitive MOS structures," *Sensors and Actuators*, **1**, 403–426.

Lundström, I. *et al.* (1992). "Electronic nose based upon field effect structures," in J. W. Gardner and P. N. Bartlett, eds., *Sensors and Sensory Systems for an Electronic Nose*, NATO ASI Series, 212, Kluwer Academic Publishers, Dordrecht, pp. 303–319.

Lutes, O. S., Nussbaum, P. S. and Aadland, O. S. (1980). "Sensitivity limits in SOS magnetodiodes," *IEEE Trans. Electron Devices*, **27**, 2156–2157.

Madou, M. J. (1997). *Fundamentals of microfabrication*, CRC Press, Boca Raton, p. 589.

Madou, M. J. and Morrison, S. R. (1989). *Chemical Sensing with Solid-State Devices*, Academic Press, New York, 1–556.

Makinwa, K. A. and Huijsing, J. H. (2000). A wind-sensor interface based on thermal sigma-delta modulation, *Proc. of Eurosensors XIV*, Copenhagen, Denmark, August, pp. 27–30.

McNie, *et al.* (1998). "Design, fabrication and testing of a silicon ring gyroscope," *J. Micromech Microeng.*, **8**, 284–292.

Meijer, G. C. M. and van Herwaarden, A. W. (1994). *Thermal Sensors*, IOP Publishing, Bristol.

Middelhoek, S. and Audet, S. A. (1989). *Silicon Sensors*, Academic Press, London.

MIRC (1990). *New and merging markets for automotive sensors in North America*, MIRC, Mountain View, Calif., USA.

Moseley, P. T. and Tofield, B. C. eds. (1987). *Solid-State Gas Sensors*, Adam-Hilger, Bristol.

Noltingk, B. E. ed. (1995). *Instrumentation Reference Book*, 2nd ed., Butterworth-Heinemann, Boston.

Norton, H. N. (1989). *Handbook of Transducers*, Prentice Hall, New York.

Oudheisen, B. W and Huijsing, J. H. (1990). "An electronic wind meter based on a silicon flow sensor," *Sensors and Actuators A*, pp. **21–24**, 420–424.

Pike, A. C. (1996). "Design of chemoresistive silicon sensors for application in gas monitoring," Ph.D. Thesis, University of Warwick (Advisor J. W. Gardner), Coventry, UK.

Pike, A. C. and Gardner, J. W. (1997). "Thermal modelling of micropower chemoresistive silicon sensors," *Sensors and Actuators B*, **45**, 19–26.

Popovic, R. S., Baltes, H. P. and Rudolf F. (1984). "An integrated silicon magnetic field sensor using the magnetodiode principle," *IEEE Trans. Electron Devices*, **31**, 286–291.

Savage, S. *et al.* (2000). SiC based gas sensors and their applications, *Proc. of ECSCRM 2000*, Erlangen, September, pp. 3–7.

Scheller, F. W. *et al.* (1991). "Biosensors: fundamentals, applications and trends," *Sensors and Actuators B*, **4**, 197–206.

Stemme, G. N. (1986). "A monolithic gas flow sensor with polyimide as thermal insulator," *IEEE Trans. Electron Devices*, **33**, 170–1474.

Sullivan, F. A. (1993). World emerging sensor technologies high growth markets uncovered, Sullivan & Frost Report No. 915-40, Mountain View, Calif., USA.

Svenningstorp, H. *et al.* (2000). MISiC schottky diodes and transistors as ammonia sensors in diesel exhausts to control SCR, *Proc. of Eurosensors XIV*, Copenhagen, Denmark, August 27–30, pp. 933–936.

Svenningstorp, H. *et al.* (2001). Detection of HC in exhaust gases by an array of MISiC sensors, *Sensors and Actuators B*, **77**, 177–185.

Tabid-Azar, M. ed. (1995). *Integrated Optics, Microstructures and Sensors*, Kluwer Academic Publishers, Dordrecht.

Taylor, R., Little, A. D. and Schultz, J. S. (1996). *Handbook of Chemical and Biological Sensors*, IOP Publishing, Bristol, p. 604.

Udd, E. ed. (1991). *Fiber-Optic Sensors*, John Wiley & Sons, New York.

Welham, C. J. (1996). "A silicon micromachined lateral resonant strain gauge pressure sensor," Ph.D. Thesis (Advisor J. W. Gardner), University of Warwick, Coventry, UK.

Welham, C. J., Gardner, J. W. and Greenwood, J. W. (1996). "A laterally driven micromachined resonant pressure sensor," *Sensors and Actuators A*, **52**, 86–91.

Welham, C. J., Greenwood, J. and Bertioli, M. (2000). A high accuracy resonator pressure sensor by fusion bonding and trench etching, *Proc. of Eurosensors XII*, August Copenhagen, Denmark.

Williams, D. E. (1987). In P. T. Moseley and B. C. Tofield, eds., *Solid-State Gas Sensors*, Adam-Hilger, Bristol, pp. 71–123.

Wolf, H. F. (1969). *Silicon Semiconductor Data*, Pergammon Press, Oxford, UK.

Wolffenbuttel, R. F., ed. (1996). *Silicon Sensors and Circuits*, Chapman and Hall, London.

Yamazoe, N., Kurokawa, Y. and Seiyama, T. (1983). "Effects of additives on semiconductor gas sensors," *Sensors and Actuators*, **4**, 283–289.

9

Introduction to SAW Devices

9.1 INTRODUCTION

Physical and chemical sensors have and will continue to play an ever-increasing role in the measurement of both physical (pressure, temperature, acceleration, strain etc.) and chemical (ion or gas concentrations, chemical potential, etc.) properties. Sensors are important for many processes, and it is impossible to imagine a world without their impact on some part of our daily lives.

Sensors based on surface acoustic waves (SAWs) form an important part of the sensor family, and in recent years, these have seen diverse applications ranging from gas and vapour detection to strain measurement (Campbell 1998). A new generation of SAW-based actuators modeled on microactuators based on microelectromechanical systems (MEMS) have also been recently announced (Campbell 1998). The advantages of using SAW-based devices in microsensors are that SAW devices are amenable to wireless interrogation, and by the application of a suitable modification in the form of an onboard antenna, they can be converted into sensors for use in remote and inaccessible locations. As the sensor frequency increases, its physical size scales down, resulting in smaller devices permitting a greater variety of applications. Consequently, we have dedicated several chapters in this book to the topic of SAW devices and their sensing applications.

9.2 SAW DEVICE DEVELOPMENT AND HISTORY

Pierre and Jacques Curie discovered the piezoelectric nature of quartz in 1880, when they observed that some crystals[1] they were studying would electrically polarise when deformed by an applied force (Smith 1976). For 30 years, this phenomenon remained a scientific curiosity. In 1910, Voigt published a book on the properties and theory of piezo-electricity. The first practical applications using piezoelectric devices began during the First World War of 1914 to 1918 (Grate *et al.* 1993a). In 1917, motivated by submarine problems, P.l Langevin of France and A. M. Nicholson of United States worked independently on using piezoelectric devices as echo-detectors for detecting compressional waves in seawater. Although too late for the war effort, these devices proved useful in shipping as detectors in depth-sounding equipment (Smith 1976).

Acoustic sensors offer a rugged and relatively inexpensive means for the development of wide-ranging sensing applications. A valuable feature of acoustic sensors is their

---

[1] Crystalline materials and lattice structures are described in Chapter 3.

direct response to physical and chemical parameters, including surface mass, stress, strain, liquid density, viscosity, permittivity, and conductivity (Grate *et al.* 1993a). Furthermore, the anisotropic nature of piezoelectric crystals allows for various angles of cut, each cut having different properties. Applications such as a SAW-based microaccelerometer, for example, utilise a quartz crystal having a stable temperature (ST) cut because the resonance frequency is almost independent of temperature (Bechmann *et al.* 1962). Again, depending on the orientation of the crystal cut, various SAW sensors having different acoustic modes may be constructed, which have a mode ideally suited toward a particular application. Other advantageous attributes include very low internal loss, uniform material density, and elastic constants (Bechmann *et al.* 1962). Owing to these properties, many different sensors can be designed and optimised to meet the needs of specific sensing applications, leading to their increasing role as chemical and physical sensors (Grate *et al.* 1993a,b). Since the early 1960s, research and development in the acoustic sensor field has increased significantly and has shown increased diversity.

The principal means of detection of a change of physical property follows from the transduction mechanism of a SAW device, which involves the conversion of signals from the physical (acoustic wave) domain to the electrical domain. Small perturbations affecting the acoustic wave manifest themselves as large-scale changes when converted to the electromagnetic (EM) domain because of the enormous difference in their velocities (Varadan and Varadan 1997). This can be understood from the following calculations:

The SAW wavelength λ is given by the ratio v/f_0. The velocity v of a SAW wave on a piezoelectric substrate depends on the material and is typically[2] 3490 ms^{-1}, whereas the synchronous frequency f_0 is set by the AC voltage applied to the interdigital transducer (IDT) and is typically 1 GHz. Thus, the SAW wavelength λ is $3490/10^9$ or about 3.5 μm. The EM wavelength λ_c is given by the ratio c/f_0, where c is the velocity of light, that is, 3×10^8 m/s; in this case, λ_c is 0.3 m. The ratio of the two wavelengths is λ/λ_c, which takes a value of 1.1×10^{-5} here. The sensing action of such transducers involves any influences that will alter the acoustic wave velocity v and, consequently, the associated properties of the wave, such as frequency and time to travel between the sensor and the detector. The slower acoustic velocity enables the use of simple, low-cost IC circuitry to transduce the sensing signals that have a high level of precision as demonstrated in later chapters.

The attributes of an ideal sensor should include the following (d'Amico and Verona 1989):

1. High sensitivity

2. Fast and linear response

3. Fully reversible behavior

4. High reliability

5. Selectivity

6. Compact

7. High signal-to-noise ratio

8. Insensitive to surrounding environmental conditions

---

[2] Value for lithium niobate.

9. Easy to calibrate

10. Low cost

11. No moving parts that can suffer from wear mechanisms

However, in practice, sensors are less than ideal, and suffer from problems associated with cross-sensitivity caused by poor selectivity.

To understand the fundamental sensing mechanism of a SAW-based acoustic sensor better, a brief description follows (Morgan 1978): The sensing mechanism of a SAW sensor relies on a change in the acoustic velocity of a bulk (or surface) wave within (or on) the surface of the piezoelectric substrate. After transduction from acoustic to the electrical domain, the frequency of the acoustic wave is measured. Another parameter that can also be measured is the amplitude of the acoustic wave, or to be more precise, the amplitude of the electrical oscillator frequency. This measurement is usually associated with a decrease in acoustic wave amplitude, and this reduction is a measure of crystal damping, thus giving an indication of viscous-damping effects.

In the development of acoustic wave sensors and their corresponding oscillators, complementary instrumentation or systems must be developed in order to measure both the dynamic and steady state sensor response. Specialised and expensive instrumentation, such as network analysers or vector voltmeters, can give a very detailed analysis of acoustic transducers or their sensor configurations. However, this type of instrumentation is far too sophisticated for general-purpose routine laboratory use. Moreover, it not only requires highly qualified personnel for its operation but also is very limited in that only one acoustic sensor can be monitored at a time. In addition, analysis of the data is extremely time consuming, thus making it impractical to use such instrumentation for measuring the signals from a microsensor array. Consequently, the vast majority of research papers have based measurements of the frequency of acoustic wave sensor oscillators on the use of frequency counters and chart recorders.

Other important design goals that concern advanced measurement systems are sensor response and data analysis. In the present system design, this involves the transfer of data directly to disk file, where postprocessing and analysis is carried out using commercial software packages, such as MathCAD or Matlab. In the case of data from an oscilloscope, the data are downloaded, and standard conversion software is invoked to convert the resulting data from a waveform format to a form that can be used as suitable input for the software packages mentioned earlier. As experimental data can be saved on disk, this will contribute toward the ease of documentation. Future directions for the system are aimed at processing and analysis being completed in real time. These can be achieved by the inclusion of the necessary software classes into the existing system software program or through dynamic data linked with other commercial software packages. In addition, system calibration has to meet the requirements of the specific application. This can be achieved by making use of the mathematical models of the sensor response. A calibrated system will aid the user, as both the graphical and numerical display will present data in the appropriate and more meaningful physical units rather than frequency (Campbell 1998).

Before any attempt is made to initiate a proposed system design for acoustic microsensors, it is mandatory to thoroughly understand the sensing mechanisms and limitations of acoustic devices for each particular sensing application. Using equivalent electronic circuit models, a greater understanding of these transducers can be achieved. External factors that cause changes to the circuit model also have to be considered. Some equivalent circuit models can be run with commercial SPICE software, which helps in the design of better acoustic microtransducers.

9.3 THE PIEZOELECTRIC EFFECT

Piezoelectric crystals play a dominant role in the communications and electronics industry where they are commonly used as filters, precision timers, and for frequency control in oscillator circuits (Mason 1942). The piezoelectric effect can be demonstrated by applying either a compressive or tensile stress to the opposite faces of a piezoelectric crystal. Figure 9.1 shows that when the equal and opposite forces, $F1$ and $F2$, (generating tensile stresses) are applied, the resulting deformation of the crystal lattice produces a separation of the centres of gravity of positive and negative charges[3].

This effect results in electrical charge appearing on the surface of the electrodes. When the force (tensile or compressive) is removed, the strain within the crystal lattice is released causing charge (and hence current) to flow, thus reestablishing a zero potential difference between the electrodes. If a sinusoidal stress alternating between the tensile and compressive forces is applied to opposite crystal faces, a sinusoidal piezoelectric voltage will appear across the electrodes. In this case, electrical energy is produced from mechanical energy (generator action). This process of crystal deformation can be reversed. In other words, when an external voltage is applied to the electrodes, the crystal lattice will deform by an amount proportional to the applied voltage. In this case, electrical energy is transformed into mechanical energy (motor action) (Campbell 1998). Therefore, it is effectively a piezoelectric microactuator.

There are essentially three important properties of piezoelectric transducers that justify their use for sensing applications (Morgan 1978):

1. An ideal coupling mechanism between the electric circuit and the mechanical properties of the crystal, ensuring that the frequency of the mechanical acoustic wave is identically equal to the electrical frequency, that is, a distortion-free interface having extremely low dissipation.

2. The anisotropic nature of piezoelectric crystals allows for different angles of cut with respect to the crystallographic axis, which, therefore, allows the use of crystals with a range of frequencies.

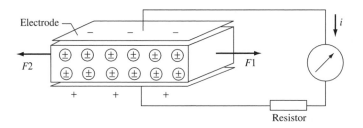

Figure 9.1   Transformation of mechanical energy into electrical energy

---

[3] Note that this effect does not occur in crystalline silicon because its lattice structure is centro-symmetric.

3. The model of the acoustic wave can be changed, for example, from compressional to shear, or the angle of propagation can be varied.

9.3.1 Interdigital Transducers in SAW Devices

Acoustic waves propagating along the surface of a piezoelectric material provide a means of implementing a variety of signal-processing devices at frequencies ranging from several MHz to a few GHz (Campbell 1998). The IDT provides the cornerstone of SAW technology. Its function is to convert electrical energy into mechanical energy, and vice versa, for generating and detecting the SAW.

An IDT consists of two metal comb-shaped electrodes placed on a piezoelectric substrate (Figure 9.2). An electric field created by a voltage V applied to the electrodes induces dynamic strains in the piezoelectric substrate, which, in turn, launch elastic waves. These waves contain, among others, the Rayleigh waves that run perpendicular to the electrodes with velocity v_R.

When an AC voltage $V(t)$ $[V_o \exp(j\omega t)]$ is applied across the electrodes, the stress wave induced by the finger pair travels along the surface of the crystal in both directions. To ensure constructive interference and in-phase stress, the distance d between two neighbouring fingers should be equal to half the elastic wavelength λ_R whence

$$d = \lambda_R/2 \tag{9.1}$$

The associated frequency is known as *synchronous frequency*, f_0, and is given by (d'Amico and Verona 1989)

$$f_0 = v_R/\lambda_R \tag{9.2}$$

At this frequency, the transducer efficiency in converting electrical energy to acoustical, or vice versa, is maximised. The exact calculation of the piezoelectric field driven by the IDT is rather elaborate (Smith 1976). For simplicity, the analysis of an IDT is carried out by means of various numerical models. The frequency response of a single IDT can be simplified by the delta-function model (Campbell 1998).

The simplest SAW device is the nondispersive delay line depicted in Figure 9.3. One IDT is connected to an electrical source and the other to a detector. The source IDT sets

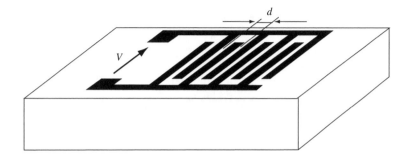

Figure 9.2   Finger-spacings and their role in the determination of the acoustic wavelength

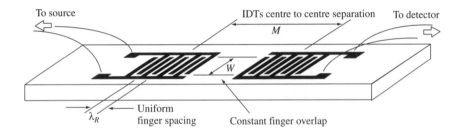

Figure 9.3   Schematic of a SAW device with IDTs metallised onto the surface

up an electric field in the substrate that launches a SAW by means of the piezoelectric effect, and the receiving transducer converts the surface wave to an electrical signal.

9.4 ACOUSTIC WAVES

The propagation of acoustic waves in solids will be described in the next chapter. It is seen that the type of acoustic wave generated in a piezoelectric material depends mainly on the substrate material properties, the crystal cut, and the structure of the electrodes used to transform the electrical energy into mechanical energy. The possibilities of various types of acoustic devices for sensor applications are explored in this section, which focuses primarily on Rayleigh and shear horizontal surface acoustic wave (SH-SAW), Love wave mode, acoustic plate mode (APM), and flexural plate wave (FPW).

9.4.1 Rayleigh Surface Acoustic Waves

SAWs were first described by Lord Rayleigh, and the Rayleigh Wave has by far become the most widely known and used (Viktorov 1967). The propagation of a Rayleigh SAW is described in detail in Section 10.5. Ripples in calm water caused by a physical disturbance can be used as an excellent analogy to describe the Rayleigh wave. The disturbance in quartz originates from the IDTs and was first described by White and Voltmer (1965). The radio frequency (RF) signal generates the SAW propagation through selective deformation of the quartz surface, as illustrated in Figure 9.4.

The elastic Rayleigh wave has both a surface-normal component and a surface-parallel component with respect to the direction of propagation. The Rayleigh wave has two particle displacement components in the sagittal plane. Surface particles move in elliptical paths having a surface-normal and a surface-parallel component. The EM field associated with the acoustic wave travels in the same direction. The wave velocity is determined by the substrate material and the crystal cut. The energy of the SAW is confined to a zone close to the surface and is of a few wavelengths thick (Morgan 1978)[4]. A SAW delay line consists of two IDTs on the surface of a piezoelectric substrate – one to launch the SAW and the other to detect it. If the SAW delay is connected to an RF amplifier in a closed loop, the system oscillates at a frequency determined by the wave velocity v_R and

---

[4] That is why they are called surface acoustic waves rather than bulk acoustic waves.

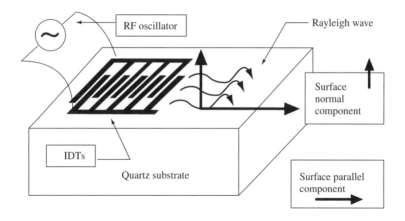

Figure 9.4 Surface acoustic waves generated in quartz by IDTs

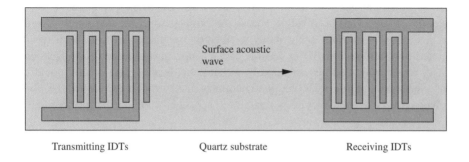

Figure 9.5 Example of a SAW device having a feedback amplifier

the geometry of the electrodes of the IDTs. The output of the amplifier can be sampled using a frequency counter or a voltage meter (Figure 9.5).

A SAW can also be generated by a network analyser sweeping across a range of signals in the region of the oscillation frequency of the SAW. Changes in the physical nature of the wave due to perturbations in its velocity can be manifested as a phase shift detectable in the output of the network analyser. The disadvantage of the output of the network analyser is that the minimum resolution possible in the sensing of a dynamic signal is limited by the minimum time taken by the network analyser to sweep across a range of frequencies defined for oscillation. The present common practice is to use real-time dynamic data-acquisition systems, for example, Labview (National Instruments, USA), in conjunction with a network analyser response to obtain the dynamic data that may be useful in measurements involving a change in the dynamic phase of the signal. This has been demonstrated with the Love wave ice sensor[5] in which the oscillator connected to the device measured the peak frequency of oscillation after and before the perturbation, whereas the device parameters such as the insertion loss and amplitude-attenuation were obtained from the network analyser results.

---

[5] This is a sensor that detects the formation of ice on the substrate surface.

Using a second set of matched IDTs, the electromechanical wave can be received and converted back into an electrical signal. Therefore, using a transmitting and receiving set of electrodes connected in a feedback loop of a suitable feedback amplifier, sustainable oscillation can be achieved (Avramov 1989). The IDT finger spacing d relates to the frequency of the device (see Section 9.2).

The electrode structures are usually made from a combination of an ultrathin 10–20 nm layer of chromium (Cr), followed by a thin layer of approximately 150 nm of gold (Au) or a thin layer of aluminum (Al) of a standard thickness of 150 nm (Atashbar 1999). In the case of a Cr–Au electrode, the chromium not only forms an ohmic contact with the quartz but also acts as an adhesion layer for the gold. Gold is used as the final interface to the environment because of its relatively inert characteristics compared with chromium and other metals. In the case of the aluminum electrode, the adhesion between aluminum and quartz is sufficiently high to prevent the need for a chromium layer to link the substrate to the metallisation layer.

SAW devices typically operate in the range of 30 MHz to 1 GHz on ST-quartz and various cuts of lithium niobate. There are two ways to perturb the device. One is to perturb the electrical field above the propagation path length and the other is to perturb the actual mechanical wave traveling across the propagation path. The latter is followed in this case. The perturbing agent tends to alter the characteristics of the wave that is propagated between the two IDTs, leading to a change in the phase or the frequency of the wave, which may then be measured by suitable application of a measurement device such as a network analyser or a frequency counter or meter (Figure 9.6).

A change of the phase velocity is related to a change in the oscillation frequency by

$$\frac{\Delta v}{v_R} = \frac{\Delta f}{f_0} \tag{9.3}$$

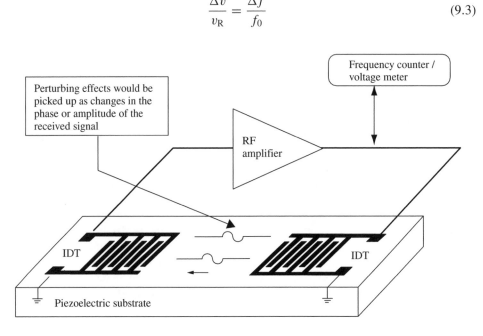

Figure 9.6 Elements of a SAW delay line–based sensor

where Δv is the change in the velocity of the wave and Δf is the corresponding frequency change. v_R is the propagation velocity of the Rayleigh wave and f_0 is the design frequency.

The particle displacement components and the quasi-static electrical potential of the SAW are described in Section 10.4 for both Rayleigh waves and Love waves.

9.4.2 Shear Horizontal Acoustic Waves

The selection of a different crystal cut can yield SH surface waves instead of vertical Rayleigh waves (Nakamura *et al.* 1977). The particle displacements of this type of wave are transverse to the wave propagation direction and parallel to the plane of the surface. The frequency of operation is determined by the IDT finger-spacing and the SAW velocity for the particular substrate material (Table 9.1). These crystals have shown considerable promise in their application as microsensors in liquid media (Kovacs and Venema 1992; Shiokawa and Moriizumi 1987). In general, the SH-SAW is sensitive to mass-loading, viscosity, conductivity, and permittivity of the adjacent liquid. The configuration of SH-APM devices is similar to the Rayleigh SAW devices, but the wafer is thinner, typically a few acoustic wavelengths. The IDTs generate SH waves that propagate in the bulk at angles to the surface. These waves reflect between the plate surfaces as they travel in the plate between the IDTs. The frequency of operation is determined by the thickness of the plate and the IDT finger-spacing. SH-APM devices are mainly used in liquid-sensing and offer the advantage of using the back surface of the plate as the sensing active area. In this manner, the IDTs can be isolated from the liquid media, and so the potential problem of a chemical attack can be avoided. Applications of these microdevices are directed mainly toward biosensor implementation (Kondoh *et al.* 1993). The particle motion is transverse, relative to the direction of wave propagation. Displacements at the surface are almost entirely in-plane (Atashbar 1999). Figure 9.7 illustrates the wave propagation through the material and the reflections at the air and liquid boundaries. It is important that these

Table 9.1 Wave velocity, propagation direction, electromechanical coupling (K^2) coefficient, and major applications for important substrates used for SAW sensors

Substrate	Propagation	Propagation velocity (m/s)	K^2 (%)	Applications
Y-cut ST quartz	X	4990	1.89	Precision oscillators. Temperature-stable narrow-band filters. Low loss RF resonator
Y-cut lithium niobate	Z	3158	4.5	Wideband midloss IF filters. Applications requiring a high electromechanical coupling
128° cut lithium niobate	X	3992	5.3	Long delay lines and wideband midloss IF filters. Disadvantage of signal drift due to temperature effects.

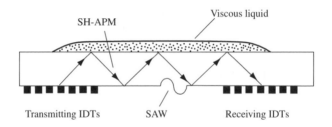

Figure 9.7   A shear horizontal acoustic plate mode device

interfaces be kept undamaged and to a sufficient grade of polish because irregularities will introduce additional noise signals owing to multiple and nonsymmetrical reflections (Shiokawa and Moriizumi 1987).

The sensitive areas of such devices are on both faces. This allows the use of the nonelectrode face for sensing and isolates the transducers from the sensing media. As the SH-APM wave does not have any surface-normal wave components on the sensitive area of either face of the quartz, it will oscillate in the feedback loop with minimal interferences in both the liquid and gas media (Atashbar 1999; Shiokawa and Moriizumi 1987). This is important when considering the sensitivity (Q-factor) of resonant oscillators. By the very nature of the interface-sensitive mode, interface parameter changes can be sensed. If conditions at these interfaces are changed, a change in the frequency will result. The increase of mass at the interface still causes the expected shift in the device's resonant frequency, and variations in the interface properties such as viscosity and density are also detectable.

9.4.3 Love Surface Acoustic Waves

Love waves may be considered as SAWs that propagate in a waveguide made of a layer of a given material M2 (Figure 9.8) deposited on a substrate made of another material M1 with different acoustic properties and infinite thickness when compared with the original layer (Love 1934). These waves are transverse and they bring only shear stresses into action. The displacement vector of the volume element is perpendicular to the propagation X-direction and is oriented in the direction of the Z-axis. Because the Love wave is a surface wave, the propagating energy is located in the M2 layer and in the part of the substrate that is close to the interface. Its amplitude decreases exponentially with depth (Ewing *et al.* 1957). In comparison, SH waves are limited by high noise levels and diffraction of the acoustic signal into the crystal bulk and background interference from the reflection off the lower surface. The layer of M2 (usually SiO_2) has the effect of confining and guiding the wave over the top layer of the device and hence circumvents the disadvantages (usually reduced sensitivity at the interface) associated with a similar SH device (Gizeli *et al.* 1992).

Figure 9.9 shows the geometry that supports a Love wave and the polarisation usually associated with a Love wave device. A necessary condition for the existence of a Love wave is that the shear acoustic velocity in the layer (v_2) must be smaller than the same in the substrate (v_1). This condition is proven in Chapter 10 (Tournois and Lardat 1969). The larger the difference, the larger the guiding effect (Du *et al.* 1996).

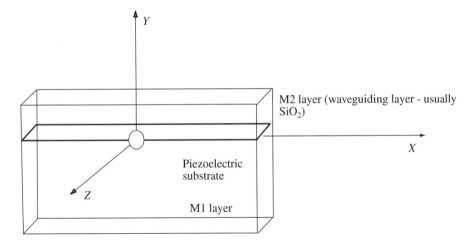

Figure 9.8   Schematic of a Love wave propagation region and relevant layers

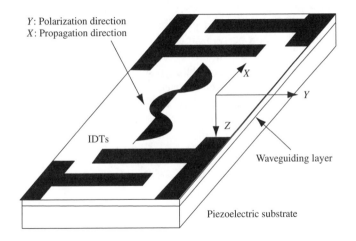

Figure 9.9   Wave generation on Love wave mode devices

The basic principle behind the generation of the waves is quite similar to that presented in the description of an SH-SAW sensor. The only difference would be the fact that the Love wave mode would be the same SH-SAW mode propagating in a layer that was deposited on top of the IDTs. This layer helps to propagate and guide the horizontally polarised waves that were originally excited by the IDTs deposited at the interface between the guiding layer and the piezoelectric material beneath (Du *et al*. 1996). The particle displacements of this wave would be transverse to the wave-propagation direction, that is, parallel to the plane of the surface-guiding layer. The frequency of operation is determined by the IDT finger-spacing and the shear wave velocity in the guiding layer. These SAW devices have shown considerable promise in their application as microsensors in liquid media (Haueis *et al*. 1994; Hoummady *et al*. 1991).

In general, the Love wave is sensitive to the conductivity and permittivity of the adjacent liquid or solid medium (Kondoh and Shiokawa 1995). The IDTs generate waves that are coupled into the guiding layer and then propagate in the waveguide at angles to the surface. These waves reflect between the waveguide (which is usually deposited from a material whose density would be lower than that of the material underneath) surfaces as they travel in the guide above the IDTs. The frequency of operation is determined by the thickness of the guide and the IDT finger-spacing (Tournois and Lardat 1969). Love wave devices are mainly used in liquid-sensing and offer the advantage of using the same surface of the device as the sensing active area. In this manner, the loading is directly on top of the IDTs, but the IDTs can be isolated from the sensing medium that could, as stated previously, negatively affect the performance of the device (Du *et al.* 1996). It is again important that interfaces (guiding layer, substrate) be kept undamaged and care taken to see that the deposition process used gives a fairly uniform film at a constant density over the thickness (Kovacs *et al.* 1993).

Love wave sensors have been put to diverse applications, ranging from chemical microsensors for the measurement of the concentration of a selected chemical compound in a gaseous or liquid environment (Kovacs *et al.* 1993; Haueis *et al.* 1994; Gizeli *et al.* 1995) to the measurement of protein composition of biologic fluids (Kovacs *et al.* 1993; Kovacs and Venema 1992; Grate *et al.* 1993a,b). Polymer (e.g. PMMA) layer–based Love wave sensors (Du *et al.* 1996) are used to assess experimentally the surface mass-sensitivity of the adsorption of certain proteins from chemical compounds. It has also been shown recently that a properly designed Love wave sensor is very promising for (bio)chemical sensing in gases and liquids because of its high sensitivity (relative change of oscillation frequency due to a mass-loading); some of the sensors with the aforementioned characteristics have already been realised (Kovacs *et al.* 1993). As is discussed in the next chapter, the main advantage of shear Love modes applied to chemical-sensing in liquids derives from the horizontal polarisation, so that they have no elastic interactions with an ideal liquid. It is also sometimes noticed that viscous liquid loading causes a small frequency-shift that increases the insertion loss of the device (Du *et al.* 1996).

9.5 CONCLUDING REMARKS

This chapter should provide the reader with the necessary background to the basic principles governing waves and SAW devices[7].

Figure 9.10 summarises the different types of waves that can propagate through a medium. These are waves that travel through the bulk of the material (Figure 9.10 (a) and (b)). The compressive (P) wave is sometimes called a *longitudinal wave* and is well known for the way in which sound travels through air. On the other hand, the S wave is a transverse bulk wave and looks like a wave traveling down a piece of string. In contrast, waves can travel along the surface of a media, (Figure 9.10 (c) and (d)). These waves are named after the people who discovered them. The Rayleigh wave is a transverse wave that travels along the surface and the classic example is the ripples created on the surface of water by a boat moving along. The Love wave is again a surface wave, but this time the waves are SH or vertical. This mode of oscillation is not supported in gases and liquids, and so produces a poor coupling constant. However, this phenomenon can be used to a great advantage in sensor applications in which poor coupling to air results in low loss (high Q-factor) and hence a resonant device with a low power consumption.

---

[7] Some of the material presented here may also be found in Gangadharan (1999).

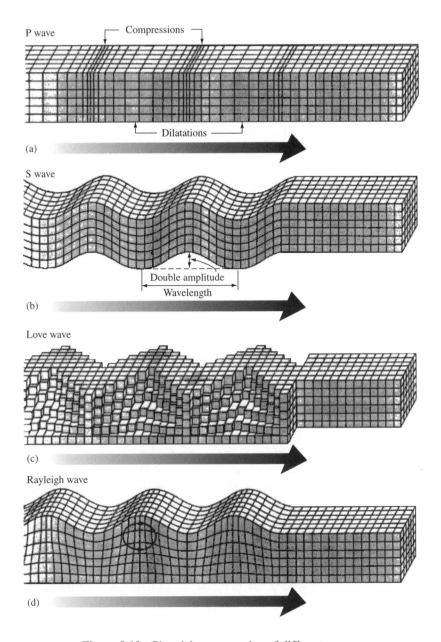

Figure 9.10 Pictorial representation of different waves

From these fundamental properties of waves, it should be noted that for applications considered here, such as ice-detection, there are a variety of possible options. Because ice-detection primarily involves sensing the presence of a liquid (e.g. water), it is obvious that Rayleigh wave modes and flexural plate (S) modes cannot be used because of their attenuative characteristics. Therefore, it is imperative that only those wave modes are used whose longitudinal component is small or negligible compared with its surface-parallel

Table 9.2 Structures of Love, Rayleigh SAW, SH-SAW, SH-APM and FPW devices and comparison of their operation

Device type	Love SAW	Rayleigh SAW	SH-SAW	SH-APM	FPW
Substrate	ST-quartz	ST-quartz	LiTaO$_3$	ST-quartz	Si$_x$N$_y$/ZnO
Typical frequency (MHz)	95–130	80–1000	90–150	160	1–6
U^a	Transverse	Transverse parallel	Transverse	Transverse	Transverse parallel
U_t^b	Parallel	Normal	Parallel	Parallel	Normal
Media	Ice to liquid chemosensors	Strain	Gas liquid	Gas liquid chemosensors	Gas liquid

$^a U$ is the particle displacement relative to wave propagation
$^b U_t$ is the transverse component relative to sensing surface

components. For this reason, either a Love SAW or an SH wave–based APM device could be used. However, because the ratio of the volume of the guiding layer to the total energy density is the largest for a Love wave device, it is natural to choose a Love wave device for the higher sensitivity toward any perturbation at the liquid interface.

Finally, Table 9.2 summarises the different types of SAW devices described in this chapter. This reference table also gives the typical operating frequencies of the devices, along with the wave mode and application area.

REFERENCES

Atashbar, M. Z. (1999). Development and fabrication of surface acoustic wave (SAW) oxygen sensors based on nanosized TiO2 thin film, PhD thesis, RMIT, Australia.

Avramov, I. D. (1989). Analysis and design aspects of SAW-delay-line-stabilized oscillators, *Proceedings of the Second International Conference on Frequency Synthesis and Control*, London, April 10–13, pp. 36–40.

Bechmann, R., Ballato, A. D. and Lukaszek, T. J. (1962). "Higher order temperature coefficients of the elastic stiffnesses and compliances of alpha-quartz," *Proc. IRE*, p. 1812.

Cambell, C. (1989). *Surface Acoustic Wave Devices and their Signal Processing Applications*, Academic Press, London, p. 470.

Campbell, C. (1998). *Surface Acoustic Wave Devices and their Signal Processing Applications*, Academic Press, London.

d'Amico, A. and Verona, E. (1989). "SAW sensors," *Sensors and Actuators*, **17**, 55–66.

Du, J. *et al.* (1996). "A study of love wave acoustic sensors," *Sensors and Actuators A*, **56**, 211–219.

Ewing, W. M., Jardetsky, W.S., and Press, F. (1957). *Elastic Waves in Layered Media*, McGraw-Hill, New York.

Gangadharan, S. (1999). Design, development and fabrication of a conformal love wave ice sensor, MS thesis (Advisor V.K. Varadan), Pennsylvania State University, USA.

Gizeli, E., Goddard, N. J., Lowe, C. R. and Stevenson, A. C. (1992). "A love plate biosensor utilising a polymer layer," *Sensors and Actuators A*, **6**, 131–137.

Gizeli, E., Liley, M. and Lowe, C. R. (1995). Detection of supported lipid layers by utilising the acoustic love waveguide device: applications to biosensing, *Technical Digest of Transducers'95*, vol. 2, IEEE, pp. 521–523.

Grate, J. W., Martin, S. J. and White, R. M. (1993a). "Acoustic wave microsensors, Part I," *Analytical Chem.*, **65**, 940–948.

Grate, J. W., Martin, S. J. and White, R. W. (1993b). "Acoustic wave microsensors. Part II," *Analytical Chem.*, **65**, 987–996.

Haueis, R. *et al.* (1994). A love wave based oscillator for sensing in fluids, *Proceedings of the 5th International Meeting of Chemical Sensors* (Rome, Italy), **1**, 126–129.

Hoummady, M., Hauden, D. and Bastien, F. (1991). "Shear horizontal wave sensors for analysis of physical parameters of liquids and their mixtures," *Proc. IEEE Ultrasonics Symp.*, **1**, 303–306.

Kondoh, J. and Shiokawa, S. (1995). Liquid identification using SH-SAW sensors, Technical Digest of Transducers'95, vol. 2, IEEE, pp. 716–719.

Kondoh, J., Matsui, Y. and Shiokawa, S. (1993). "New biosensor using shear horizontal surface acoustic wave device," *Jpn. J. Appl. Phys.*, **32**, 2376–2379.

Kovacs, G. and Venema, A. (1992). "Theoretical comparison of sensitivities of acoustic shear wave modes for (bio)chemical sensing in liquids," *Appl. Phys. Lett.*, **61**, 639–641.

Kovacs, G., Vellekoop, M. J., Lubking, G. W. and Venema, A. (1993). A love wave sensor for (bio)chemical sensing in liquids, *Proceedings of the 7th International Conference on Solid-State Sensors and Actuators*, Yokohama, Japan, pp. 510–513.

Love, A. E. H. (1934). *Theory of Elasticity*, Cambridge University Press, England.

Mason, W. P. (1942). *Electromechanical Transducers and Wave Filters*, Van Nostrand, New York.

Morgan, D. P. (1978). *Surface-Wave Devices for Signal Processing*, Elsevier, The Netherlands.

Nakamura, N., Kazumi, M. and Shimizu, H. (1977). "SH-type and Rayleigh-type surface waves on rotated Y-cut LiTaO3," *Proc. IEEE Ultrasonics Symp.*, **2**, 819–822.

Shiokawa, S. and Moriizumi, T. (1987). Design of SAW sensor in liquid, *Proceedings of the 8th Symposium on Ultrasonic Electronics*, July, Tokyo.

Smith, W. R. (1976). "Basics of the SAW interdigital transducer," *Wave Electronics*, **2**, 25–63.

Tournois, P. and Lardat, C. (1969). "Love wave dispersive delay lines for wide band pulse compression," *Trans. Sonics Ultrasonics*, **SU-16**, 107–117.

Varadan, V. K. and Varadan, V. V. (1997). "IDT, SAW and MEMS sensors for measuring deflection, acceleration and ice detection of aircraft," *SPIE*, **3046**, 209–219.

Viktorov, I. A. (1967). *Rayleigh and Lamb Waves*: *Physical Theory and Applications*, Plenum Press, New York.

White, R. W. and Voltmer, F. W. (1965). "Direct piezoelectric coupling to surface elastic waves," *Appl. Phys. Lett.*, **7**, 314–316.

10

Surface Acoustic Waves in Solids

10.1 INTRODUCTION

Acoustics is the study of sound or the time-varying deformations, or vibrations, in a gas, liquid, or solid.

Some nonconducting crystalline materials become electrically polarised when they are strained. A basic explanation is that the atoms in the crystal lattice are displaced when it is placed under an external load. This microscopic displacement produces electrical dipoles within the crystal and, in some materials, these dipole moments combine to give an average macroscopic moment or electrical polarisation. This phenomenon is approximately linear and is known as the *direct piezoelectric* (PE) effect (Auld 1973a). The direct PE effect is always accompanied by the inverse PE effect in which the same material will become strained when it is placed in an external electric field.

A basic understanding of the generation and propagation of acoustic waves (sound) in PE media is needed to understand the theory of surface acoustic wave (SAW) sensors. Unfortunately, most textbooks on acoustic wave propagation contain advanced mathematics (Auld 1973a) and that makes it harder to comprehend. Therefore, in this chapter, we set out the basic underlying principles that describe the general problem of acoustic wave propagation in solids and derive the basic equations required to describe the propagation of SAWs.

The different ways of representing acoustic wave propagation are outlined in Sections 10.2 and 10.3. The concepts behind stress and strain over an elastic continuum are discussed in Section 10.4, along with the general equations and concepts of the piezoelectric effect. These equations together with the quasi-static approximation of the electromagnetic field are solved in Section 10.5 in order to derive the generalised expressions for acoustic wave propagation in a PE solid. The boundary conditions that restrict the propagation of acoustic waves to a semi-infinite solid are included, and the general solution for a SAW is presented. An overview of the displacement modes in Love, Rayleigh, and SH-SAW waves are finally presented in Section 10.5. Consequently, this chapter is only intended to serve as an introduction to the displacement modes of Love, Rayleigh, and SH waves.

The components of displacements have been shown only for an isotropic elastic solid. The equations for the complex reciprocity and the assumptions used to derive the perturbation theory are elaborated in Appendix I.

More advanced readers may wish to omit this chapter or refer to specialised textbooks published elsewhere (Love 1934; Ewing *et al.* 1957; Viktorov 1967; Auld 1973a,b;

Slobodnik 1976). This chapter on the basic understanding of wave theory, together with the next chapter on measurement theory, should provide all readers with the necessary background to understand the application of interdigital transducer (IDT) microsensors and microelectromechanical system (MEMS) devices presented later in Chapters 13 and 14.

10.2 ACOUSTIC WAVE PROPAGATION

The most general type of acoustic wave is the plane wave that propagates in an infinite homogeneous medium. As briefly summarised at the end of Chapter 9 for those readers omitting that chapter, there are two types of plane waves, *longitudinal* and *shear* waves, depending on the polarisation and direction of propagation of the vibrating atoms within the medium (Auld 1973a). Figure 10.1 shows the particle displacement profiles for these two types of plane waves[1]. For longitudinal waves, the particles vibrate in the propagation direction (y-direction in Figure 10.1(a)), whereas for shear waves, they vibrate in a plane normal to the direction of propagation, that is, the x- and z-directions as seen in Figure 10.1(b) and (c).

When boundary restrictions are placed on the propagation medium, it is no longer an infinite medium, and the nature of the waves changes. Different types of acoustic waves may be supported within a bounded medium, as the equations given below demonstrate. *Surface Acoustic Waves* (SAWs) are of great interest here; in these waves, the traveling

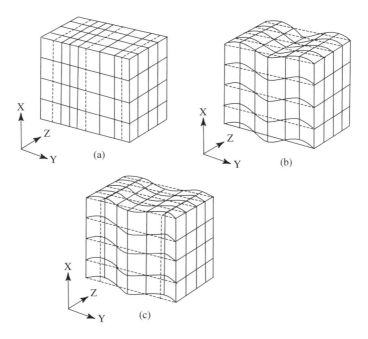

Figure 10.1 Particle displacement profiles for (a) longitudinal, and (b,c) shear uniform plane waves. Particle propagation is in the y-direction

---

[1] Also see Figure 9.10 in Chapter 9 on the introduction to SAW devices.

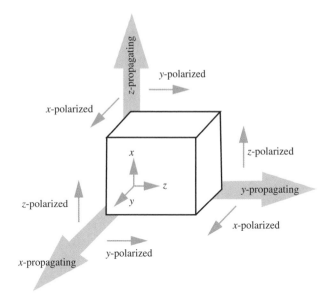

Figure 10.2   Acoustic shear waves in a cubic crystal medium

wave is guided along the surface with its amplitude decaying exponentially away from the surface into the medium. Surface waves were introduced in the last chapter and include the Love wave mode, which is important for one class of IDT microsensor.

10.3 ACOUSTIC WAVE PROPAGATION REPRESENTATION

Before a more detailed analysis of the propagation of uniform plane waves in piezoelectric materials in the following sections, a pictorial representation of the concept of shear wave propagation is presented. Figure 10.2 illustrates shear wave propagation in an arbitrary cubic crystal medium. An acoustic wave can be described in terms of both its propagation and polarisation directions. With reference to the X, Y, Z (x, y, z) coordinate system, propagating SAWs are associated with a corresponding polarisation, as illustrated in the figure.

10.4 INTRODUCTION TO ACOUSTICS

10.4.1 Particle Displacement and Strain

As stated earlier, acoustics is the study of the time-varying deformations or vibrations within a given material medium. In a solid, an acoustic wave is the result of a deformation of the material. The deformation occurs when atoms within the material move from their equilibrium positions, resulting in internal restoring forces that return the material back to equilibrium (Auld 1973a). If we assume that the deformation is time-variant, then

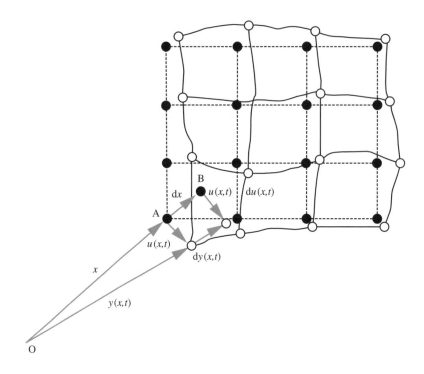

Figure 10.3   Equilibrium and deformed states of particles in a solid body

these restoring forces together with the inertia of the particles result in the net effect of propagating wave motion, where each atom oscillates about its equilibrium point. Generally, the material is described as being *elastic* and the associated waves are called *elastic* or *acoustic waves*. Figure 10.3 shows the equilibrium and deformed states of particles in an arbitrary solid body – the equilibrium state is shown by the solid dots and the deformed state is shown by the circles.

Each particle is assigned an equilibrium vector x and a corresponding displaced position vector $y(x, t)$, which is time-variant and is a function of x. These continuous position vectors can now be related to find the displacement of the particle at x (the equilibrium state) through the expression

$$u(x, t) = y(x, t) - x \tag{10.1}$$

Hence, the particle vector-displacement field $u(x, t)$ is a continuous variable that describes the vibrational motion of all particles within a medium.

The deformation or *strain* of the material occurs only when particles of a medium are displaced relative to each other. When particles of a certain body maintain their relative positions, as is the case for rigid translations and rotations[2], there is no deformation of the material. However, as a measure of material deformation, we refer back to Figure 10.3 and extend the analysis to include two particles, A and B, that lie on the position vector x and $x + dx$, respectively. The relationship that describes the deformation of the particles

---

[2] Only at constant velocity because acceleration induces strain.

between these two points after a force has been applied may be written as

$$\Delta' = dy^2(x, t) - (dx)^2 \tag{10.2}$$

$$\Delta'(x, t) = 2S_{ij}(x, t)dx_i\, dx_j \tag{10.3}$$

where $S_{ij}(x, t)$ is the second-order strain tensor[3] defined by

$$S_{ij} = \frac{1}{2}\left[\frac{\partial u_i}{\partial x_j} + \frac{\partial u_j}{\partial x_i} + \frac{\partial u_k}{\partial x_i}\frac{\partial u_k}{\partial x_j}\right] \tag{10.4}$$

with the subscripts of i, j, and k being x, y, or z.

For rigid materials, the deformation gradient expressed in Equation (10.4) must be kept small to avoid permanent damage to the structure; hence, the last term in the above expression is assumed to be negligible, and so the expression for the strain-displacement tensor is rewritten as

$$S_{ij}(x, t) = \frac{1}{2}\left[\frac{\partial u_i(x, t)}{\partial x_j} + \frac{\partial u_j(x, t)}{\partial x_i}\right] \tag{10.5}$$

10.4.2 Stress

When a body vibrates acoustically, elastic restoring forces, or *stresses,* develop between neighbouring particles. For a body that is freely vibrating, these forces are the only ones present. However, if the vibration is caused by the influence of external forces, two types of excitation forces (*body* and *surface forces*) must be considered. Body forces affect the particles in the interior of the body directly, whereas surface forces are applied to material boundaries to generate acoustic vibration. In the latter case, the applied excitation does not directly influence the particles within the body but it is rather transmitted to them through elastic restoring forces, or stresses, acting between neighbouring particles. Stresses within a vibrating medium are defined by taking the material particles to be volume elements, with reference to some orthogonal coordinate system (Auld 1973a). In order to obtain a clearer understanding of stress, we make the use of the following simple example. Let us assume a small surface area ΔA on an arbitrary solid body with a unit normal n, which is subjected to a surface force ΔF with uniform components ΔF_i. The surface ΔA may be expressed as a function of its surface components ΔA_j and the unit normal components n_j as follows:

$$\Delta A_j = n_j \cdot \Delta A \tag{10.6}$$

with the subscript j taking a value of 1, 2, or 3.

The stress tensor, T_{ij}, is then related to the surface force and the surface area through

$$T_{ij} = \frac{\Delta F_i}{\Delta A_j} \tag{10.7}$$

with the subscripts i and j taking a value of 1, 2, or 3.

---

[3] A tensor is a matrix in which the elements are vectors.

Moreover, if we consider the stress tensor T_{ij} to be time-dependent and acting upon a unit cube (assumed free body), the stress analysis may be extended to deduce the dynamical equations of motion through the sum of the acting forces. Thus,

$$\frac{\partial T_{ij}}{\partial x_j} + \rho F_i = \rho \frac{\partial^2 u_i}{\partial t^2} \tag{10.8}$$

where ρ is the mass density, F_i are the forces acting on the body per unit mass, and u_i represents the components of particle displacements along the i-direction.

10.4.3 The Piezoelectric Effect

Within a solid medium, the mechanical forces are described by the components of the stress field T_{ij}, whereas the mechanical deformations are described by the components of the strain field S_{ij}. For small static deformations of nonpiezoelectric elastic solids, the mechanical stress and strain fields are related according to Hooke's Law (Slobodnik 1976):

$$T_{ij} = c_{ijkl} S_{kl} \tag{10.9}$$

where T_{ij} are the mechanical stress second-rank tensor components (units of N/m^2), S_{kl} are the strain second-rank tensor components (dimensionless), and c_{ijkl} is the elastic stiffness constant (N/m^2) represented by a fourth-rank tensor. Taking into account the symmetry of the tensors, the previous equation can be reduced to a matrix equation using a single suffix. Thus, the tensor components of T, S, and c are reduced according to the following scheme of replacement (Auld 1973a; Slobodnik 1976):

$$\begin{array}{llll} (11) \Leftrightarrow 1; & (22) \Leftrightarrow 2; & (33) \Leftrightarrow 3 \\ (32) = (23) \Leftrightarrow 4; & (13) = (31) \Leftrightarrow 5; & (21) = (12) \Leftrightarrow 6 \end{array} \tag{10.10}$$

Therefore, the elastic stiffness constant can be reduced to a 6×6 matrix. Depending on the crystal symmetry, these 36 constants can be reduced to a maximum of 21 independent constants. For example, quartz and lithium niobate, which present trigonal symmetry, have their number of independent constants reduced to just 6 (Auld 1973b):

$$\begin{pmatrix} c_{11} & c_{12} & c_{12} & c_{14} & 0 & 0 \\ c_{12} & c_{11} & c_{13} & -c_{14} & 0 & 0 \\ c_{12} & c_{13} & c_{33} & 0 & 0 & 0 \\ c_{14} & -c_{14} & 0 & c_{44} & 0 & 0 \\ 0 & 0 & 0 & 0 & c_{44} & c_{14} \\ 0 & 0 & 0 & 0 & c_{14} & \frac{1}{2}(c_{11} - c_{12}) \end{pmatrix} \tag{10.11}$$

In piezoelectric materials, the relation given by Equation (10.8) no longer holds true. Coupling between the electrical and mechanical parameters gives rise to mechanical deformation and vice versa upon the application of an electric field. The mechanical stress relationship is thus extended to

$$T_{ij} = c_{ijkl}^E S_{kl} - e_{kij} E_k \tag{10.12}$$

where e_{kij} is the piezoelectric constant in units of C/m^2, E_k is the kth component of the electric field, and c_{ijkl}^E is measured either under a zero or a constant electric field.

In nonpiezoelectric materials, the electrical displacement D is related to the electric field applied by

$$D = \varepsilon_r \varepsilon_o E \tag{10.13}$$

where ε_r is the relative permittivity, formerly called the *dielectric constant*, and ε_o is the permittivity of vacuum, now known as the *electric constant*. For piezoelectric materials, the electrical displacement is extended to:

$$D_i = e_{ikl} S_{kl} + \varepsilon_{ik}^S E_k \tag{10.14}$$

where ε_{ik}^S is measured at constant or zero strain.

Equations (10.12) and (10.14) are often referred to as *piezoelectric constitutive equations*. In matrix notation, Equations (10.12) and (10.14) can be written as (Auld 1973b):

$$[T] = [c][S] - [e^T]E$$

$$D = [e][S] + [\varepsilon]E \tag{10.15}$$

where, $[e]$ is a 3×6 matrix with its elements depending on the symmetry of the piezoelectric crystal and $[e^T]$ is the transpose of the matrix $[e]$. For quartz having a trigonal crystal classification, the $[e]$ matrices are

$$[e] = \begin{pmatrix} e_{11} & -e_{11} & 0 & e_{14} & 0 & 0 \\ 0 & 0 & 0 & 0 & -e_{14} & -e_{11} \\ 0 & 0 & 0 & 0 & 0 & 0 \end{pmatrix} \tag{10.16}$$

The difference between poled and naturally piezoelectric materials is that in the former, the presence of a large number of grain boundaries and its anisotropic nature would lead to a loss of acoustic signal fidelity at high frequencies. This is one of the reasons SAW devices are, usually, only fabricated out of single-crystal piezoelectrics.

10.5 ACOUSTIC WAVE PROPAGATION

10.5.1 Uniform Plane Waves in a Piezoelectric Solid: Quasi-Static Approximation

For the numerical calculations of acoustic wave propagation, the starting point is the equation of motion in a piezoelectric material (Auld 1973a)

$$\rho \ddot{u}_i = T_{ij,j} \quad i, j = 1, 2, 3 \tag{10.17}$$

where, ρ is the mass density, and u_i is the particle displacement.

In tensor notation, the two dots over a symbol denotes $\partial^2/\partial t^2$ and a subscript i preceded by a comma denotes $\partial/\partial x_i$. The piezoelectric constitutive equations in (10.15) are rewritten in tensor notation:

$$T_{ij} = c_{ijkl}^E S_{kl} - e_{kij} E_k \tag{10.18}$$

$$D_i = e_{ikl} S_{kl} + \varepsilon_{ik}^S E_k \tag{10.19}$$

with i, j, k, and l taking the values of 1, 2, or 3.

The strain-mechanical displacement relation is:

$$c_{ijkl} S_{kl} = c_{ijkl} u_{l,k} \tag{10.20}$$

The absence of intrinsic charge in the materials is assumed; therefore,

$$D_{j,j} = 0 \tag{10.21}$$

The quasi-static approximation is valid because the wavelength of the elastic waves is much smaller than that of the electromagnetic waves, and the magnetic effects generated by the electric field can be neglected (Auld 1973a):

$$E_k = -\phi_{,k} \tag{10.22}$$

where ϕ is the electric potential associated with the acoustic wave.

The problem of acoustic wave propagation is fully described in Equations (10.17) to (10.22). These equations can be reduced through substitution to

$$\rho \ddot{u}_i = c_{ijkl}^E u_{l,jk} + e_{kij} \phi_{,jk} \tag{10.23}$$

$$0 = e_{jkl} u_{l,jk} - \varepsilon_{jk}^S \phi_{,jk} \tag{10.24}$$

The geometry for the problem of SAW wave propagation is shown in Figure 10.4. It has a traction-free surface ($x_3 = 0$) separating an infinitely deep solid from the free space.

The traction-free boundary conditions are (Viktorov 1967; Varadan and Varadan 1999)

$$T_{i3} = 0 \quad \text{for } x_3 = 0 \tag{10.25}$$

where i takes a value of 1, 2, or 3.

The solutions of the coupled wave Equations (10.23) and (10.24) must satisfy the mechanical boundary conditions of Equation (10.25). The solutions of interest here are

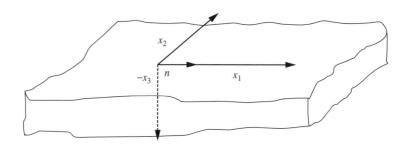

Figure 10.4   Coordinate system for SAW waves showing the propagation vector

SAWs that propagate parallel to the surface with a phase velocity v_R and whose displacement and potential amplitudes decay with distance away from the surface ($x_3 > 0$). The direction of propagation can be taken as the x_1-axis, and the (x_1, x_3) plane can be defined as the sagittal plane.

Note that the propagation geometry axes depicted in Figure 10.4 do not always correspond to the axes in which the material property tensors are expressed. There are transformation formulae that can be applied to the property tensors so that all the above equations hold for the new axes. The elastic constants (c_{ijkl}), the piezoelectric constants (e_{ijkl}), and the dielectric constants (ε_{ij}) can be substituted by c'_{ijkl}, e'_{ijkl}, ε'_{ij}. The primed parameters refer to a rotated coordinate system through the Euler transformation matrix (Auld 1973a).

The solutions for Equations (10.23) and (10.24) have the form of running waves: the surface wave solution is in the form of a linear combination of partial waves of the form (Auld 1973a)

$$u_i = A_i \exp(-kx_3) \exp\left[-j\omega\left(t - \frac{x_1}{V_R}\right)\right] \tag{10.26}$$

$$\phi = B \exp(-kx_3) \exp\left[-j\omega\left(t - \frac{x_1}{V_R}\right)\right] \text{ and } x > 0 \tag{10.27}$$

Here, ω is the angular frequency of the electrical signal, k is the wave number, given by $2\pi/\lambda$, and λ is the wavelength, given by $2\pi v_R/\omega$.

When the three particle displacement components exist, the solutions are called *generalised Rayleigh waves*. The crystal symmetry and additional boundary conditions (electrical and mechanical) impose further constraints on the partial wave solutions. If the sagittal plane is a plane-of-mirror symmetry of the crystal, x_1 is a pure-mode axis for the surface wave, which involves only the potential and the sagittal-plane components of displacement.

Because the Rayleigh wave has no variation in the x_2-direction, the displacement vectors have no component in the x_2-direction and the solution is given as follows (Varadan and Varadan 1999):

Assume displacements u_1 and u_3 to be of the form $A \exp(-bx_3) \exp[jk(x_1 - ct)]$ and $B \exp(-bx_3) \exp[jk(x_1 - ct)]$, and u_2 equal to zero, where the elastic half-space that exists for x_3 is less than or equal to zero, B and A are unknown amplitudes, k is the wave number for propagation along the boundary (x_1-axis) and c is the phase velocity of the wave. Physical consideration requires that b can, in general, be complex with a positive real part. Substitution of the assumed displacement into Navier–Stokes equation gives (Varadan and Varadan 1999)

$$\nabla \cdot \tau - \rho \frac{\partial^2 u}{\partial t^2} = 0 \tag{10.28}$$

and use of the generalised Hooke's law for an isotropic elastic solid yields two homogeneous equations in A and B. For a nontrivial solution, the determinant of the coefficient matrix vanishes, giving two roots for b in terms of the longitudinal and transverse velocities. Substitution of the roots of b obtained, as shown earlier, into the homogeneous equations in A and B gives the amplitude ratios. Thus, we obtain the general displacement solution (Equation 10.28) (Varadan and Varadan 1999).

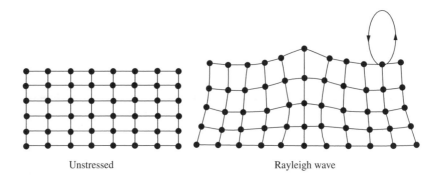

Unstressed	Rayleigh wave

Figure 10.5   Particle displacement on the sagittal plane for the Rayleigh wave

These displacements are as shown in Figure 10.5. It is seen that as u_3 is in phase quadrature with u_1, the motion of each particle is an ellipse. Because of the change in sign in u_1 at a depth of about 0.2 wavelengths, the ellipse is described in different directions above and below this point. At the surface, the motion is retrograde, whereas lower down it is prograde.

$$u_1 = [A_1 \exp(-b_1 x_3) + A_2 \exp(-b_2 x_3)] \exp[jk(x_1 - ct)]$$

$$u_3 = (-b_1/jk)A_1 \exp(-b_1 x_3) + (jk/b_2)A_2 \exp(-b_2 x_3) \exp[jk(x_1 - ct)] \quad (10.29)$$

where $b_1 = k(1 - c^2/v_l^2)^{1/2}$ and $b_2 = k(1 - c^2/v_t^2)^{1/2}$

The longitudinal and transverse velocities, v_l and v_t, are given by

$$v_l = \sqrt{\frac{\lambda + 2G}{\rho}} \quad v_t = \sqrt{\frac{G}{\rho}}$$

where the Lames' constants G is given by $E_m/2(1 + v)$ and λ is given by $vE_m/[(1 + v)(1 - 2v)]$ with v being Poisson's ratio and E_m being Young's modulus.

10.5.2 Shear Horizontal or Acoustic Plate Modes

Acoustic plate modes (APW) or shear horizontal (SH) waves in a half-space utilise single-crystal quartz substrates. These act as an acoustic waveguide by confining the acoustic energy between the upper and lower surfaces of the plate. Such a mechanism is used to confine waves traveling between an input and output IDT. SH modes may be thought of as those waves with a superposition of SH plane waves, which are multiply reflected at some angle between the upper and lower surfaces of the quartz plate. These upper and lower faces impose a transverse resonance condition, which results in each SH mode having the displacement maxima at the surfaces, with sinusoidal variation between the surfaces.

The solution is simply a plane shear wave propagating parallel to the surface, with its amplitude independent of x_3 within the material. The phase velocity is equal to v_t. The particle displacement associated with the nth order SH plate mode (propagating in the

x_1-direction) has only an x_2-component and is given by the following equation (see Auld 1973a,b):

$$u_2 = u_0 \cos\left[\frac{n\pi}{b}\left(x_2 + \frac{b}{2}\right)\right] \exp\left[j\left(\omega t - \beta_n x_1\right)\right] \tag{10.30}$$

where b is the plate thickness, u_2 is the particle displacement at the surface, n is the transverse modal index $(0,1,2,3\ldots)$, and t is time. The exponential term in the equation describes the propagation of the displacement profile down the length of the waveguide (along the x_1-direction) with angular frequency ω and wave number β_n given by

$$\beta_n = \sqrt{\left(\frac{\omega}{v_0}\right)^2 - \left(\frac{n\pi}{b}\right)^2} \tag{10.31}$$

where v_0 is the unperturbed propagation velocity of the lowest-order mode.

The cross-sectional displacement profiles (in the $x_2 - x_3$ plane) for the four lowest-order isotropic SH plate modes are shown in Figure 10.6. It is also noticed that each mode has equal displacements on both sides of the acoustic plate mode (APM) sensor, allowing the use of either side for sensing measurements.

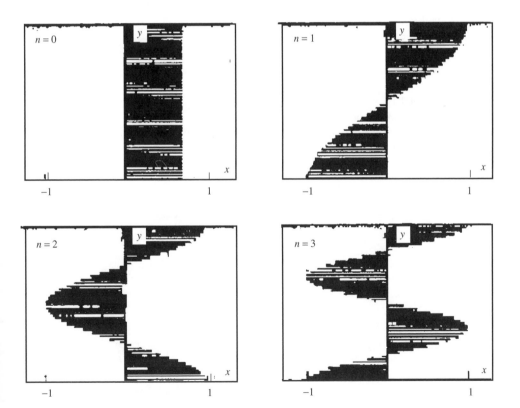

Figure 10.6 Displacement modes for $(n = 0, 1, 2, 3)$ SH–APM modes $(z::x_1, x::x_2, y::x_3)$

10.5.3 Love Modes

Ewing and co-workers (1957) were one of the first to point out from long-period seismographs that in addition to measuring the characteristic horizontal motion during the main disturbance of the earthquake, the seismographs also showed a large amount of transverse components. This early established fact in seismology was explained in 1911 by Love, and he easily showed that there could be no SH surface wave on the free surface of a homogeneous elastic half-space (Love 1934). Hence, this simple model could not explain the measurements. Love, however, showed subsequently that the waves involved were SH waves, confined to a superficial layer of an elastic half-space and the layer having a different set of properties from the rest of the half-space. Following Love's treatment here, Love waves can be considered as SAWs that propagate along a waveguide made of a layer of a given material M_2 (e.g. glass) deposited on a substrate made of another material M_1, (e.g. stable temperature (ST)-cut quartz), with different acoustic properties and, effectively, an infinite thickness when compared with the original layer.

These waves are transverse and they bring only shear stresses into action. The displacement vector of the volume element is perpendicular to the propagation direction O-x_1 and is oriented in the direction of the O-x_2 axis. Because the Love wave is a surface wave, the propagating energy is located in the layer and in that part of the substrate that is close to the interface. Its amplitude decreases exponentially with depth. However, it should also be noted that materials should have appropriate properties to propagate and carry a Love wave, as shall be discussed in the section hereby.

10.5.3.1 Existence conditions of Love waves dispersion equation

The case in which the two propagating media are isotropic is examined first. The coordinate origin is chosen on the interface; the O-x_1 axis is oriented in the direction of propagation and the x_3-axis is oriented vertically upwards (see Figure 10.7). The plane

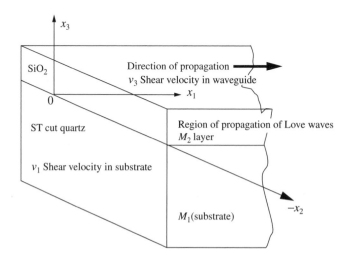

Figure 10.7   Structure of a Love waveguide: M_1 is the substrate; M_2 is the guiding layer

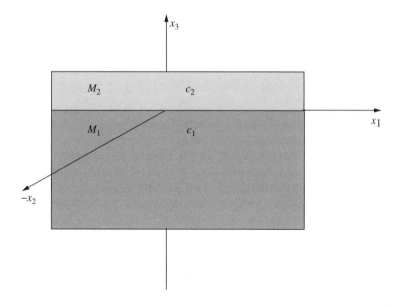

Figure 10.8   Schematic of Love wave device used for calculations

$x_3 = h$ represents the free boundary of the layer. Let us assume that displacements are oriented along the x_2-axis and are independent of x_1. Then, let us consider a monochromatic progressive wave of frequency ω propagating along the x_1-axis.

Using the symbols ρ_1, G_1, u_1 and ρ_2, G_2, and u_2 for the density, the shear modulus and the displacement vector of the volume elements for the substrate and the layer, respectively; v_{T1} and k_1 (equal to ω/v_{T1}), the phase velocity of the transverse waves and the wave number in medium M_1; and v_{T2} and k_2 (ω/v_{T2}), the same quantities in medium M_2, let us finally call c and k (equal to ω/c) the phase velocity and the wave number of the Love wave, whose existence is postulated (see Figure 10.8).

The solutions of the propagation's Equation (10.29) can now be written in the following way (Varadan and Varadan 1999):

$$u_{1\times2} = A \exp(j\omega t - jkx_1 + \alpha_1 x_3)$$
$$u_{2\times2} = (B_1 + B_2) \exp (j\omega t - jkx_1 + \alpha_2 x_3) \qquad (10.32)$$

where

$$\alpha_1 = -k\sqrt{1 - c^2/v_{T1}^2} \quad \alpha_2 = -k\sqrt{1 - c^2/v_{T2}^2} \qquad (10.33)$$

It can be verified that the above equation satisfies the Naviers Equation (10.28) in the two media and further that $u_3 \to 0$ as $x_3 \to -\infty$ (Varadan and Varadan 1999). v_{T1} and v_{T2} are the transverse wave velocities, as defined earlier by Equation (10.28).

The three constants A, B_1, and B_2 are determined by the boundary conditions that require not only that the tangential stresses σ_{23} cancel out in the plane $x_3 = h$ but also that they are continuous as well as the displacements $u_{1\times2}$ and $u_{2\times2}$ in the plane $x_3 = 0$ (Ewing *et al.* 1957; Slobodnik 1976).

The first of these conditions leads to

$$B_1 \exp(-\alpha_2 h) - B_2 \exp(+\alpha_2 h) = 0 \qquad (10.34)$$

and the two other conditions lead to $A = B_1 + B_2$ and

$$\alpha_1 \, G_1 A = \alpha_2 \, G_2(B_1 - B_2) \qquad (10.35)$$

This system of three linear equations has a solution different from zero (Tournois and Lardat 1969) if

$$\tan(\alpha_2 h) = \frac{G_1 \alpha_1}{G_2 \alpha_2} \qquad (10.36)$$

The roots of this equation have a real value when $k_1 < k < k_2$, that is, when $c_2 < c < c_1$. Therefore, a necessary condition of existence of Love waves is that the propagation velocity of transverse waves in the layer must be smaller than the propagation velocity of the transverse waves in the substrate.

It is easy to deduce from Equation (10.36) that there are infinite modes owing to the periodicity of the tangential function; when c tends to c_1, $\tan(\alpha_2 h)$ tends toward the value of $n\pi$ (with n being $0, 1, 2, \ldots$), and for the first mode, the wavelength becomes infinite compared with the thickness.

The particle displacements occurring during the propagation of Love waves are easily obtained from Equation (10.35).

$$u_{1\times 2} = A \exp(\alpha_1 x_3) \exp[j(\omega t - k x_1)]$$

$$u_{2\times 2} = A \frac{\cos[\alpha_2(h - x_3)]}{\cos \alpha_2 h} \exp[j(\omega t - k x_1)] \qquad (10.37)$$

where A is a propagation constant determined by the excitation signal.

The equations in (10.37) show that the displacement amplitude $u_{1\times 2}$ decreases exponentially in the substrate. It also shows that the different modes $u_{2\times 2}$ correspond to $0, 1, 2, \ldots$ nodal planes in the layer. Figure 10.9(a) gives the shape of particle displacements in the layer and the substrate for the first three modes.

The displacement amplitude also depends on the frequency, and Figure 10.9(b) shows its variation for the first mode. Therefore, it could be noticed that the energy is entirely located in the substrate for very low frequencies and that the Love wave propagates at a velocity c_1 as if the layer does not exist. Its thickness is, in fact, negligible when compared with the wavelength. Conversely, the acoustic energy is concentrated in the layer for very high frequencies, and the phase velocity of the Love waves tends toward c_2, the wavelength being very small with respect to the thickness of the layer. Between these two limits, the energy progressively transfers from the substrate to the layer, whereas the phase velocity varies between c_1 and c_2 (Tournois and Lardat 1969).

Having obtained the nature of the displacement and the similarity between the SH-SAW waves and Love modes, we can derive an expression for the change in the velocity and frequency shift for a Love wave device using perturbation theory. The derivation for the frequency shift and the corresponding change in velocity have been presented in Appendix I using the basic equations derived in this chapter.

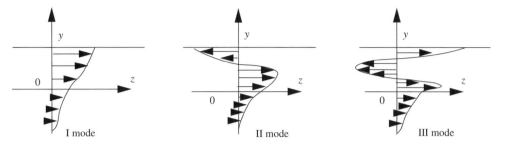

Particle displacement for the first three Love modes

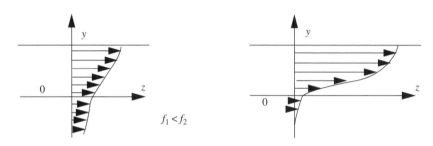

Frequency variation of the particle displacement for the first three modes

Figure 10.9 Displacement modes for Love wave devices (note that z corresponds to x_2 and y corresponds to x_3 in Figure 10.8)

10.5.3.2 Discussions of the characteristics of the Love waveguiding materials

It may be noticed from earlier discussions, that the two most important parts of a Love wave sensor are the overlayer material and the piezoelectric substrate. Our discussion now focuses on the salient points of the waveguide, particularly with respect to the properties of the material.

Love waves propagate near the surface of a suitable substrate material when the surface is overlaid by a thin film with appropriate properties for a guiding layer. An essential condition for the propagation of a Love wave is that the shear velocity in the film is less than that in the substrate. Sensitivity to mass-loading is enhanced by the low density of the film as well as a large difference between the shear velocities. For a particular guiding-layer material, an optimum layer thickness exists, which results in maximum acoustic energy density close to the surface and maximum sensitivity to mass-loading.

Love wave devices incorporating guiding layers of poly(methyl methacrylate) (PMMA) and sputtered SiO_2 overlaid on single-crystal quartz have been successfully demonstrated (Du *et al.* 1996). PMMA has a density of about 1.18 kg/m^3 and has a shear acoustic velocity of 1100 m/s (Kovacs *et al.* 1993; Jakoby and Vellekoop 1998; Du *et al.* 1996), whereas sputtered silicon dioxide has a density of about 2.3 kg/m^3 and a shear acoustic velocity[4] of 2850 m/s (Auld 1973a). Gizeli and coworkers (1995) utilised PMMA layers of thickness up to 5.6 µm spun onto Y-cut quartz with IDTs of periodicity of 45 µm at the quartz–PMMA interface. A network analyser was used to monitor the phase of the wave. The maximum thickness reported by Gizeli and coworkers (1995) (∼1.6 µm) is considerably less than the estimated optimum thickness of approximately 3 µm of PMMA (Shiokawa and Moriizumi 1988). Kovacs and co-workers (1993) have utilised sputtered silicon dioxide on ST-cut quartz. Acoustic losses in SiO_2 are low when compared with polymers such as PMMA. SiO_2 is more resistant to most chemicals and, when sputtered under optimal conditions, has excellent wear resistance. Because of technical reasons, it was reported that the maximum thickness of SiO_2 that was utilised was 5.46 µm – considerably less than the optimum value of approximately 6 µm (for devices of wavelength 40 µm).

Another criterion for the choice of a suitable waveguiding material would be the absorption coefficient. It essentially depends on the material structure, which can be polycrystalline, crystalline, or amorphous. In polycrystalline materials, when the wavelength becomes comparable to the grain size because of the phenomenon of Rayleigh scattering (Rayleigh 1924), the energy absorption increases proportionally to frequency to the fourth power (Tournois and Lardat 1969). At higher frequencies, it is obvious that the materials employed will have to be without loss-inducing grain boundaries, that is, either single-crystalline or amorphous. Amorphous bodies, such as certain glasses and fused silica, will allow propagation with a limited absorption at frequencies much higher than 100 MHz.

10.6 CONCLUDING REMARKS

In this chapter, the basic equations that describe the propagation of different types of waves in an elastic solid have been presented and expressions for the displacement of particles therein[5] have also been obtained. The emphasis has been directed toward the fundamental differences between the Rayleigh and SH modes and SH of vibration. The SH and Love wave modes have been examined from the point of view of waveguide structure, that is, the nature of the overlayer and the substrate. This mathematical discourse should help readers to understand the nature and application of SAW microsensors and MEMS devices in other chapters.

REFERENCES

Auld, B. A. (1973a). *Acoustic Fields and Waves in Solids I*, John Wiley and Sons, New York.
Auld, B. A. (1973b). *Acoustic Fields and Waves in Solids II*, John Wiley and Sons, New York.
Du, J. *et al*. (1996). "A study of Love wave acoustic sensors," *Sensors and Actuators A*, **56**, 211–219.
Ewing, W. M., Jardetsky, W. S., and Press, F. (1957). *Elastic Waves in Layered Media*, McGraw-Hill, New York.
Gangadharan, S. (1999). Design, development and fabrication of a conformal love wave ice sensor, MS thesis, Pennsylvania State University, USA.

---

[4] This value is sensitive to the deposition conditions.
[5] Some of the material presented here may also be found in Gangadharan (1999).

Gizeli, E., Liley, M. and Lowe, C. R. (1995). Detection of supported lipid layers by utilizing the acoustic Love waveguide device: application to bioengineering, *Technical Digest of Transducers '95*, pp. 521–523.

Jakoby, B. and Vellekoop, M. J. (1998). "Analysis and optimisation of Love wave sensors," *IEEE Trans. Ultrasonics*, Ferroelectrics and Frequency control, **45**, 1293–1302.

Kovacs, G., Vellekoop, M. J., Lubking, G. W. and Venema, A. (1993). A Love wave sensor for (bio)chemical sensing in liquids, *Sensors and Actuators*, **43**, 38–43.

Love, A. E. H. (1934). *Theory of Elasticity*, Cambridge University Press, England.

Rayleigh, R. (1924). *Theory of Sound*, Macmillan, New York.

Shiokawa, S. and Moriizumi, T. (1988). Design of SAW sensor in liquid, *Proc. of 8th Symp. on Ultrasonic Electronics*, Tokyo, pp. 142–144.

Slobodnik, A. J. (1976). "Surface acoustic waves and materials," *Proc. IEEE*, **64**, 581–595.

Tournois, P. and Lardat, C. (1969). "Love wave dispersive delay lines for wide band pulse compression," *Trans. Sonics Ultrasonics*, **SU-16**, 107–117.

Varadan, V. V. and Varadan, V. K. (1999). Elastic wave propagation and scattering, *Engineering Science and Mechanics*, Pennsylvania State University, USA.

Viktorov, I. A. (1967). *Rayleigh and Lamb Waves: Physical Theory and Applications*, Plenum Press, New York.

11

IDT Microsensor Parameter Measurement

11.1 INTRODUCTION TO IDT SAW SENSOR INSTRUMENTATION

There is no specific design procedure for interdigital transducer (IDT) microsensors based on surface acoustic wave (SAW) delay lines. Although substantial work has been done on delay line designs for filtering and signal-processing applications, the requirements for SAW-based devices are essentially different from those for commercial non-SAW oscillator-based sensors (Avramov 1989). The SAW device should not only have the appropriate frequency-transfer characteristics, but its physical dimensions should also allow for miniaturisation and remote-sensing of a variety of physical and chemical media.

This chapter deals with the instrumentation and measurement aspects of a typical IDT-SAW sensor during the course of its operation and so covers the different measurement techniques available and makes a comparison between them.

Specifically, Section 11.3 describes the basic principles of a 'network analyser,' whereas subsequent sections describe its use to measure the amplitude (Section 11.4), phase (Section 11.5), and frequency (Section 11.6) of signals. Finally, a brief overview of a network analyser system is given with particular emphasis on the topics of forward-matching, reverse-matching, and transmission (see Gangadharan 1999).

11.2 ACOUSTIC WAVE SENSOR INSTRUMENTATION

11.2.1 Introduction

Acoustic wave sensors convert the physical or chemical property of interest into a signal suitable for measurement. Ultimately, the measured sensor data must be processed so that they can be presented to the user in both a sensible and meaningful way. The role of system instrumentation is to implement this task.

Ever since the early work by Sauerbrey and King (Avramov 1989), and many of those that have followed, the vast majority of acoustic-sensing applications has involved acoustic sensors being used as the active frequency-control elements in oscillator circuits. The oscillator output frequency is then used as the (desired) measured parameter. This is not so surprising, as acoustic sensors are digital by nature[1] with a good dynamic range and linearity. Therefore, state-of-the-art system instrumentation can readily exploit the digital acoustic microsensor. One of the main advantages digital instrumentation has over the earlier system technologies is that there is no requirement to perform any frequency to voltage (for output) conversion – a process that often causes a considerable loss of resolution. In the state-of-the-art systems, frequency to digital conversion takes place without loss[2] through the use of simple electronic counters. However, other measuring system instrumentation must be considered and compared in terms of its viability, functionality, and suitability. Having accepted that either *frequency* or multiple-period measurements (timing signal) is the most appropriate parameter to measure, consideration will now be given to past and current forms of such instrumentation (Campbell 1998).

There are three electronic configurations that may be used to measure the response of a SAW (two-port) microsensor:

1. Where the microsensor is connected to a network analyser (or vector voltmeter) and scanned by a narrow radio frequency (RF) band on either side of its fundamental resonant frequency (see Section 11.3).

2. Where the microsensor is used as the *passive* element, being driven from a fixed RF source (amplitude or phase measurements) (see Sections 11.4 and 11.5).

3. Where the microsensor is used as the *active* feedback-determining element, controlling the frequency of an oscillator circuit (frequency measurement) (see Section 11.6).

11.3 NETWORK ANALYSER AND VECTOR VOLTMETER

Primarily, the network analyser and vector voltmeters are used by radio and electronic engineers for network analysis and design. Although SAW resonators have equivalent electric circuit analogues, network analysis is usually performed on the basis of the idea of electrically matching resonators with their corresponding oscillator circuits. This is particularly true for the SAW-sensor oscillator design. From either the network analysers or the vector voltmeters, the characteristic admittance Y or impedance Z can be measured. These measurements will produce a locus of the admittance or impedance parameters over a set of discrete frequency data points. The admittance (or impedance) can be expressed in either rectangular (real and imaginary) or polar (magnitude and phase) coordinates. Analysis from such instruments will deliver the most information possible about an acoustic transducer or the sensor that is under load, detailing shifts in resonant frequency and any changes in the quality factor Q (Subramanian 1998; Piscotty 1998).

Despite the advantages that either the network analyser (or vector voltmeter) has in delivering very detailed information on a number of important parameters, both these instruments are impractical as general-purpose laboratory tools. The reasons for this are that the resonator setup is:

---

[1] Digital in the sense that they are active frequency elements in oscillator circuits where frequency counters are used to measure the output frequency.
[2] High resolutions are achieved through long counting periods.

1. Far too time-consuming

2. Restricted, as only one acoustic sensor can be measured at a time

3. Allows only steady state measurements; no dynamic response measurements are possible

4. Very expensive to operate, in terms of both the instrument price and the need for highly qualified personnel

5. Unable to offer remote-sensing

6. Too cumbersome and not portable

Nevertheless, there are instances in which it is useful to use a network analyser to understand an IDT-SAW problem. For example, most of the work described here on the device referred to previously as an ice microsensor has been more of a characterisation study, so it is sensible to employ a network analyser in order to measure the relevant phase difference parameters[3]. This is despite the fact that the network analyser carries with it all the relevant disadvantages just listed.

Before choosing the configuration to employ, the following question should always be asked:

'What is the most practical and appropriate form of instrumentation or system to be used for measuring and analyzing acoustic-based microsensors?'

Wohltjen and Dessy (1979) answered the question by describing three general methods for measuring the response of a SAW microsensor – amplitude, phase, or frequency measurements. The following three sections provide typical examples of such measuring systems (Wohltjen and Dessy 1979) based on these methods.

11.4 ANALOGUE (AMPLITUDE) MEASURING SYSTEM

Figure 11.1 shows a block diagram for an amplitude measurement system.

The output from a common RF source is split with zero phase shift and is used to excite an acoustic device and RF step attenuator. Diode detectors rectify the incoming signal,

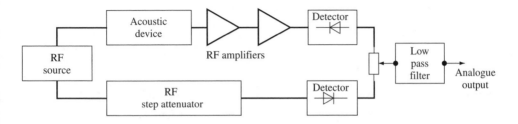

Figure 11.1   Amplitude measurement system

3 See Chapter 8 for details of the SAW-based ice microsensor.

producing a negative direct current (DC) level from the acoustic device and a positive DC level from the attenuator arm. A potentiometer is used to set the signal to zero for the unperturbed acoustic resonator. A subsequent low-pass filter restricts the bandwidth of the resultant analogue output signal.

11.5 PHASE MEASUREMENT SYSTEM

In a phase measurement system, the initial stages are the same as in an amplitude measurement system, but a double-balanced mixer (Figure 11.2) replaces the diode detectors and low-pass filter. The two RF signals from the acoustic and attenuator branches are fed into the mixer and a potential difference appears at the mixer intermediate frequency (IF) port, which corresponds to the phase difference. The output voltage is then applied to an instrument amplifier through a simple resistance capacitance (RC) filter. Under normal operation, the RF attenuator is adjusted to match the RF output level from the acoustic device. Wohltjen and Dessy (1979) reported that changes in the amplitude of the acoustic wave of several percent did not introduce any significant error; however, substantial attenuation occurs, which makes it difficult to interpret measurements.

Unlike the vector voltmeter and network analyser, the amplitude and phase measurement systems are portable. They also have the advantage of being driven from an external fixed RF source, where the amplitude of the source can be adjusted to compensate for heavy damping conditions. However, a number of reasons make amplitude (or phase measurements) a poor choice for monitoring acoustic sensor response. The most significant is the much-reduced dynamic range, typically between 10 000:1 and 1000:1. It is well known that the use of amplitude signals by instruments, such as a voltmeter or chart recorder, is neither as precise nor as accurate as measurements based on frequency or the time domain. The dynamic range for *frequency* or *time* measurements is on the order of 10 million to 1. Furthermore, the use of a potentiometer to preadjust for a zero baseline unnecessarily complicates the amplitude system, as it will most likely require readjustment from time to time because of the drift. Grate and coworkers (1993) describe several mechanisms that will contribute to shifts in the unperturbed resonance frequency, namely, environmental contaminants, for example, oxidation, changes in the piezoelectric material, mounting stresses, thermal gradients, hysteresis, and deterioration through aging. Both systems require the user to adjust the RF step attenuator in order to match the RF level from the acoustic device. In our opinion, good engineering design should at best eliminate the need for preset adjustments or, at worst, limit the requirement, in order to allow system instrumentation to be used by nonexperts.

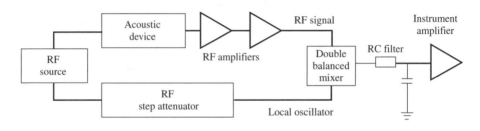

Figure 11.2 Phase measurement system

11.6 FREQUENCY MEASUREMENT SYSTEM

Figure 11.3 shows a typical configuration in which a SAW resonator is used as the feedback element in an oscillator circuit. Oscillation will be sustained, provided the gain of the RF amplifiers exceeds the insertion loss of the resonator. Unfortunately, measuring the acoustic oscillator frequency alone provides no information regarding signal amplitude because it is a measure of mechanical-loading (damping effects) and gives an estimate of the quality factor Q. Furthermore, under heavy loading conditions, oscillation may cease. To partially overcome this limitation, some researchers have used oscillator circuits using automatic gain control (AGC) techniques. Measurement of the AGC feedback permits an estimate of the oscillator frequency amplitude (damping) levels to be made. The sole purpose of AGC is to maintain oscillation under heavy loading by maintaining the magnitude of acoustic wave within the crystal substrate constant (Smith and Gerard 1971). AGC, however, will not restore any loss of circuit quality factor.

The authors believe that the acoustic resonator oscillator design should be optimised so as to match the nominal environmental quiescent feedback characteristics of the acoustic sensor. By adhering to this design philosophy, the simple frequency-measurement system should be more than adequate to meet most sensing application needs (Grate *et al*. 1993).

To sum up, the measurement of frequency provides the simplest and most cost-effective solution for the processing of acoustic resonator responses because of the following reasons:

1. It has the greatest dynamic range 10^7:1.

2. It is less complex and costly compared with the amplitude or the phase measurement systems.

3. No manual preset adjustments required, making operation easier for the user.

4. There are fewer component parts and, therefore, potential sources of noise are reduced.

In the majority of applications, a SAW microsensor serves as a feedback element that controls the oscillator frequency. Oscillators constructed from transistor transistor logic (TTL), logic inverter chips, and transistors have been reported (Shiokawa and Moriizumi 1988).

Figure 11.4 is a representation of such an oscillator circuit, in which the electronic circuit not only delivers the necessary driving signal to sustain oscillation but also provides the output port in which the resonant frequency can be monitored.

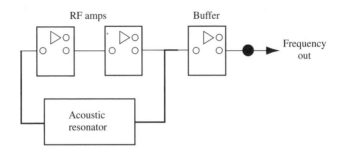

Figure 11.3   Frequency measurement system

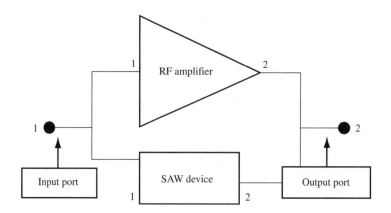

Figure 11.4   SAW device oscillator circuit

Non-AGC oscillator circuits, such as the Brukenstein and Shay (see Smith *et al.* (1969)), are a viable alternative. These circuits feature high output voltage and will drive under high viscous load conditions. Advantages of their design include fewer circuit components making less-possible noise and drift sources and are, therefore a much simpler and compact alternative.

11.7 ACOUSTIC WAVE SENSOR OUTPUT FREQUENCY TRANSLATION

Typically, the unperturbed resonant frequency of acoustic resonators used for sensing applications ranges between 30 and 300 MHz for SAW devices. As sensors, typical sensing effects constitute relatively small frequency deviations from their unperturbed resonance, from several kHz to a few MHz. However, measuring such frequencies in the very high frequency (VHF) and ultrahigh frequency (UHF) bands requires very expensive RF instrumentation. Therefore, moves from the RF spectrum down to the audio spectrum has proved a popular alternative. This is achieved by a mixing process that involves heterodyning the reference and sensing oscillator frequencies. A low-pass filter ensures that only the difference frequency is passed on. Figure 11.5 shows the schematic diagram of a typical mixing circuit.

The circuit consists of a dual-oscillator system in which the frequency of each oscillator is controlled by reference and sample resonators.

The difference frequency Δf is the output from the low-pass filter

$$\Delta f = (f_{\text{ref}} - f_{\text{sample}}) \tag{11.1}$$

Digital mixing circuits offer a viable alternative to classical analogue circuit techniques (Smith *et al.* 1969). The digital technique has advantages over analogue methods in that it removes the need for both the low-pass filter and the RF mixing transformer. This substantially lowers cost and size. However, irrespective of whether analogue or digital mixing is used, there will always exist some finite difference frequency between the

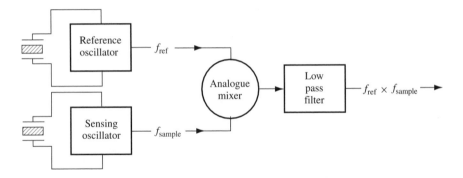

Figure 11.5 Block diagram of dual reference and sample analogue mixing circuit

sensor and reference oscillators, producing some degree of baseline offset. Although the mixing circuit technique will significantly reduce the effects of common mode interference, there is always the possibility that interference could compound and, therefore, increase measurement errors.

Another option is to use an environmentally isolated precision reference oscillator. As the frequency from this protected reference oscillator will remain fixed, the mixed frequencies from the reference and indicator sensor oscillator will not contain frequency contributions from any interfering source (Crabb and Lewis 1973).

11.8 MEASUREMENT SETUP

The vector network analyser and associated calibration techniques make it possible to accurately measure the transmission parameters of the devices under test. The measurement schematic is shown in Figure 11.6. The network analyser consists of a synthesized sweeper (10 MHz–40 GHz), test setup (45 MHz–40 GHz), HP8510B network analyser, and a display processor (Subramanian 1998; Piscotty 1998). The sweeper provides the stimulus and the test setup provides signal separation. The display panel of the HP8510B is used to define and conduct various measurements. The system bus is instrumental in controlling various other instruments. The device to be tested is connected between the test Port 1 and Port 2. The point at which the device is connected to the test setup is called the *reference plane*. All measurements are made with respect to this reference plane. The measurements are expressed in terms of the scattering parameters referred to as *S parameters* (Subramanian 1998). These describe the signal flow within the network.

S parameters are defined as ratios and are represented by $S_{\text{in}/\text{out}}$, where the subscripts in and out refer to the *input* and *output* signal, respectively. Figure 11.7 shows the energy flow in a two-port network. It can be shown that (see HP 8510B Network Analyser Manual 1987)

$$b_1 = a_1 S_{11} = a_2 S_{12} \quad \text{and} \quad b_2 = a_1 S_{21} = a_2 S_{22} \tag{11.2}$$

where S_{11} is b_1/a_1 and S_{21} is b_2/a_1 when a_2 is zero; S_{12} is b_1/a_2 and S_{22} is b_2/a_2 when a_1 is zero. S_{11} and S_{21} (S_{12} and S_{22}) are the reflection and transmission coefficients for Port 1(2), respectively.

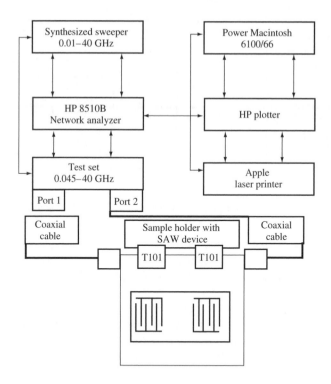

Figure 11.6   Schematic of measurement setup

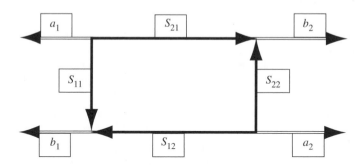

Figure 11.7   Signal flow of a two-port network

11.9 CALIBRATION

Calibration of any measurement is essential in order to ensure the accuracy of the system. The errors that exist in systems may be random or systematic. Systemic errors are the most significant source of measurement uncertainty. These errors are repeatable and can be measured by the network analyser. Correction terms can then be computed from these measurements. This process is known as *calibration*. Random errors are not repeatable and are caused by variations due to noise, temperature, and other environmental factors that surround the measurement system.

A series of known standards are connected to the system during calibration. The systemic effects are determined as the difference between the measurand and the known response of the standards. These errors can be mathematically related by solving the signal-flow graph (Subramanian 1998). The frequency response is the vector sum of all test setup variations in magnitude and phase and the frequency. This is inclusive of all signal-separation devices, such as test setup and cabling.

The mathematical process of removing errors is called *error correction*. Ideally, using perfectly known standards, these errors should be completely characterised. The measurement system is calibrated using the full two-port calibration method. The four standards that are commonly used are shielded open circuit, short circuit, load, and through. This method provides full correction of directivity, source match, reflection and transmission-signal path, frequency response, load match, and isolation for S_{11}, S_{12}, S_{21}, and S_{22}. The procedure involves taking a reflection, transmission, and isolation measurement.

For the reflection measurement (S_{11}, S_{22}), the open, short, and load standards are connected to each port in turn and the frequency response is measured. These six measurements result in the calculation of the reflection error coefficients for both ports.

For the transmission measurement, the two ports are connected and the following measurements are carried out forward through transmission (S_{21}-frequency response), forward through match (S_{21}-load), reverse through transmission (S_{12}-frequency response), and reverse through match (S_{12}-load). The transmission error coefficients are computed from these four measurements.

Loads are connected to the two ports and the S_{12} and S_{21} noise floor level is measured. From these measurements, the forward and reverse-isolation error coefficients are computed. The calibration is saved in the memory of the network analyser and the correction is turned on to correct systemic errors that may occur.

By making these measurements, it is possible to identify the critical acoustic parameters and thus design the optimal IDT-SAW microsensor. The SAW microsensor may now be fabricated, and the process is provided in the following chapter.

REFERENCES

Avramov, I. D. (1989). Analysis and design aspects of SAW-delay-line-stabilised oscillators, *Proceedings of the 2nd Int. Conf. on Frequency Synthesis and Control*, London, April 10–13, pp. 36–40.

Campbell, C. (1998). *Surface Acoustic Wave Devices and their Signal Processing Applications*, Academic Press, London.

Crabb, J. and Lewis, M. F. (1973). "Surface acoustic wave oscillators: mode selection and frequency modulation," *Electronics Lett.*, **9**, 195–197.

Gangadharan, S. (1999). Design, development and fabrication of a conformal Love wave ice sensor, MS thesis, Pennsylvania State University, USA.

Grate, J. W., Martin, S. J. and White, R. M. (1993). "Acoustic wave microsensors, Parts I and II," *Anal. Chem.*, **65**, 940–948, 987–996.

HP 8510B Network Analyzer Manual (1987). Hewlett-Packard Company, Santa Rosa, Calif.

Piscotty, D. J. (1998). 150 MHz wireless detection of a ST-cut quartz substrate surface acoustic wave device, MS thesis, Pennsylvania State University, USA.

Shiokawa, S. and Moriizumi, T. (1988). Design of SAW sensor in liquid, *Proc. of 8th Symp. on Ultrasonic Electronics*, Tokyo, July, pp. 142–144.

Smith, W. R. and Gerard, H. M. (1971). "Differences between in-line and cross-field three-port circuit models for integrated transducers," *IEEE Trans. Microw. Theory Techniques*, **19**, 416–417.

Smith, W. R. *et al.* (1969). "Analysis of interdigital surface wave transducers by use of an equivalent circuit model," *IEEE Trans. Microw. Theory Techniques*, **16**, 856–864.

Subramanian, H. (1998). Experimental validation and design of wireless microaccelerometer, MS thesis, Pennsylvania State University, USA.

Wohltjen, H. and Dessy, R. (1979). "Surface acoustic wave probe for chemical analysis," *Anal. Chem.*, **51**, 471–477.

12
IDT Microsensor Fabrication

12.1 INTRODUCTION

Surface acoustic wave (SAW) devices are fabricated using processes that have been primarily developed for integrated circuit (IC) technology in the microelectronics industry.

In this chapter, we describe all the steps required to fabricate an interdigital transducer (IDT) SAW microsensor from a stable temperature (ST) cut quartz wafer. A basic overview of this process is given in Figure 12.1. Specifically, there are two processes that are commonly used to define the IDTs: *etching* and *lift-off* (Hatzakis *et al.* 1980). Both methods[1] are suitable for the fabrication of IDT-SAW delay-line sensors, but the ultimate choice of either the etching or the lift-off process mainly depends on the minimum feature size (resolution and accuracy) of the patterned structure required. Although the etching procedure is relatively easy to realise and acceptable resolution is achievable, it is more susceptible to electrical shorts between features than that of the lift-off process. This is a major concern, especially for minimum feature sizes approaching 1–2 µm, where the influence of contaminants, such as large dust particles, becomes more significant (Vellekoop 1994). However, for larger minimum feature sizes, of 5 µm or greater, it is recognised that the etching process is acceptable and comparable in terms of device fabrication yield and quality to that of the lift-off process.

Section 12.2 provides full details of the steps required to make an IDT microsensor through either an etching process or a lift-off technique. The process given here is meant to serve as an example, and variations in the precise choice of materials and equipment used will vary from laboratory to laboratory.

Next, the steps required to make a Rayleigh-SAW microsensor from the IDTs are shown, together with a waveguiding layer of SiO_2 (Section 12.3) to fabricate a Love wave microsensor.

Finally, in Section 12.4, we provide tables that summarise the etching and lift-off processes and present their relative merits.

12.2 SAW-IDT MICROSENSOR FABRICATION

12.2.1 Mask Generation

SAW-IDT designs are written onto square, low-expansion glass plates using a process of electron-beam (E-beam) lithography. The SAW designs are first created using a

---

[1] Pattern transfer and etching methods were introduced in Chapter 2.

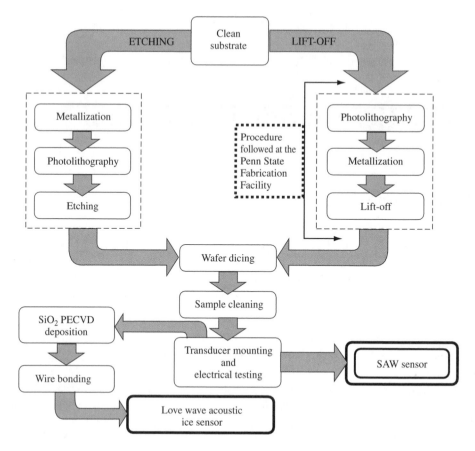

Figure 12.1   Overview of process required to fabricate Rayleigh wave and Love wave IDT microsensors

computer-aided design (CAD) system (e.g. L-Edit from Tanner Tools Inc.) and then the electronic design files are exported in a standard format (e.g. GDS II) that offers compatibility with the E-beam writer. The IDT structures are thus written on a positive resist material that coats the mask plate on which a thin chromium layer has already been deposited. The resist is developed and the chrome is etched away to leave the desired IDT structures. It is common practice to make an inverse mask, or negative, from the master positive mask plates using a quicker and more inexpensive ultraviolet (UV) optical lithographic process. It is these copies that are then used in the silicon run and, if damaged, can be replaced immediately. Figure 12.2 shows a typical IDT design that would be written onto the positive and negative mask plates.

12.2.2 Wafer Preparation

Effective cleaning of the quartz wafers is a vital procedure, which is an essential requirement for the successful fabrication of IDT microsensors. In order to obtain good adhesion and a uniform coating of the metallic film used to make the IDTs, a thorough cleaning of

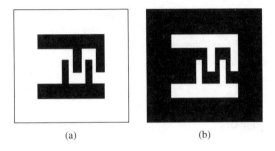

(a)   (b)

Figure 12.2   Basic layout of a photolithographic mask plate showing an IDT structure: (a) positive and (b) negative fields

the wafer surface is essential. The cleaning of the wafers should be performed in a fume cupboard (in a clean room) to allow the safe and fast removal of any possible harmful fumes produced during the cleaning process (Campbell 1998; Atashbar 1999).

The wafers are initially cleaned of any surface contaminants, such as dust, grease, or any other soluble organic particles, by immersion in trichloroethylene[2] at 60 °C for 10 minutes, followed by an acetone bath at 60 °C for 10 minutes. The wafers are then rinsed with methanol and finally with deionised water. It is best to avoid the use of nitrogen gas for drying the sample during the aforementioned procedure so as to minimise further surface contaminants. Instead, a slow evaporation in a protected fume cupboard is employed. Further cleaning is then undertaken for the removal of the more obstinate contaminants. The wafers are immersed in a mixture of three parts of deionised water (3H_2O), one part ammonium hydroxide (NH_4OH), and one part of 30 percent unstabilised hydrogen peroxide (H_2O_2) at 75 °C for 10 minutes. Caution is required because the mixture is harmful, and it is recommended that the hydrogen peroxide is added last so as to minimise any reaction side effects. Next, the wafers are placed in a solution of industrial grade detergent and subjected to ultrasonic agitation at 60 °C for ten minutes. Following a rinse in deionised water, the wafers are placed in a circulating deionised water bath for 30 minutes. The wafers are then dried using compressed filtered nitrogen and stored in an appropriate container and environment.

12.2.3 Metallisation

A metal layer now needs to be deposited, from which IDT structures are to be formed. In general, aluminum is evaporated using, for example, a Kurt Lesker™ E-beam evaporator. Aluminum is employed because it is commonly used in IC foundries and exhibits chemical resistance to many different liquids[3].

Typically, a 100 to 150 nm layer of aluminum is deposited on the clean surface of a quartz wafer. For example, the beam voltage of an E-beam evaporator is set to 6 keV during the deposition of 150 nm of aluminum, the pre-evaporation pressure is set at 10^{-6} torr, and the beam current is set to almost 100 mA. This gives an evaporation rate of 0.2 nm/s. It is to be noted that aluminum could have also been evaporated onto the

---

[2] Caution needs to be exercised since trichloroethylene fumes are toxic.
[3] Clearly, strong acids attack aluminum and should be avoided.

device using thermal evaporation instead of using the E-beam technique. The E-beam technique, however, allows more control over the deposition rate, and the films tend to be more uniform and to possess fewer stacking faults and dislocations.

E-beam evaporation of aluminum is, indeed, compatible with both the etching and the lift-off processes used later on.

12.2.4 Photolithography

The photolithography process is conducted in a clean room environment at a constant temperature of, typically, 25 °C ± 1 °C and at a relative humidity of 40 ± 5 percent.

The IDT structures need to be oriented correctly with respect to the quartz wafer in order to generate the required Rayleigh (or Love) waves. Figure 12.3 shows the correct orientation of the wafer and the SAW-IDTs[4].

12.2.4.1 Etching process

The etching process begins with the initial cleaning of the metallised wafers, followed by the deposition of a positive photoresist. The wafers are first rinsed in a bath of acetone and then in isopropanol to remove any possible loose surface contaminants that could have appeared during storage since the initial wafer-cleaning procedure. Next, the wafer is thoroughly rinsed in a deionised water bath for 5 minutes, followed by an oven bake at 75 °C for 20 minutes. This removes any moisture from the surface of the wafer. Using a Headway Research Inc.® spinner, hexamethyl disilazane (HMDS) is spun on the wafer at 3000 rpm for 60 seconds to improve the adhesion of the resist to the wafers. After allowing the HMDS thin film to sit for 2 minutes, AZ-1512® positive photoresist (Hoechst) is then spun at 3000 rpm for 30 seconds. A photoresist layer, approximately 1.2 μm thick, is formed. The wafer is then baked in an oven at 90 °C for 30 minutes to

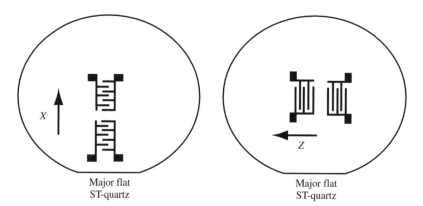

Major flat
ST-quartz

Major flat
ST-quartz

Figure 12.3 Orientation of an ST-quartz wafer and the SAW-IDT structures to fabricate Love and Rayleigh wave sensors

---

[4] The relationship between wafer flats and crystal orientation is defined in Section 4.2.

remove any excess solvents from the photoresist. Then, it is cooled to room temperature for approximately 15 minutes before exposing it to UV light in the mask aligner.

A contact mask aligner (Karl Suss MRK-3) is used to align the *positive* chrome mask plate with the quartz that is wafer-coated with the photoresist. A UV light exposure of 6 seconds is subsequently required. The exposed wafer is then developed in a mixture of (ratio 1:4) AZ-450® developer (Hoechst) and deionised water for 40 seconds. Great care should be taken at this stage because under or overdeveloping the photoresist layer will degrade the fabrication success. It is strongly recommended that an immersion style is adopted, so that the wafer is slowly agitated during the developing process at 10 second intervals, followed by a deionised water rinse and a close inspection using a microscope. This will provide for greater control in the important developing stage of fabrication. A 'soft' post bake is then performed at 75 °C for 10 minutes, which assists in the hardening and formation of sharp features of the photoresist. The wafer is then allowed to cool to room temperature for approximately 15 minutes. At this stage, the IDT pattern should have been successfully transferred to the wafer; if not, the photoresist can be stripped off in acetone and the entire procedure repeated before the etching of the wafer.

Chemical wet-etching of the unwanted aluminum is then performed. The aluminum layer is first etched in a solution of a commercial etchant and deionised water (3.25 g of etchant in 50 ml of deionised water) at room temperature for approximately 60 seconds. The etching time is extremely critical because undercutting of the structure walls may occur if prolonged times are employed. It is strongly recommended that etching is performed at 10 second intervals, followed by a deionised water rinse and close inspection with a microscope.

The temperatures of the etchant solutions, together with the thickness of the metal layers, are important factors that have a significant influence on the etching times. It is recommended that the etching procedure is inspected for assurance before the processing of valuable quartz wafers.

Once the IDT design has been successfully transferred to the metallised wafer via the etching process, the wafer is ready for dicing (Campbell 1996, 1998). The dicing process is described briefly in Section 12.2.5.

12.2.4.2 Lift-off process

The lift-off process begins with an initial cleaning of the wafers, followed by the deposition of a positive photoresist. A similar cleaning procedure to that used for the etching process is used to remove any possible loose surface contaminants that may have appeared during storage since the initial wafer-cleaning procedure. Similarly, HMDS is spun on to the wafer using, for example, a Headway Research Inc.® spinner at 3000 rpm for 60 seconds to improve the adhesion of the resist to the wafer. After allowing the HMDS thin film to sit for 2 minutes, AZ-1512® positive photoresist (Hoechst) is then spun at 3000 rpm for 30 seconds. A photoresist layer of approximately 1.2 μm is formed. The wafers are then baked in an oven at 75 °C for 30 minutes to remove any excess solvents from the photoresist. The wafers are cooled to room temperature for approximately 15 minutes before UV light exposure.

After aligning the *negative* IDT chrome mask plate with the photoresist-coated wafer having a similar orientation to that used in the etching process, the wafer is exposed to

UV light for 6 seconds. To improve the lift-off capability, the wafer is then immersed in a chlorobenzene bath at room temperature for 3 to 3.5 minutes. It is important to note that this time varies depending on the intensity of the UV exposure lamp; typically, for an intensity of 21 W/cm^2, the characteristic time in chlorobenzene ranges from about 220 to 280 seconds. This is an extremely critical step, and the procedure should be validated before it is applied to the set of SAW wafers.

Chlorobenzene modifies the surface of the photoresist by developing a characteristic 'lip' in the developed pattern. This creates a discontinuity at the edges of the patterned photoresist when a metal is evaporated on the surface of the wafer; thus, unwanted metal is subsequently removed more easily. The wafer is then baked in an oven at 75 °C for 30 minutes to remove any excess solvents from the photoresist and allowed to cool to room temperature for approximately 15 minutes. The wafer is then developed in a mixture (ratio 1:4) of AZ-450® developer (Hoechst) and deionised water for 40 seconds.

Again, great care should be taken at this stage, as under or overdeveloping the photoresist layer will degrade the fabrication success. As in the etching process, an immersion method is strongly recommended, whereby the wafer is slowly agitated during the developing process at 10 second intervals, followed by a deionised water rinse and then close inspection with a microscope. It is important to prevent damage to the photoresist-patterned structures at this stage, so extremely gentle agitation is required in the immersion step, and the use of compressed filtered nitrogen for drying the wafer should be avoided.

Close inspection of the wafer surface using an optical microscope is then performed to examine the transferred SAW-IDT pattern. The edges of the photoresist patterns should be well defined and sharp to facilitate the lift-off process. Again, if found unacceptable, the photoresist can be removed using acetone and the entire procedure can be repeated before the metallisation of the wafer.

After the metallisation of the photoresist-patterned wafer using the metal evaporation technique (Section 12.2.4), the photoresist is removed by immersing the wafers in an acetone bath at room temperature for 30 minutes. Ultrasonic agitation may be used to assist in the removal process but caution is advised as damage to small patterned structures (feature sizes ≤2 µm) may occur.

Once the IDT designs have been successfully transferred to the wafers via the lift-off process, the wafers are then ready for dicing (Section 12.2.5).

A summary of the photolithography process for both the etching and lift-off procedures is shown in Figure 12.4.

12.2.5 Wafer Dicing

The wafers are finally cut into small, individual chips using, for example, a Deckel™ wire saw, together with a diamond impregnated wire and slurry. The slurry is made from a mixture of silicon, glycerol, and deionised water (3:5:1) and has a particle size of 25 µm.

Before cutting the wafer, a thick layer of AZ-4562® positive photoresist (Hoechst) is spun at 2000 rpm for 30 seconds, following the deposition of a thin HMDS layer spun on to improve the photoresist adherence on the wafer. The wafer is then baked in an oven at 75 °C for 30 minutes and then allowed to cool to room temperature. The resulting thick layer protects the delicately patterned IDT structures during the debris cutting.

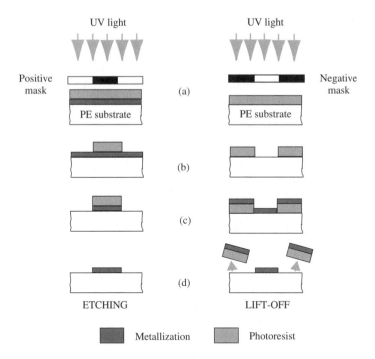

Figure 12.4   Basic steps involved in two lithographic processes used to make IDT structure: etching (left) and lift-off (right) on a piezoelectric (PE) substrate

12.3 DEPOSITION OF WAVEGUIDE LAYER

12.3.1 Introduction

Love wave sensors require the deposition of a guiding layer made from an acoustic material that has a shear wave velocity less than that of the quartz wafer. Described next are the process conditions and steps that should be followed to deposit SiO_2 as a guiding layer on top of a quartz wafer.

Steps that occur during a typical chemical deposition process include (Campbell 1996) the following:

1. The transport of precursors from the chamber inlet to the proximity of the wafer

2. Reaction of these gases to form a range of daughter molecules

3. Transport of these reactants to the surface of the wafer

4. Surface reaction to release the SiO_2

5. Desorption of the gaseous by-products

6. Transport of the by-products away from the surface of the wafer

7. Transport of the by-products away from the reactor

12.3.2 TMS PECVD Process and Conditions

One of the necessary conditions for the deposition of SiO_2 is that the temperature of deposition should be as low as possible. This is desirable because higher temperatures can adversely affect the poling characteristics of quartz (in spite of the fact that quartz is a naturally piezoelectric material) and because the melting point of the metallisation layer (aluminum is 650 °C) should not be exceeded.

We should therefore choose SiO_2 that is either sputtered or deposited by plasma-enhanced chemical vapour deposition (PECVD) from silane gas. The sputtering process provides better step-coverage than evaporation and far less radiation damage than E-beam evaporation (Campbell 1996). A simple sputtering system consists of a parallel-plate plasma reactor in a vacuum chamber and the target material (SiO_2) placed on the electrode such that it receives the maximum ion flux. An inert gas (at a pressure of 0.1 torr) is usually used to supply the chamber with high-energy ions that strike the target at high velocities and dislodge the SiO_2 molecules, which deposit conformal to the wafer (the SAW-IDT device). The only disadvantage in this process is that on account of the physical nature of the process, sputtering could also bombard and damage the delicate IDT fingers on the surface of the quartz. Sputtering can also introduce a variety of contaminants from the substrate holder because of the physical nature of the process. Hence, sputtering is not the ideal means of depositing SiO_2, despite the fact that the process can be carried out under conditions of low temperature.

An alternative approach is to use chemical vapour deposition (CVD). A simple CVD process is shown in Figure 12.5. The reactor consists of a tube with a rectangular cross section, and the walls of the tube are maintained at a temperature T_w. A single wafer rests on a heated susceptor in the centre of the tube.

This susceptor is maintained at a temperature T_s (where $T_s \gg T_w$). The obvious choice is to use oxidised silane gas (SiH_4) (also referred to as tetraethoxysilane TEOS) to form SiO_2 in the presence of an oxidising agent, such as O_2, and an inert carrier gas, such as H_2 (to improve the uniformity of deposition). Excessive homogeneous reactions occurring spontaneously in the gas above the wafer will result in the deposition of large Si particles in the gas phase, and their subsequent deposition on the wafer will cause poor surface morphology and inconsistent film properties.

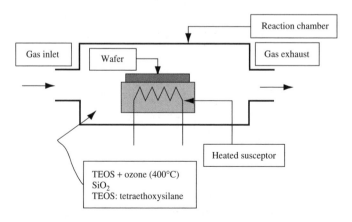

Figure 12.5   A simple CVD process flow system

Some of the other problems associated with PECVD (TEOS) are that (*1*) quality plasma-enhanced chemical vapour–deposited tetraethoxysilane (PETEOS) SiO_2 films are difficult to achieve at temperatures below 250 °C (Alaonso *et al*. 1992; Itani and Fukuyama 1997) and (*2*) TEOS has a low vapour pressure of approximately 2 mTorr (25 °C and 1 atm), which necessitates the heating of all delivery lines and chamber surfaces to prevent TEOS condensation and prevents gas metering with conventional mass-flow controllers, thus rendering the resulting process prohibitively expensive (Ballantine *et al*. 1997). Conventional mass-flow controllers, on the other hand, easily meter silane gas, but great care must be used because silane is a toxic and pyrophoric gas and constitutes an explosion hazard at high SiH_4 concentrations. These limitations add to the cost and complexity of TEOS and silane-based silicon deposition equipment. To achieve a low temperature, good quality oxide, and for the circumvention of the safety issues associated with silane-based oxides and the manufacturing complexities inherent with TEOS, an alternative precursor needs to be employed.

Potential organo-silicon precursors are compiled and their critical physical and chemical properties are tabulated for comparison with the properties of silane and TEOS. Of all the precursors listed in Table 12.1, tetramethylsilane (TMS) can be chosen as the best precursor for the current low-temperature application for several reasons.

TMS is known to be nontoxic and nonpyrophoric, and its high vapour pressure (580 mTorr) allows for the use of conventional mass-flow controllers at room temperature.

Table 12.1 Tabulation of relevant parameters for feasible PECVD precursors (Gangadharan 1999)

Precursor	Silane	TEOS	TMS	MS	TMCTS	LTO-410, DES
Chemical Name	Silane	Tetraethoxy silane	Tetramethyl silane	Methyl silane	1,3,5,7 Tetra methylcyclo tetrasiloxane	Diethyl- silane
Formula	SiH_4	$Si(C_2H_5O)_4$	$Si(CH_3)_4$	CH_3SiH_3	$C_4H_{16}O_4Si_4$	$SiH_2(C_2H_5)_2$
MW	32	208.3	88.2	46	240.5	88.2
State @ 20 °C	Gas	Liquid	Liquid	Gas	Liquid	Liquid
Best Assay (%)	**	>99.99	99.90	**	99.90	>99.70
VP@20 °C (mTorr)	Gas	1.5	589	**	6	207
Use stand. MFC	Yes	No	Yes	Yes	Not sure	Yes
Stability	Unstable	Stable	Stable	**	**	Stable
Flammable	Yes	Yes	Yes	Yes	Yes	Yes
Pyrophoric	Yes	No	**	**	Not sure	**
Toxicity (ppm)	Toxic (0.5)	Nontoxic (100)	**	**	**	**

** Values not known.

Also, each parent TMS molecule (Si(CH$_3$)$_4$) contains half as much carbon and three-fifths as much hydrogen as a TEOS molecule (Si(OC$_2$H$_5$)$_4$), and it is hypothesised that carbon and hydrogen-free films will be obtainable at lower temperatures from this precursor. Additionally, the lower molecular weight of TMS might allow for higher surface mobility than TEOS at any given temperature, thereby resulting in better-quality films at temperatures lower than those obtainable by PETEOS ($\sim$250 °C). Finally, it is thought that PECVD TMS oxide (PETMS-O$_x$) deposition conditions could mimic very closely those conditions found to produce high-quality PETEOS and silane oxides in the semiconductor industry (Campbell 1996; Ghandi 1994). Such deposition is carried out using a cluster tool that is specifically fabricated for this process, and the four-chamber showerhead Vactronics PDS-5000 S cluster tool PECVD reactor (Figure 12.6) is used.

The deposition procedures and conditions involve units 3 and 4 as follows: initially, in the deposition chambers, TMS, O$_2$, and He gas lines are evacuated of residual gas and then a sample is placed in a load-lock chamber (unit 3), which is evacuated from atmosphere to a low pressure (typically 10^{-5}–10^{-6} torr). This preinsertion vacuum time is held at 30 minutes. The SAW-IDT wafer is then placed on the preheated sample stage (unit 4) in the deposition chamber, which is maintained at 10^{-6} to 10^{-7} torr, by the robotic loading mechanism. A period of 1 hour is allotted for the sample to come to temperature, after which O$_2$ and He gases are input via the gas-dispersion showerhead and a period of 5 minutes is allotted for the flows to stabilise. A plasma is struck with the same pressure, RF power, and gas flow rates. This 10-minute preclean plasma purge serves three purposes:

1. It removes any residual carbonaceous matter left on the SAW device

2. It helps to form a stable interface oxide

3. It provides a high flow, stable plasma into which a miniscule flow of TMS can be injected

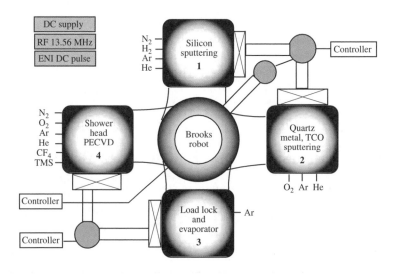

Figure 12.6 Schematic representation of a PECVD unit

Table 12.2   Main steps involved in the etching process

Step	Description
(a)	Exposure of photoresist metallised wafer with positive IDT mask plate.
(b)	Develop photoresist patterned structures.
(c)	Removal of unwanted metallisation layer via chemical wet-etching.
(d)	Removal of photoresist layer.

Table 12.3   Main steps involved in the lift-off process

Step	Description
(a)	Exposure of photoresist bare wafer with negative IDT mask plate.
(b)	Develop photoresist and formation of the characteristic lip.
(c)	Deposition of metal layer onto the wafer.
(d)	Removal of unwanted metallisation via acetone rinse.

Table 12.4   Summary of the main advantages and disadvantages of the etching and lift-off procedures

Etching procedure	
Advantages:	Disadvantages:
Simple and reproducible	Process parameters must be characterised
Good resolution achievable	Compatibility of chemical etchants with substrates
Fast realisation	Loss of feature resolution due to overetching, tendency to undercut
Ideal for small batch processing	Susceptible to electrical shorts
Lift-off procedure	
Advantages:	Disadvantages:
Capable of higher resolution	Extreme care in handling before metallisation
Occurrence of electrical shorts minimised	Intimate-contact photolithography required to achieve vertical sidewalls on patterned photoresist structures
Photolithography process is independent of pattern resolution	
Etchants not required	Poor 'lift-off' possible because of incorrect formation of characteristic 'lip'
Ability to reprocess patterned photoresist structures before metal layer deposition	

At the end of this preclean, the plasma remains and TMS vapour is introduced. It is metered using a conventional 10 cubic centimeter per second (cc/s) mass-flow controller (MFC) to the desired volumetric flow rate. The oxide deposition begins at this point and is continued for a predetermined time to achieve an oxide film of the desired thickness. Following the deposition, the TMS gas is turned off, but the O_2–He plasma is kept on for postdeposition cleaning. Once the 15 minute postdeposition cleaning is completed, the plasma is extinguished, the gases are turned off, and the chamber is evacuated for 3 minutes. Finally, a robotic transporter can shuttle the SAW-IDT wafers to the load-lock chamber (unit 3), where they are subsequently removed.

12.4 CONCLUDING REMARKS

This chapter has described in detail two process runs that can be followed to fabricate a Rayleigh wave IDT microsensor or a Love wave IDT microsensor (Gangadharan 1999).

The main steps for the etching process are given in Table 12.2 and for the lift-off process in Table 12.3. These tables provide the reader with a list of the key steps of the two processes described in Section 12.2.4 earlier.

A summary of the main advantages and disadvantages of the etching and lift-off processes is given in Table 12.4. They are relevant to the fabrication of SAW-IDT microsensors and are taken from a number of sources (Campbell 1996, 1998; Atashbar 1999).

The next chapter describes the use of SAW-IDT devices in a number of different sensing applications.

REFERENCES

Alaonso, J. C., Ortiz, A. and Falcony, C. (1992). "Low temperature SiO_2 films deposited by plasma enhanced techniques," *Vacuum*, **43**, 843–847.

Atashbar, M. Z. (1999). Development and fabrication of surface acoustic wave (SAW) oxygen sensors based on nanosized TiO_2 thin film, PhD Thesis, RMIT, Australia.

Ballantine, D. S. *et al.* (1997). *Acoustic Wave Sensors: Theory, Design and Physico-Chemical Applications*, Academic Press, London.

Campbell, A.S. (1996). *The Science and Engineering of Microelectronic Fabrication*, Oxford University Press, Oxford, England.

Campbell, C. (1998). *Surface Acoustic Wave Devices and their Signal Processing Applications*, Academic Press, New York.

Gangadharan, S. (1999). Design, development and fabrication of a conformal Love wave ice sensor, MS thesis (advisor V. J. Varadan), Pennsylvania State University, USA.

Ghandi, S. K. (1994). *VLSI Fabrication Principles: Silicon and Gallium Arsenide*, John Wiley and Sons, New York.

Hatzakis, M., Canavello, B. J. and Shaw, J. M. (1980). "Single-step optical lift-off process," *IBM J. Res. Develop.*, **24**, 452–460.

Itani, T. and Fukuyama, F. (1997). "Low temperature synthesis of plasma TEOS SiO_2," *Mat. Res. Soc. Symp.*, **446**, p. 255.

Vellekoop, M. J. (1994). A smart Lamb-wave sensor system for the determination of fluid properties, PhD Thesis, Delft University, The Netherlands.

13

IDT Microsensors

13.1 INTRODUCTION

Surface acoustic wave (SAW) devices possess several properties such as high reliability, crystal stability, good reproducibility, and relatively small size that make them suitable for many sensing applications. They can be used to sense many different properties, for example, strain, stress, force, pressure, temperature, gas concentration, electric voltage, and so forth. Readers are referred to a recent article by Hommady *et al.* (1997) for a review of their applications.

One attractive feature of some types of SAW sensor is that they can be read remotely. The operating frequency of a SAW device typically ranges from 10 MHz to a few GHz, which corresponds to the operating frequency range of radio and radar communication systems, respectively. Thus, when an interdigital transducer (IDT) sensor is directly connected to an antenna, the electromagnetic waves received by wireless transmission can excite SAW in the piezoelectric material. The fundamentals of both SAW devices and acoustic waves in solids were considered in Chapters 9 and 10, and it was evident that passive, wireless (or remotely operable) SAW devices can be made. The latter is an attractive proposition when low-power sensors are needed and are even more attractive for use in remote, inaccessible locations, for example, when buried in concrete or in the ground. Wireless SAW-based microsensors are described in detail in Section 13.3.

The sensing mechanism of SAW- IDT microsensors is based on a change in the properties of the SAW (e.g. amplitude, phase, frequency, or velocity) when the measurand changes. Basic descriptions of the acoustic parameters that can be used in a generalised measurement system have been given in Chapter 11.

In this chapter, we present a number of different applications of SAW microsensors together with the equations that govern their behaviour. For example, in chemical sensors, the SAW couple into a thin chemically sensitive coating and its properties perturb the nature of the waves. Several different properties of the film coatings can affect the acoustic waves, namely, mass, density, conductivity, electrical permittivity, strain, and viscoelasticity. In general, the change in acoustic velocity v_a can be related by the total differential theorem to the change in any property or properties. The following equation applies for changes in mass, electrical, mechanical, and environmental parameters (Hommady *et al.* 1997).

$$\frac{\Delta v_a}{v_a} \approx \frac{1}{v_a} \left[\frac{\partial v_a}{\partial_{mass}} \Delta_{mass} + \frac{\partial v_a}{\partial_{elec}} \Delta_{elec} + \frac{\partial v_a}{\partial_{mech}} \Delta_{mech} + \frac{\partial v_a}{\partial_{env}} \Delta_{env} \right] \qquad (13.1)$$

Because the change in acoustic velocity of a SAW microsensor is a combination of these different parameters, care must be taken in the choice of IDT design and signal processing techniques so that only changes in the desired parameter, such as mass, are measured and not the cross-interfering signals from, for example, mechanical strain or environmental temperature. The coupled-mode theory of SAW devices helps us to understand the nature of these types of microsensors.

13.2 SAW DEVICE MODELING VIA COUPLED-MODE THEORY

The use of coupled-mode theory on SAW devices for different geometric designs and choice of piezoelectric material is clearly described by Pierce (1954) and Campbell (1998). The benefit of this approach is that a SAW device can be represented by a set of transfer matrices corresponding to its basic elements.

There are generally three elements of a SAW device: IDT, spacing, and reflector. These can be described by the transfer matrices of **T**, **D**, and **G**, respectively. **T** matrix is a 3×3 matrix, whereas **D** and **G** are 2×2 matrices. The **T** matrix describes the IDT input and output of SAWs as well as the electromechanical conversion between the electrical signal and the SAW. Thus, the **T** matrix has three ports of which two are acoustical ports and one is an electrical port. The transfer matrix **D** describes a SAW propagation path between two representative sections, while matrix **G** represents a reflector array. Detailed mathematical forms of these transfer matrices are given in Appendix J.

Depending on the precise configuration of a SAW device, any number of **T**, **D**, and **G** matrices can be used, but their basic forms remain the same. For example, a SAW microsensor comprising an IDT and a reflector (Figure 13.1) can be modeled simply by using three transfer matrices $\mathbf{T}_1$, $\mathbf{D}_2$, and $\mathbf{G}_3$, as illustrated in Figure 13.2.

Figure 13.1   Basic elements of a SAW-IDT microsensor: IDT (left), spacing and reflector (right)

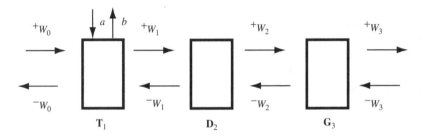

Figure 13.2   Schematic representation of a SAW device using transfer matrix elements

Figure 13.1 shows the actual device layout that has a metallic IDT, metallic reflector, and spacing in between on top of a piezoelectric substrate[1]. Thus, they can be represented, as shown in Figure 13.2, with transfer matrices **T**, **D**, and **G** for each element of the SAW device (the numbers 1, 2, and 3 are shown for bookkeeping purposes when dual devices or even array devices are modeled).

The electrical signals passing in and out of the IDT are represented by the scalars a and b. The SAWs coming in and out of each representative element are described by the symbols ^{+}W and ^{-}W – one for each propagation direction. Thus, any $(n-1)$th SAW amplitude coming in and out of the nth section (**T**, **D**, or **G**) has the following relation, where the components of the transfer matrices are represented by italic typeface.

$$\begin{bmatrix} ^{+}W_{n-1} \\ ^{-}W_{n-1} \end{bmatrix} = [T, D, G]_n \cdot \begin{bmatrix} ^{+}W_{n} \\ ^{-}W_{n} \end{bmatrix} \tag{13.2}$$

This matrix representation of a lumped system model of a SAW device allows other SAW structures to be modeled as well. As long as the SAW device is a combination of IDT, reflectors, and spacings, corresponding transfer matrices can be used in the same order as the actual device layout. Figures 13.3 and 13.4 show the structures and models of an IDT–IDT pair and a two-port SAW resonator. More complex structures of SAW devices can also be modeled by just adding more transfer matrices at appropriate locations.

The acoustic part and the electrical part of the signals W_i can be conveniently separated for the IDT equations and hence solved to determine the SAW amplitudes. Then, the overall acoustic part can be represented by the simple product of each acoustic transfer matrix in turn. For example, the overall acoustic matrix for the resonator shown

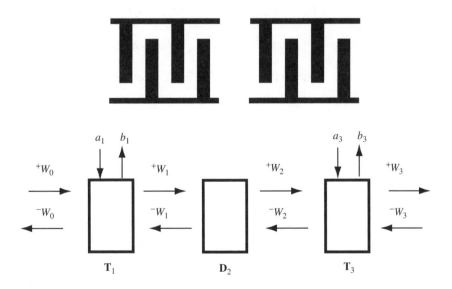

Figure 13.3 Schematic representation of an IDT–IDT pair and its transfer matrix model

[1] Fabrication details of IDT microsensors are given in Chapter 12.

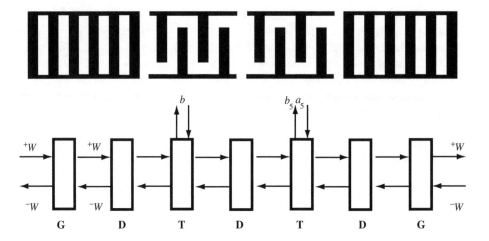

Figure 13.4 Schematic representation of a two-port SAW resonator and its transfer matrix model

in Figure 13.4 may be described by

$$\begin{bmatrix} +W_0 \\ -W_0 \end{bmatrix} = [G_1] \cdot [D_2] \cdot [T_3] \cdot [D_4] \cdot [T_5] \cdot [D_6] \cdot [G_7] \cdot \begin{bmatrix} +W_7 \\ -W_7 \end{bmatrix} \qquad (13.3)$$

where $[T_3]$ and $[T_5]$ are 2×2 acoustic submatrices of a 3×3 **T** matrix, and an overall acoustic matrix $[M]$ (or **M**) can be defined as

$$[M] = [G_1] \cdot [D_2] \cdot [T_3] \cdot [D_4] \cdot [T_5] \cdot [D_6] \cdot [G_7] \qquad (13.4)$$

Likewise, the acoustic part of other SAW devices can also be modeled in a straightforward manner.

The SAW amplitudes associated with an IDT have an electrical part as an input or output power. For example, from Figure 13.4,

$$\begin{bmatrix} +W_2 \\ -W_2 \end{bmatrix} = [T_3] \cdot \begin{bmatrix} +W_3 \\ -W_3 \end{bmatrix} + a_3 \cdot [\tau_3] \qquad (13.5)$$

where a_3 is the scalar input power to IDT 3 and $[\tau_3]$ is a 2×1 submatrix of the 3×3 **T** matrix.

Knowing that

$$\begin{bmatrix} +W_0 \\ -W_0 \end{bmatrix} = [G_1] \cdot [D_2] \cdot \begin{bmatrix} +W_2 \\ -W_2 \end{bmatrix} \qquad (13.6)$$

and

$$\begin{bmatrix} +W_3 \\ -W_3 \end{bmatrix} = [D_4] \cdot [T_5] \cdot [D_6] \cdot [G_7] \cdot \begin{bmatrix} +W_7 \\ -W_7 \end{bmatrix} \qquad (13.7)$$

substituting Equations (13.6) and (13.7) into Equation (13.5) gives an overall transfer matrix of the SAW device in terms of W_0's and W_7's for a given input a_3, as shown in the following equation:

$$\begin{bmatrix} +W_0 \\ -W_0 \end{bmatrix} = [M] \cdot \begin{bmatrix} +W_7 \\ -W_7 \end{bmatrix} + a_3 \cdot [G_1] \cdot [D_2] \cdot [\tau_3] \tag{13.8}$$

By applying the appropriate boundary conditions, Equation (13.8) becomes soluble with two subequations and two unknown parameters. Usually, the boundary conditions are $+W_0 = 0$ and $-W_7 = 0$ because there are no external sources to SAWs, that is, from outside the device. Any reflections of the SAWs from the substrate edges, or other structures outside the SAW device, are suppressed by using an acoustic absorber and/or serrated (or slanted) edges.

The basic form of the transfer matrices remains the same for other devices, whereas some of the parameters inside the transfer matrix are changed according to the choice of material and geometric constants. For example, a SAW gyroscope is a combination of a SAW resonator (Figure 13.4) and a SAW sensor (Figure 13.1) placed orthogonal to each other, as shown in Figure 13.5.

By providing a known power to an IDT of the resonator, the response of the resonator part can be solved in just the same way as before. The only difference in solving the sensor part is the boundary condition on each IDT because secondary waves are generated upon device rotation and they become an input SAW to the passive IDT that acts as a *Coriolis* sensing element. The secondary SAWs are $+W_2$ and $-W_1$ and $+W_0$ and $-W_3$ and are again zero, provided there are no external SAW sources. Outputs b_1 and b_3 are the resultant electrical signals because of the secondary SAW (Figure 13.6). Again, different SAW devices can be modeled in similar ways and solved by applying the appropriate boundary conditions.

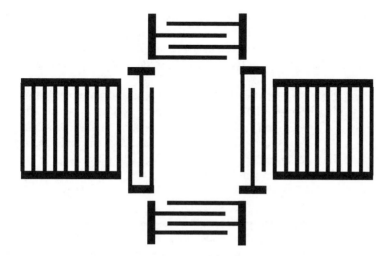

Figure 13.5 Basic layout of a SAW-IDT gyroscope: a pair of IDTs and a SAW resonator

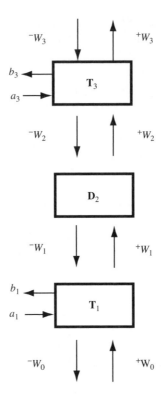

Figure 13.6 Model of a SAW sensor with secondary SAW as boundary conditions

13.3 WIRELESS SAW-BASED MICROSENSORS

In order to obtain a high sensitivity, SAW microsensors are usually constructed as electric oscillators[2] using the SAW device as the frequency control component. By accurately measuring the oscillation frequency, a small change in the physical variables can be detected by the sensors. A typical SAW oscillator sensor schematic is shown in Figure 13.7.

Briefly, an amplifier connects two IDTs on a piezoelectric wafer so that oscillations of the SAW propagating from one IDT to the other are set up by feedback. The oscillation frequency satisfies the condition that the total phase shift of the loop equals $2n\pi$ and varies with the SAW velocity or the distance (spacing) between the IDTs. The oscillator includes an amplifier and so requires an external electrical power supply and, therefore, *cannot* be operated in a passive wireless mode.

As stated earlier, the operating frequency of SAW devices ranges from 10 MHz to a few GHz. When an IDT is directly connected to an antenna, the SAW can be excited remotely by electromagnetic waves. Thus, it is possible to construct *passive*, wireless, remotely operable SAW devices. The applications of remote sensors was first reported by Bao *et al.* (1987). The temperature of a passive SAW device with a small antenna can be remotely read out by a microwave communications system (Suh *et al.* 2000).

---

[2] Acoustic wave oscillators are described in Section 11.7.

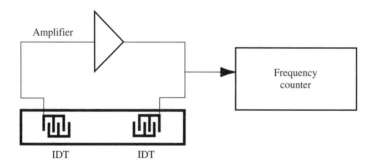

Figure 13.7   Schematic diagram of an oscillator SAW sensor with a SAW resonator

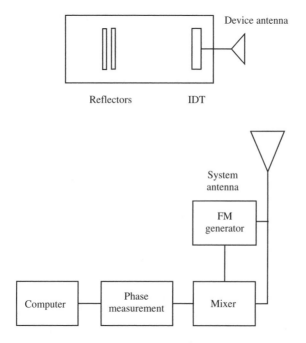

Figure 13.8   Schematic diagram of a remote reading sensor system with passive SAW sensor

The schematic diagram of IDT and reflectors in Figure 13.8 shows the basic operating principle of an IDT with a wireless communication interface. One IDT and two reflectors are fabricated on the surface of a piezoelectric crystal wafer. These micro IDT and SAW sensors can be fabricated using the microlithographic process described in the previous chapter.

The IDT connects directly to a small antenna called the *device antenna*. This antenna-IDT configuration is able to convert the microwave signal from air to SAW signal on the wafer surface and vice versa. The reading system has a linear frequency-modulated (FM) signal generator with a system antenna that transmits these FM signals. The signals are then received by the device antenna and converted by the antenna-IDT to SAWs that propagate along the surface of the piezoelectric wafer.

The echoes from the two reflectors are picked up by the antenna-IDT and sent back to the system antenna. The echo signals are delayed copies of the transmitted FM signal. The delay times mainly depend on the velocity of the SAW and the distance between the IDT and the reflectors. A mixer, which takes the transmitted FM as a reference signal, outputs the signals of frequency difference between the reflected and the transmitted signals. Because the transmitted signal is linearly FM, the frequency difference is proportional to the time delay. By using a spectrum analysis technique, such as a fast fourier transform (FFT), the two echo signals can be separated in the frequency domain because the delay times are different.

Figure 13.9 shows the layout of a transceiver telemetry system developed by a small US company (HVS Technologies).

This system operates in the range of 905 to 925 MHz. The circuit operates as follows: The input signal is pulsed FM. A pulser synchronises the direct current (DC) voltage ramp circuit, voltage controlled oscillator (VCO) output, and the A/D converter during pulses of typically 16 ms duration. During the pulse, the DC voltage ramp circuit linearly tunes the VCO from 905 to 925 MHz. The VCO output is controlled by a diode switch and then amplified to 50 mW by a high isolation amplifier. A coupler diverts a sample of the signal to the LO input of the mixer. A circulator sends the transmitted signal to the antenna and also the reflected signal, through an automatic gain control amplifier, to the radio frequency (RF) input of the mixer. Then, a low-pass filter removes any high-frequency noise and signals and then the signal is digitised at 10 M samples per second at 10 bit resolution. Finally, a programmable digital signal processing (DSP) chip, such as the TI TMS320C3X, is used to extract the delay information and compute the desired parameter. This value is then shown on a liquid crystal display (LCD).

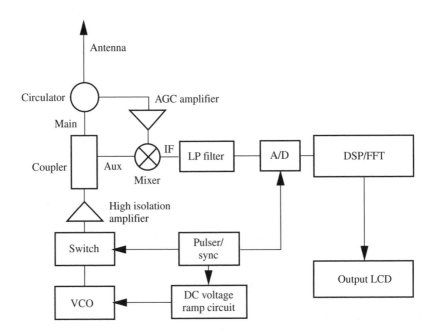

Figure 13.9   System for remote sensing application

13.4 APPLICATIONS

In this section, we present in detail some examples of the applications of SAW-IDT devices as temperature, strain, pressure, torque, rotation rate (gyroscope), humidity, and so forth sensors. In the next chapter, the applications are extended to include micro-electromechanical system (MEMS) IDT structures along with IDTs for remote sensing of acceleration.

13.4.1 Strain Sensor

In this section, a remote MEMS-IDT strain sensor system is employed to study the deflection and strain of a 'flex-beam' type structure of a helicopter rotor (Varadan *et al.* (1997)). The system is based on the fact that the phase delay is changed because of the strain in the sensor substrate. The system consists of a remote passive SAW sensor read by a fixed microwave system station.

The FM signal sent by the system antenna is expressed as

$$S(t) = A \cos(\omega_0 + \mu t/2)t \tag{13.9}$$

where ω_0 is the initial frequency of the FM signal, μ is 2π times the rate of modulation, and t is time.

The echoes from the two reflectors, $S_1(t)$ and $S_2(t)$, are the same as the transmitted signal $S(t)$ but with different amplitudes and time delays t_1 and t_2, respectively. These may be written as

$$S_1(t) = A_1 \cos(\omega_0 + \mu t/2)(t - t_1) \tag{13.10}$$

and

$$S_2(t) = A_2 \cos(\omega_0 + \mu t/2)(t - t_2) \tag{13.11}$$

with

$$t_1 = 2d_1/v + \tau_e \tag{13.12}$$

$$t_2 = 2d_2/v + \tau_e \tag{13.13}$$

where v is the SAW velocity, d_1 and d_2 are the distances from the IDT transducer to the two reflectors, and τ_e is the total of other delays (such as the delay in the electronic circuit and devices and the traveling time of the electromagnetic wave[3]) that is the same for both echoes.

Through the mixer that uses the transmitted signal as a reference and low-pass filter, frequency differential signals are obtained as

$$E_1(t) = B_1 \cos[\mu t_1 t + (\omega_0 t_1 - \mu t_1^2)] = B_1 \cos[\omega_1 t + \varphi_1] \tag{13.14}$$

and

$$E_2(t) = B_2 \cos[\mu t_2 t + (\omega_0 t_2 - \mu t_2^2)] = B_2 \cos[\omega_2 t + \varphi_2] \tag{13.15}$$

---

[3] For short distances, this time is negligible.

with different amplitudes B and angular frequencies. The frequencies and phases of these two signals depend on the delay times. The two signals can be separated in the frequency domain. Because ω_0 is usually much greater than μ, the phase shift is more sensitive to the variation of the delay time than the frequency.

The difference of the two phases can be written as

$$\varphi = (\varphi_1 - \varphi_2) = [\omega_0 - \mu(t_1 + t_2)/2][t_2 - t_1] \tag{13.16}$$

where the extra delay time of the second echo compared with the first is equal to the round trip time of the acoustic wave traveling from the first reflector to the second and is given by

$$\tau = (t_2 - t_1) = \frac{2d}{v} \tag{13.17}$$

where d is the distance between the two reflectors.

The phase difference is sensitive to the change in the delay time. The variation of the phase difference because of the change in the delay time is expressed as

$$\Delta(\varphi_1 - \varphi_2) = \Delta\varphi = [\omega_0 - \mu(t_1 + t_2)/2]\Delta\tau \quad \text{and} \quad \Delta\tau = 2\Delta d/v \tag{13.18}$$

Because ω_0 is usually much larger than the other term, the equation reduces to

$$\Delta\varphi \approx \omega_0 \Delta\tau \tag{13.19}$$

The wave traveling time τ is proportional to the distance between the two reflectors and is inversely proportional to the velocity. If the possible velocity variation of the SAW under stress is neglected and we take into account only the direct effect on the distance, we have

$$\Delta\varphi \approx \omega_0 \frac{2\varepsilon d}{v} = \omega_0 \varepsilon \tau_0 \tag{13.20}$$

where ε is the mechanical strain $(\Delta d/d)$ and τ_0 is the traveling time when the strain is zero.

Thus, the sensitivity of this wireless sensor system depends on the operating frequency and the round traveling time of the SAW between the two reflectors.

13.4.1.1 Sensor dimensions and reading system

A strain sensor was developed by Varadan et al. (1997), wherein YZ-cut lithium niobate (LiNbO$_3$) crystal of dimensions 3.7 mm × 6.7 mm × 0.5 mm was used. An IDT was designed with a central frequency of 912.5 MHz. The IDT is connected directly to a small antenna of dimensions 25 mm × 75 mm × 0.3 mm. The distance between the two reflectors was 0.15 mm and the round trip traveling time of the surface waves was 0.1 μs at zero stress. The wireless reading system had a patch antenna to send out FM microwave signals and to receive the echoes coming from the reflectors through the antenna-IDT. The transmitted signal was linearly modulated from 905 to 920 MHz in a pulse lasting 1/60th of a second. The resolution of the phase detector of the system is 1°. According to Equation (13.20), the strain sensitivity of the system is estimated to be an impressive 2×10^{-5} per phase degree.

13.4.1.2 Calibration

The calibration was done on an uniform cantilever beam. The sensor was glued to a fibreglass cantilever beam of length 172 mm, width 25 mm and thickness 1.5 mm. The centre of the piezoelectric wafer is 53.3 mm from the root of the beam and the SAW propagating direction is along the beam length. In order to reduce the effect of the substrate in the deformation of the structure, a relatively soft plastic epoxy was used to mount the sensor substrate. The softness of the bonding layer and the wall of the package of the sensor make the strain on the crystal wafer surface less than the strain on the beam surface. Therefore, a calibration is required to determine the actual sensitivity of the sensor to the strain on the beam surface.

If we neglect the effect of the bonded piezoelectric wafer, the strain on the surface of the beam can be calculated by simple beam theory and written as (Gere and Timoshenko 1990),

$$\varepsilon_x = -d_s \frac{3t_b(l - x)}{2l^3} \tag{13.21}$$

where l is the length of the beam, x is the distance of the location of the piezoelectric wafer from the root, t_b is the thickness of the beam, and d_s is the displacement at the tip of the beam. In this experiment, a force at the tip bends the beam and a dial indicator measures the tip displacement.

The wireless IDT microsensor system successfully detects the shift of the phase difference ($\Delta\varphi$) between the two echoes with respect to the variation of the tip displacement. The strains ε_x are calculated, using Equation (13.21), from the tip displacements, and the results are plotted in Figure 13.10.

The data show that the phase shift varies almost linearly with the strain when the strain is less than 0.0012. For larger strains, nonlinearity and larger fluctuations appear. The nonlinearity may be due to the glue used to bond the sensor substrate, the large deformation of the beam, or some other cause.

13.4.1.3 Dynamic strain measurement test

In a dynamic strain measurement test, the beam was extended to 750 mm in length. A mass of 200 g was mounted at the tip to reduce the vibration frequency. The beam was tilted off the vertical axis because of the inclination of the beam support itself, as shown in Figure 13.11. A static strain was developed on the surface of the beam from gravitational forces. The end of the original beam was shifted a distance w of 6 mm from the neutral line. The theoretical strain at this sensor position was calculated from Equation (13.21) to be about -224 microstrains. The sensor detected an actual strain of -276 microstrains. The difference between the theoretical and experimental values is less than the resolution of the sensor system, that is, about 51 microstrains.

Then, a force applied at the tip further deformed the beam, resulting in an additional deflection of the beam by 15 mm. A free vibration was established by releasing the tip.

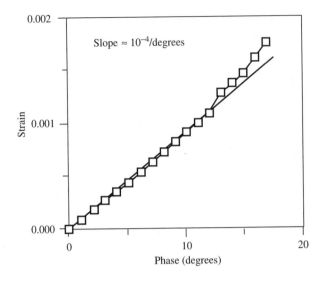

Figure 13.10   Experimental results of a strain guage SAW-IDT sensor

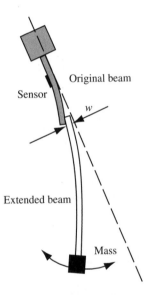

Figure 13.11   Test setup for dynamic strain sensing

The dynamic variation of the strain in the vibrating beam was measured by the use of a wireless passive detector. The results are presented in Figure 13.12. The surface strain of the beam was found to vary periodically corresponding to the vibration time period of 1.4 second and decayed because of damping effects. The initial maximum strain was found to be -1100 microstrains, which agrees with the calculated value of strain. The sensor wirelessly measured the strain caused by both the static and dynamic loads.

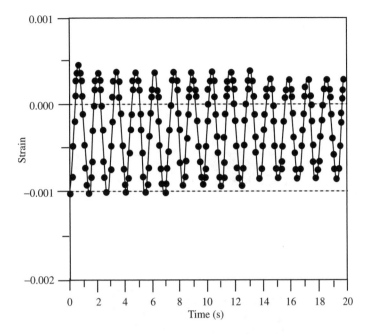

Figure 13.12   Measured strain on the surface of a vibrating beam

13.4.2 Temperature Sensor

An IDT and two reflectors are patterned onto the surface of a YZ-cut lithium niobate wafer, as shown in Figure 13.13. This transducer connects directly to a small antenna. In the remote reading system, an FM generator sends a linearly FM signal to an antenna and to a mixer. The signal transmitted by the system antenna is received by the small antenna connected to the LiNbO$_3$ wafer and converted into a SAW by the transducer. The echoes from the reflectors are received by the IDT and are transmitted back to the system antenna and mixed with the original FM signal in the mixer. The echoes are delayed copies of the original FM signal. The time delays depend on the SAW velocity, which is a sensitive function of the ambient temperature. The difference in frequency signals, usually called the *intermediate frequencies* (IFs), appear at the output of the mixer. The frequencies and the phase shifts of the IFs vary with the time delays. Because the changes in time delay with temperature are very small, the phase shifts, which are more sensitive than the frequency, are preferred. In order to avoid the effects of time delay variations other than temperature changes (e.g. the changes of distance between the two antennae), the temperature is determined by the difference in the phase shifts of the two IFs corresponding to the two reflectors (Bao *et al.* 1987, 1994).

As stated earlier, the initial FM signal is expressed as

$$S(t) = A\cos[\theta(t)] = A\cos[(\omega_0 + \mu t/2)t + \theta_0] \tag{13.22}$$

The echo from the first reflector input (S_1) to the mixer is the same as the original FM signal, but with a time delay t_1 and different amplitude, so that it is written as

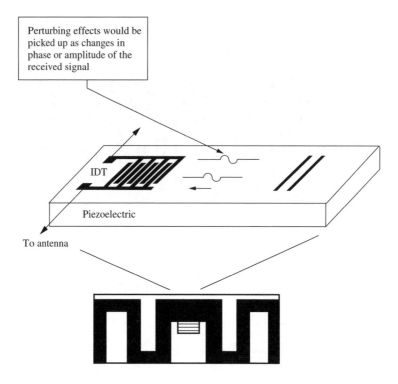

Figure 13.13   Schematic diagram of the antenna integrated with IDTs

$$S_1(t) = K_1 S(t - t_1) = A_1 \cos\{[\omega_0 + \mu(t - t_1)/2](t - t_1) + \theta_0\} = A_1 \cos[\theta_1(t)] \quad (13.23)$$

where

$$t_1 = \tau_e + \tau_1 \qquad (13.24)$$

and τ_1 is the time delay corresponding to the surface wave traveling from the transducer to the first reflector and back. This delay can be written as

$$\tau_1 = 2d_1/v \qquad (13.25)$$

where d_1 is the distance between the IDT and the first reflector, v is the SAW velocity, and τ_e is the time delay due to the electronic circuit and signal propagation.

The IF corresponding to $S_1(t)$ is expressed as

$$I_1(t) = B_1 \cos[\theta(t) - \theta_1(t)] = B_1 \cos[+\mu t_1 t + \omega_0 t_1 - \mu_1 t^2/2] \qquad (13.26)$$

Both the frequency μt_1 and the phase shift $\varphi_1 = \omega_0 t_1 - \mu_1 t^2/2$ depend on the time delay t_1. Because ω_0 is usually much greater than μt_1, the phase shift, as stated earlier, is more sensitive than the variation of the frequency.

From Equation (13.24), we know that the total delay t_1 depends not only on the traveling time of the surface wave that is a function of the temperature but also on the microwave propagation path. The latter varies with the distance between the transmitted excitation and the SAW device. To eliminate the error from the variation of τ_e, a second reflector is put on the wafer. The corresponding time delay is τ_2. Similar to the first reflector, we have the IF corresponding to the second reflector as

$$I_2(t) = B_2 \cos[\theta(t) - \theta_2(t)] = B_2 \cos[+\mu t_2 t + \omega_0 t_2 - \mu_2 t^2/2] \tag{13.27}$$

where

$$t_2 = \tau_e + \tau_2 \quad \text{and} \quad \tau_2 = 2d_2/v \tag{13.28}$$

and d_2 is the distance between the IDT and the second reflector. The difference between the two phase shifts can now be written as

$$\varphi_d = (\varphi_2 - \varphi_1) = [\omega_0 - \mu/2(t_2 + t_1)](t_2 - t_1) = K\tau \tag{13.29}$$

where

$$K = \omega_0 - \mu/2(t_2 + t_1) \approx \omega_o \tag{13.30}$$

Since $\omega_0 \gg \mu/2(t_2 + t_1)$ as can be seen from the values calculated and

$$\tau_0 = 2d/v \tag{13.31}$$

where τ_0 is the total travel time of the surface wave from the first reflector to the second and back. This time being inversely proportional to the surface wave speed is very sensitive to the temperature in the vicinity of the SAW device and we propose the following relationship between the travel time τ and the temperature T

$$\tau = \tau_0[1 + \alpha(T - T_0)] \tag{13.32}$$

where α is the temperature coefficient of the time delay of the SAW device and T_0 is the ambient temperature.

From Equation (13.29),

$$\varphi_d = K\tau_0[1 + \alpha(T - T_0)]$$
$$= \alpha K\tau_0 T + K\tau_0(1 - T_0) \tag{13.33}$$
$$= aT + b$$

$$a = K\alpha\tau_0 \quad \text{and} \quad b = K\tau_0(1 - T_0) \tag{13.34}$$

If the resolution of phase shift difference in degrees is $\Delta\varphi$, then the resolution of the temperature reading will be

$$\Delta T = \Delta\varphi/a \tag{13.35}$$

The wafer is made of YZ-cut lithium niobate with

$$\alpha = 94 \times 10^{-6}/°C \tag{13.36}$$

The two reflectors are located such that the time delay at room temperature T_0 is

$$\tau_1 = 1 \ \mu s \quad \text{and} \quad \tau_2 = 1.1 \ \mu s \tag{13.37}$$

Then,

$$\tau_0 = 0.1 \ \mu s \tag{13.38}$$

The transmitted FM signal is pulse-modulated with a time duration of $1/60^{th}$ of a second. The carrier frequency varies linearly from 905 MHz to 925 MHz during the period. The parameters in Equation (13.22) for the FM signal are

$$\omega_0/2\pi = 905 \ \text{MHz} \tag{13.39}$$

$$\mu/2\pi = 1.2 \times 10^{-3} \ \text{MHz}/\mu s \tag{13.40}$$

In operation, the distance between the two antennas is 1 to 2 meters so that τ_e can be neglected compared with τ_1 or τ_2. The temperature variation can be in the range 0 to 200 °C. The first and second terms in Equation (13.30) can now be calculated

$$
\begin{aligned}
K_0 &= \omega_0 - \mu/2(t_1 + t_2) \\
&= 2\pi \times 905 \times 10^6 - 1.2 \times 10^{-3} \times 1.05/2 \\
&= 2\pi \times 905 \times 10^6
\end{aligned}
\tag{13.41}
$$

So, the approximation in Equation (13.30) is clearly justified. From Equations (13.34), (13.36), and (13.38), the constant a is

$$a = 3.06 \ \text{angular degrees}/°C \tag{13.42}$$

The resolution of the phase shift is $1°$ so that the resolution of the temperature reading is given by Equation (13.35) as

$$\Delta T = 0.33°C \tag{13.43}$$

The experimental calibration is done in a temperature-controlled chamber called a *Delta 9023*, (Figure 13.14).

A digital hand-held thermometer (Keruco Instruments Co.) with an accuracy of $\pm 0.2 °C$ is taken as the temperature standard. The temperature range in the experiment is from room temperature (near 20 °C) to a maximum of 140 °C. The straight line theory, as shown in Figure 13.15, fits well the experimental points of phase against temperature. The equation for the line in Figure 13.15 is obtained by the least mean square method and is given by

$$\varphi = 2.89T - 49.1 \tag{13.44}$$

The value of the slope coefficient a (2.89) is in good agreement with that (3.06) estimated from Equation (13.42). The root mean square (rms) error of the phase difference is equal to $0.78°$, which corresponds to an rms error of $0.26 °C$.

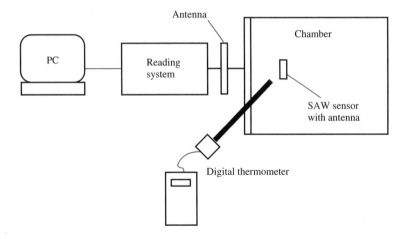

Figure 13.14 Calibration of the remote reading system

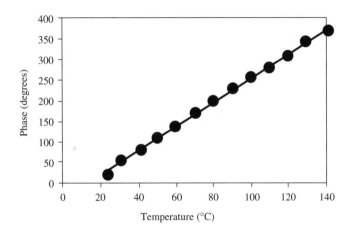

Figure 13.15 Phase shift difference versus temperature (after jump correction) of a SAW-IDT microsensor

13.4.3 Pressure Sensor

The SAW-based pressure sensors use the strain principle described in Section 13.4.1. To illustrate this principle, let us consider a circular quartz membrane with thickness h. The deflection of the plate, w, under uniformly distributed pressure (P) is assumed to be smaller than $h/5$. The differential equation describing the elastic behaviour of the middle plane of a thin plate is obtained from the elementary theory of plates

$$\nabla^4 w = \frac{P}{D} \qquad (13.45)$$

where D is the stiffness of the plate with piezoelectric properties of quartz. In the case of simply supported edge, the deflection w is given by

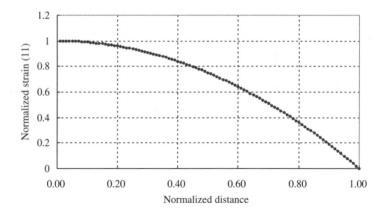

Figure 13.16 Normalised strain against the distance from the centre of the membrane

$$w = \frac{P(a^2 - r^2)}{64D}\left[\frac{5 + v}{1 + v}a^2 - r^2\right] \tag{13.46}$$

where a is the radius of the plate, r is measured in a coordinate system fixed to the centre of the plate, and v is Poisson's ratio.

The radial strain is given by

$$
\begin{aligned}
\varepsilon_r(r, z) &= -\frac{Dz}{Eh^3/12}\left(\frac{d^2w}{dr^2} + \frac{v}{r}\frac{dw}{dr}\right) \\
&= \frac{3}{8}\frac{Pa^2(h_p - h_m)(3 + v)}{E(h_p + h_m)^3}\left[1 - \frac{r^2}{a^2}\right]
\end{aligned} \tag{13.47}
$$

Figure 13.16 illustrates the calculated static average strain distribution as a function of the radial distance from the centre of the circular membrane. The strain is then calibrated with change of pressure (Vlassov *et al.* 1993).

13.4.4 Humidity Sensor

There is a need for the development of a remote, wireless, and passive sensor system for humidity measurement that is more accurate than conventional methods. Here, such a system is discussed on the basis of a SAW device.

The following sections describe a wireless sensor system that can remotely interrogate a passive SAW sensor for the measurement of relative humidity (RH) (Hollinger *et al.* 1999). The principle of operation of the wireless SAW-based sensor system was described earlier. The FM generator continuously emits pulses with duration of 16.7 ms that are linearly frequency modulated from 905 to 925 MHz.

The original FM signal is expressed as

$$S(t) = A\cos[\theta(t)] \tag{13.48}$$

where

$$\theta(t) = (\omega_0 + \mu t/2)t + \theta_0 \tag{13.49}$$

Consider a simplified SAW device with only two reflectors. The echo from the first (S_1) input to the mixer is the same as the original FM signal, but with a time delay t_1 and a different amplitude

$$S_1(t) = A_1 \cos[\theta_1(t)] \tag{13.50}$$

$$\theta_1(t) = [\omega_0 + \mu(t - t_1)/2](t - t_1) + \theta_0 \tag{13.51}$$

The IF corresponding to $S_1(t)$ is expressed as

$$I_1(t) = B_1 \cos[\mu t_1 t + \omega_0 t_1 - \mu t_1^2/2] \tag{13.52}$$

Both the frequency μt_1 and the phase shift $\varphi_1 = (\omega_n t_1 - \mu t_1^2)/2$ depend on the time delay t_1. Once again, because ω_n is usually much greater than μt_1, the phase shift is more sensitive than the variation of the frequency.

The total delay t_1 depends not only on the travel time of the surface wave that is a function of the physical parameter being sensed but also the RF propagation time, which is a function of separation between the sensor and the reader. To eliminate this error, a second reflector is used that has a different time delay t_2 and results in the following IF.

$$I_2(t) = B_2 \cos[\mu t_2 t + \omega_0 t_2 - \mu t_2^2/2] \tag{13.53}$$

The difference between the two phase shifts can be written as

$$\varphi_d = [\omega_0 - \mu/2(t_2 + t_1)](t_2 - t_1) = K\tau \tag{13.54}$$

where

$$K = \omega_0 - \mu/2(t_2 + t_1) \approx \omega_0 \tag{13.55}$$

because $\omega_0 \gg \mu/2(t_2 + t_1)$ for the values of the present system.

$$\text{Now } \tau = t_2 - t_1 = 2d/v \tag{13.56}$$

where τ is the total travel time of the surface wave from the first reflector to the second and back, d is their physical separation, and v is the SAW velocity.

For the design of the humidity sensor, the difference in delay between two reflectors (τ) is 1.40 μs. The differential phase shift can be written as

$$\varphi_d \approx \frac{2\omega_0 d}{v} \tag{13.57}$$

The change in the differential phase shift due to a change in humidity is then

$$\Delta\varphi_d = 2\omega_0 d \left(\frac{1}{v_1} - \frac{1}{v_0}\right) = \frac{2\omega_0 d}{v_0}\left(\frac{v_0 - v_1}{v_1}\right) \tag{13.58}$$

From preliminary measurements of humidity, it is seen that

$$\frac{\Delta v}{v_0} = 3.05 \times 10^{-6} \times \text{RH}(\%) \tag{13.59}$$

Although the change in velocity is small, we can replace v_1 in the denominator of the last term in Equation (13.58) by v_0. Then, using Equations (13.56), (13.58), and (13.59) we have

$$\Delta\varphi_d = \omega_0\tau_0[3.05 \times 10^{-6} \times \text{RH}(\%)] \tag{13.60}$$

Using a frequency of 905 MHz (i.e. $\omega_0 = 2\pi \times 905 \times 10^6$ rad/s), the above value (1.40 µs) for τ_0, and by converting from radians to degrees, we get

$$\Delta\varphi_d = 1.39 \times \text{RH}(\%) \text{ in degrees} \tag{13.61}$$

Therefore, because the wireless system has a phase difference measurement resolution of 1°, this system should provide a resolution of 0.72 percent RH for measurements.

The substrate material for the SAW device is made of YZ-LiNbO$_3$, which is a Y-axis cut and Z-axis-propagating lithium niobate crystal. The size of the piezoelectric substrate is approximately 4.3 mm by 8 mm and 0.5 mm thick. The IDTs and reflectors are made of aluminum and are deposited by sputtering using appropriate masks.

Because the SAW sensor response can be affected by both the temperature and humidity, a design has been developed that allows for the simultaneous measurement of temperature and humidity. As shown in Figure 13.17, four IDTs are arranged in a staggered manner along the centre of the piezoelectric substrate with reflectors on both sides of the IDTs. On one side, between the IDTs and the right set of reflectors, a moisture-sensitive coating (SiO$_2$ in this case) is deposited. The substrate between the IDTs and the left set of reflectors is left bare, as lithium niobate shows very little response to changes

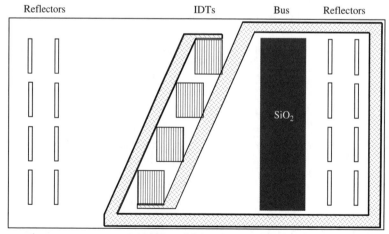

Figure 13.17 Wireless SAW sensor design for simultaneous temperature and humidity measurement at 915 MHz

in humidity. Therefore, the uncoated side is used for measuring the temperature, which is then used to compensate for temperature changes in the humidity measurement from the coated side. There are up to three phase shifters (not shown) in front of each reflector. The unique arrangement of these phase shifters gives the sensors their unique identification number, which can also be determined by the wireless system based on the reflected electromagnetic signal. The bus is connected to the two terminals of each IDT and this is inductively coupled with the sensor antenna through an air gap.

The SiO_2 coating used on the wireless humidity sensor is 40 to 50 nm thick and was deposited using plasma-enhanced chemical vapour deposition (PECVD). A glass mask was prepared to deposit the film only on the area indicated in Figure 13.17. The coating is amorphous and porous to give it a high sensitivity to water vapour.

In order to protect the IDTs and reflectors, they should be coated with a very thin (about 1 μm) layer of a passivation layer that is not affected by moisture. One candidate is amorphous silicon nitride (SiN_xH_y). SiN_xH_y has excellent passivation properties, which make it ideal for use in the semiconductor industry as an insulator and as a protective layer for silicon devices. SiN_xH_y can be deposited from the reaction of silane (SiH_4) with either ammonia (NH_3) or nitrogen (N_2). Using plasma to assist in the deposition, these films are deposited at much lower temperatures than by any other techniques (about 350 °C instead of more than 750 °C).

Figure 13.18(a) shows the experimental setup for remote humidity measurement using wireless and passive SAW sensors. The main components of the system consist of the SAW sensor, the transceiver, central interface unit (CIU), and computer. A commercial RH sensor (Omega RH82) is also placed inside the humidity chamber for calibration and verification.

SAW devices based on a $LiNbO_3$ substrate with part of the substrate coated with a thin layer of humidity-sensitive silicon dioxide are used. The delay line in the SAW sensor is now sensitive to changes in humidity as the silicon dioxide adsorbs moisture from the humid air. The transceiver emits RF pulses that are picked up by the SAW sensor; these pulses are converted to acoustic waves on the surface of the sensor, which are further reflected by the uniquely spaced reflectors. The reflected signal received back by the transceiver contains the sensor information. This signal passes through the CIU in which all the signal processing takes place and then the processed data is sent to the computer through the serial interface.

The phase difference between two reflectors along the coated part of the propagation path can be used as a measure of the humidity. The RH of the enclosed chamber is changed by pumping nitrogen into the chamber through a bubbler. The bubbler wets the nitrogen gas, which when introduced into the chamber, causes the humidity inside the chamber to increase.

The following graph, Figure 13.18(b), shows the plot between RH, as measured by the commercial humidity sensor, and the phase change in the humidity sensor, as measured by the transceiver and computer system. It can be seen that the phase change varies linearly with RH.

The straight line, shown in Figure 13.18(b), has been fitted to the experimental data using least squares. The equation of the fitted line is $\varphi = 1.4$ percent RH $+ 75.9$. Therefore, as stated earlier, because the wireless system has a phase difference measurement resolution of 1°, this system is able to provide an excellent resolution of 0.69 percent RH measurements.

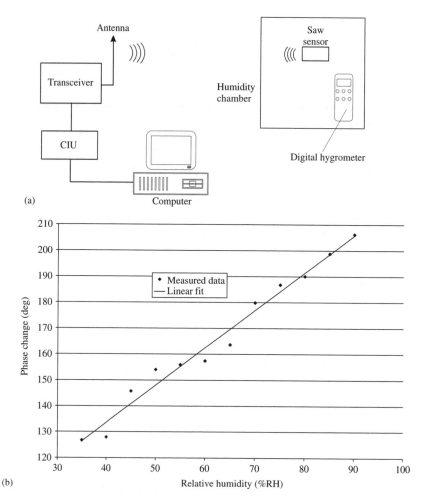

Figure 13.18   (a) Setup for humidity measurement and (b) effect of relative humidity on phase change

13.4.5 SAW-Based Gyroscope

In general, gyroscopes have a wide range of applications. They are

- Consumer electronics – picture stabilisation in three-dimensional (3D) pointers, camcorders, geostationary positioning system (GPS)

- Industrial products – robots, machine control, guided vehicles.

- Medical applications – wheel chairs, surgical tools

- Military applications – smart ammunition, guided missiles and weapon systems (Soderkvist 1994; Yazdi *et al.* 1998)

As technology matures, the market for low-cost angular sensors will extend into high volume of micromachined sensors for complete automotive pitch and roll control, which is the final step toward realisation of integrated chassis control (Eddy and Soarks 1998). This includes sensing the steering, suspension, power train and breaking systems for real-time vehicle control. All these systems require an application-specific low-cost sensor and actuator. Future air bag systems will be equipped with a low-g accelerometer and a gyroscope. Low-cost and low-g accelerometers are already in use, whereas gyroscopes that are compatible with the automotive industries' requirements, are not yet on the market.

The construction of the gyroscope is based on integration of a SAW resonator (Bell and Li 1976; Staples 1974) and a SAW sensor (White 1985; Ballantine *et al.* 1997) that operates primarily in the Rayleigh mode. The Rayleigh wave is a SAW that has its energy concentrated within one wavelength of the substrate surface (Achenbach 1973). The displacement of particles near the surface, due to the Rayleigh wave, has out-of-surface motion that traces an elliptical path (Slobodnik 1976). The Rayleigh wave can be generated at the surface of a piezoelectric material by applying a voltage to an IDT patterned on the substrate (White and Voltmer 1965). Lao (1980) derived theoretically the dependence of SAW velocity on the rotation rate of the wave-propagating medium and it is established that, for an isotropic medium, the rotation rate is a function of Poisson's ratio. Kurosowa *et al.* (1998) proposed a SAW gyroscope sensor based on an equivalent circuit simulation. He concluded that the angular velocity could not be detected because of the mismatch of resonant frequencies. Whereas Varadan *et al.* (2000a,b) presented the design, proof of concept through fabrication, and performance evaluation of a SAW gyroscope using a two-port resonator and a sensor. In this section, we present the equivalent circuit model analysis and experimental evaluation of a SAW gyroscope (Varadan *et al.* 2000b). The SAW resonator is designed and optimised using the coupled-mode theory. The gyroscope is optimised using a cross-field circuit model for numerical simulation. This gyroscope has the added capability that it can be used as a wireless gyroscope that can be easily integrated onto a SAW accelerometer (Subramanian *et al.* 1997).

13.4.5.1 Principle of SAW-based gyroscope

Any mechanical gyroscope must have a stable reference vibrating motion (V) of a mass (m) such that when subjected to a rotation, the angular rotation (Ω) perpendicular to the reference motion would cause Coriolis forces at the frequency of the reference motion. The strength of the Coriolis force F is a measure of the rotation rate and is given by

$$F = 2mV \times \Omega \qquad (13.62)$$

It is well known that in a standing Rayleigh surface wave, the particle vibration will be perpendicular to the surface. This particle vibration can be cleverly utilised for the creation of a reference vibratory motion for the gyroscope.

The concept of utilising a SAW for gyroscopic motion is illustrated in Figure 13.19. It consists of IDTs, reflectors, and a metallic dot array within the cavity, which are fabricated through microfabrication techniques on the surface of a piezoelectric substrate. The resonator IDTs create a SAW that propagates back and forth between the reflectors and forms a standing wave pattern within the cavity because of the collective reflection from the reflectors. A SAW reflection from individual metal strips adds in phase if the reflector periodicity is equal to half a wavelength. For the established standing wave pattern in the cavity, shown in Figure 13.19, a typical substrate particle at the nodes of the standing wave has no amplitude of deformation in the z-direction. However, at or near the antinodes of the standing wave pattern, such particles experience large amplitude of vibration in the z-direction, which serves as the reference vibrating motion for this gyroscope. To amplify acoustically the magnitude of the Coriolis force in phase, metallic dots (proof mass) are positioned strategically at the antinode locations. The rotation (Ω, x-direction) perpendicular to the velocity (V in $\pm z$-direction) of the oscillating masses (m) produces Coriolis force ($F = 2mV \times \Omega$, in $\pm y$-direction) in the perpendicular direction, as illustrated in Figure 13.19. This Coriolis force establishes a SAW in the y-direction with the same frequency as the reference oscillation. The metallic dot array is placed along the y-direction such that the SAW, because of the Coriolis forces, adds up coherently. The generated SAW is then sensed by the sensing IDTs placed in the y-direction.

The operating frequency of the device is determined by the separation between the reflector gratings, periodicity of reflectors, and the IDTs. The separation between reflectors was chosen as an integral number of half-wavelengths such that standing waves are created between both reflectors. The periodicity of IDT was chosen as a half-wavelength ($\lambda/2$) of the SAW. Therefore, for a given material, the SAW velocity in the material and the desired operating frequency f_0 define the periodicity of the IDT.

The substrate used for the present gyroscope is 128YX LiNbO$_3$ because of its high electromechanical coupling coefficient. For such materials, the wave velocities in the

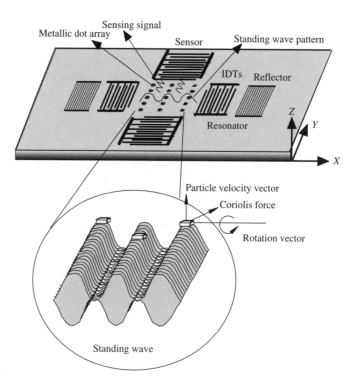

Figure 13.19 Working principle of a MEMS SAW gyroscope

Table 13.1 Comparison of measured SAW velocities in the x- and y-directions on LiNbO$_3$

Substrate	Measured velocity (m/s)	Reference velocity (m/s)
128YX-LiNbO$_3$		
x-direction	3961	3997
y-direction	3656	not available

Source: Campbell and Jones (1968).

x- and y-directions are different because of its anisotropy. The effective SAW velocities were experimentally measured in the x- and y-directions because the device utilises wave propagation in these directions and the effect of the metallic dot array on the velocities in both directions is not fully known. To measure the effective velocities of the waves in both directions, two narrow-band IDT sets with the same periodicity were placed in the x- and y-direction so that they included the dot array in the middle. Hence, the response measured using IDTs in x- and y-directions were different for the same periodicity of these IDT sets, and the velocities of both directions were measured as 3961 m/s and 3656 m/s, respectively. The difference between published values (Campbell and Jones 1968) (3980 $\sim$ 4000 m/s) and these experimental results is mainly due to the effect of metallisation and the metallic dot array. The width of IDT fingers and its spacing for the SAW gyroscope were determined by these velocities, as shown in Table 13.1.

Changes in wave velocity, frequency, or amplitude indicate physical property changes occurring at the device surface. To reduce the effect of the metallic dot array inside the SAW resonator, the size of each dot in the array was chosen such that it is sufficiently smaller than the wavelength in both directions. Because the amplitudes of the standing waves are dependent on material damping and electromechanical transduction losses, the transmitting and sensing IDTs are located at the standing wave maxima in order to reduce the transduction loss. To obtain a good resonator performance with this high-coupling coefficient substrate, the aperture of the IDTs and the number of IDT fingers were minimised but were large enough to avoid acoustic beam diffraction. Also, a larger spacing between the IDTs was chosen compared with conventional resonators because electromagnetic coupling between IDTs has to be avoided and sufficient metallic dots were accommodated within the cavity to induce the Coriolis effect. The number and aperture of IDTs, electromechanical coupling coefficient, and dielectric permittivity of the substrate determine the electrical impedance of the gyroscope. Thus, the spacing, aperture, and number of IDTs had to be chosen as a compromise between the differing requirements.

13.4.5.2 Design of a SAW gyroscope

It is important to know the characteristic impedance, admittance, bandwidth, and sensitivity near the operating frequency of the SAW gyroscope because the sensing IDTs have to be designed such that they efficiently pick up the SAW waves generated from Coriolis force. As stated earlier, the numerical simulation of the SAW resonator is done using coupling-of-mode (COM) theory because a SAW device can be easily represented by several basic elements that are transfer matrices of representative sections of a SAW device. The sensor is modeled using Mason's equivalent circuit concept. For the effective numerical calculation of the gyroscope output, a cross-field equivalent circuit model was chosen because then Coriolis force can be directly related to input force on the SAW sensor model.

Various SAW devices have been modeled using COM theory. The application of coupled-mode theory on SAW devices for given geometry and material data is well documented by Cross and Schmidt (1977), Haus and Wright (1980), and Campbell (1998).

Figures 13.20 and 13.21 show a modeling example of an IDT–IDT pair and a two-port SAW resonator, respectively. Different complex layouts of SAW devices can be modeled by adding more transfer functions at appropriate locations.

Figure 13.21 presents the two-port resonator represented in terms of its component matrices. The **G** matrix represents SAW reflectors, **T** matrix represents the relation

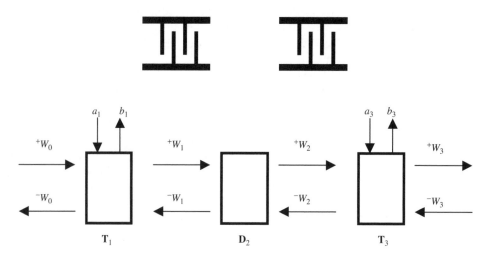

Figure 13.20   Schematic representation of an IDT–IDT pair

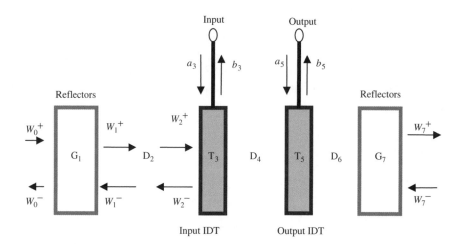

Figure 13.21   Schematic matrix representation of a two-port SAW resonator

between the electric and acoustic parameters, and **D** matrix represents the acoustic space between the IDTs and the reflectors. The boundary conditions are applied assuming that there are no incoming waves from outside the reflectors and output impedances are matched at the electrical terminal. The frequency response can be computed from transmission line matrices for appropriate sections. The amplitudes of the waves are represented by scalar electric potential Φ, which can be written in terms of SAW velocity $v(x)$ as

$$\frac{d^2\Phi}{dx^2} + \left[\frac{\omega^2}{v^2(x)}\right]\Phi = 0 \tag{13.63}$$

where ω is the radiation frequency. The wave velocity $v(x)$ is perturbed sinusoidally about v_0 while passing through the metallic strips and can be written as

$$v(x) = v_0 - \frac{\Delta v}{2}\cos(Ax) \tag{13.64}$$

where A is equal to $2\pi/L$, and L is the period.

The solution of Equation (13.63) gives a pair of coupled-wave equations that can be represented as forward and backward SAW waves with amplitudes R and S.

$$-R - j\delta R = jkS \tag{13.65}$$

$$-S - j\delta S = jkR \tag{13.66}$$

The grating matrix, **G**, for an array can be written as shown in the following equation (Campbell 1998). The equation for this matrix and that for **T** are provided in Appendix J together with a list of symbols. We repeat the equations here for the sake of clarity.

$$G = [G]$$

$$= C\begin{bmatrix} \left[\frac{\sigma}{k} + j\left(\frac{\delta - j\alpha}{k}\right)\tanh(\sigma L)\right]e^{j\beta_0 L} & je^{-j\theta}\tanh(\sigma L)e^{j\beta_0 L} \\ -je^{-j\theta}\tanh(\sigma L)e^{j\beta_0 L} & \left[\frac{\sigma}{k} - j\left(\frac{\delta - j\alpha}{k}\right)\tanh(\sigma L)\right]e^{-j\beta_0 L} \end{bmatrix}$$

$$\tag{13.67}$$

where k is the coupling coefficient, α is the attenuation constant, L is the period, β_0 is the unperturbed phase constant, and

$$\sigma = \sqrt{k^2 - (\delta - j\alpha)^2} \quad \text{and} \quad C = \frac{k}{\sigma}\cosh(\sigma L)$$

Similarly, the transmission matrix **T**, which relates the electrical and acoustic parameters of IDTs, as shown in Figure 13.22, can be obtained from the knowledge of the scattering matrix of an IDT (Cross and Schmidt 1977).

$$\mathbf{T} = [T] = \begin{bmatrix} t & \tau \\ \tau_s & t_{33} \end{bmatrix} \tag{13.68}$$

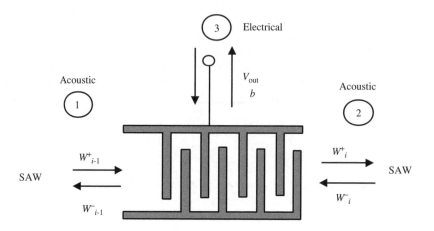

Figure 13.22   Schematic representation of electrical and acoustic ports

where

$$
t = \begin{bmatrix} s\left[1 + \dfrac{G_a(R_s + Z_e)}{1 + \theta_e i}\right]\exp(\theta_t i) & -s\left[\dfrac{G_a(R_s + Z_e)}{1 + \theta_e i}\right] \\[4mm] -s\left[\dfrac{G_a(R_s + Z_e)}{1 + \theta_e i}\right] & s\left[1 - \dfrac{G_a(R_s + Z_e)}{1 + \theta_e i}\right]\exp(-\theta_t i) \end{bmatrix} \tag{13.69}
$$

with $\theta_e = (\omega C_t + B_a)(R_s + Z_e)$ and $\theta_t = N_t A \delta$

$$
\tau = \begin{bmatrix} \dfrac{\sqrt{2G_a Z_e}}{1 + \theta_e i}\exp(\theta_t i/2) \\[4mm] \dfrac{\sqrt{2G_a Z_e}}{1 + \theta_e i}\exp(\theta_t i/2)\exp(-i\theta_t) \end{bmatrix} \tag{13.70}
$$

$$
\tau_s = \begin{bmatrix} s\dfrac{\sqrt{2G_a Z_e}}{1 + \theta_e i}\exp(\theta_t i/2) - s\dfrac{\sqrt{2G_a Z_e}}{1 + \theta_e i}\exp(\theta_t i/2)\exp(-i\theta_t) \end{bmatrix} \tag{13.71}
$$

where $G_0 = k^2 \cdot C_s \cdot f_0$ is the characteristic radiation conductance and

$$
G_a = G_0(N_t - 1)^2 \quad \text{if} \quad \omega = \omega_0; \text{ else } G_a = G_0(N_t - 1)^2\frac{\sin(\theta_t/2)^2}{(\theta_t/2)^2} \tag{13.72}
$$

$$
B_a = 0 \quad \text{if} \quad \omega = \omega_0; \text{ else } B_a = 2G_0(N_t - 1)^2\frac{\sin(\theta_t) - \theta_t}{\theta_t^2} \tag{13.73}
$$

$$
t_{33} = \left[1 - \frac{2\theta_c i}{1 + \theta_e i}\right] \tag{13.74}
$$

$$
\theta_c = \omega C_t(R_s + Z_e) \tag{13.75}
$$

$$
C = \frac{k_{12}}{\sigma}\cosh(\sigma L) \tag{13.76}
$$

The transmission matrix **T** can be related to the incoming and outgoing waves as

$$\begin{pmatrix} w_{i-1}^+ \\ w_{i-1}^- \\ b_i \end{pmatrix} = [T] \begin{pmatrix} w_i^+ \\ w_i^- \\ a_i \end{pmatrix} \tag{13.77}$$

Equation (13.77) can be further simplified using SAW amplitudes and the electrical output from the IDT.

$$[W_{i-1}] = [t_i][W_i] + a_i \cdot [\tau_i] \tag{13.78}$$

where a_i is the input electrical signal at the ith plane and the acoustic submatrix can be written as

$$[t_1] = \begin{pmatrix} t_{11} & t_{12} \\ -t_{12} & t_{22} \end{pmatrix}_i \tag{13.79}$$

and

$$[\tau_i] = \begin{pmatrix} t_{13} \\ -t_{23} \end{pmatrix}_i \tag{13.80}$$

Also, space between the IDTs and reflectors is represented by

$$D = \begin{bmatrix} e^{i\beta d} & 0 \\ 0 & e^{-i\beta d} \end{bmatrix} \tag{13.81}$$

The total acoustic matrix $[M]$ for a two-port resonator can be now obtained as

$$[M] = [G_1][D_2][t_3][D_4][t_5][D_6][G_7] \tag{13.82}$$

Here, $[G_1]$ and $[G_7]$ are related to two SAW reflectors at the end and $[D_2]$ and $[D_6]$ are the spacings between the gratings and adjacent IDTs. $[D_4]$ is the separation between the IDTs. $[t_3]$ and $[t_5]$ are the acoustic submatrices as shown in Equation (13.79). Equation (13.82) can be solved for the frequency response of the resonator by applying the boundary conditions.

$$w_0^+ = w_7^- = 0 \tag{13.83}$$

which assumes that there is no external SAW at the input reference side.

It is also noted that the source and load impedances determine the electrical properties of the transmission line. It can be seen from Figure 13.22 that

$$[W_2] = t_3[W_3] + a_3\tau_3 \tag{13.84}$$

Applying the foregoing boundary conditions for the transducer $[t_3]$, the first-order response of two-port resonator can be obtained by solving the following equations:

$$[W_5] = [D_6][G_7][W_7] \tag{13.85}$$

By applying the aforementioned boundary conditions, 2×2 matrix equation could be solved:

$$\begin{pmatrix} 0 \\ w_0^- \end{pmatrix} = [M] \begin{pmatrix} w_7^+ \\ 0 \end{pmatrix} + a_3[G_1][D_2][\tau_3] \tag{13.86}$$

The output voltage V_{out} (S_{21}) from the IDT can be written as

$$V_{\text{out}} = b_5 = [\tau_{s5}][W_5] \tag{13.87}$$

The output voltage is computed for different frequencies and is compared with the measured data from 65 to 85 MHz obtained from the SAW resonator using an HP 8510C Network Analyzer. Figure 13.23 presents the measured and computed voltages using Equation (13.87) for a resonator with five IDT pairs and a period of 52 μm.

The voltage on the transmission line, which is related to the surface potential, mainly shows the behaviour of the particle displacement at the surface. The surface electrical potential can then be computed after determining all the W vectors. The magnitude of surface potential in the cavity region of the resonator is an indication of the formation of standing waves, as shown in Figure 13.24.

There is a fixed ratio between the SAW displacement components and the surface electrical potential depending on the crystal cut. 128YX lithium niobate has the ratio of 0.2 nm z-displacement per unit surface potential (Datta 1986). For the present gyroscope, the z-displacement is of extreme importance because the SAW standing wave has the maximum z-displacement at antinodal points. Any mass at these antinode points has maximum amplitude of vibration, which is utilised as a reference vibration. When these

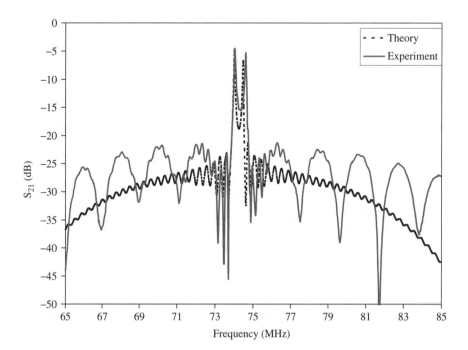

Figure 13.23 Computed and measured performance of the SAW resonator

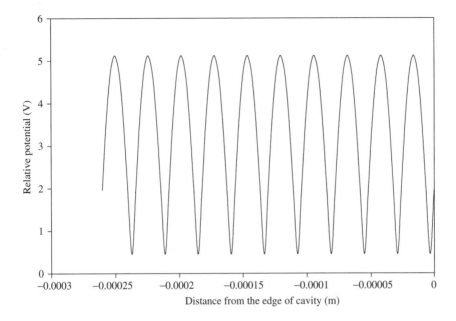

Figure 13.24   Variation of the potential inside the resonator with distance from the centre

particles are subjected to a rotation vector, a Coriolis force will be generated perpendicular to their vibration velocity vectors, as mentioned earlier. The velocity and the displacement of a particle at the antinode points are studied from the calculated surface potential using COM theory and are presented in Figures 13.25 and 13.26, respectively.

The velocity of a particle in the middle of the resonator at the antinode point is calculated for the resonant frequency. Using this velocity value and the mass at the antinodal points, the Coriolis force is calculated for different input rates. This Coriolis force is the input to the gyroscope sensor, which is discussed later.

After modeling the resonator, it is imperative to know the sensor characteristics. The cross-field model was chosen to simulate the sensor and gyroscope. Various SAW sensors have been successfully represented and simulated using the cross-field equivalent circuit model derived from the Mason circuit. In order to model the SAW gyrosensor, which generates a voltage output owing to rotation, the induced SAW due to the Coriolis force has to be modeled as an input force through the transformer ratio in the SAW equivalent circuit. In this model, the electric field distribution under the electrode is approximated as normal to the piezoelectric substrate, which is equivalent to a parallel-plate capacitor. Each IDT is represented by a three-port network, as shown in Figure 13.27. Here port 1 and 2 represent the electrical equivalent of acoustic ports and port 3 is a true electrical port.

Using an RF transmission line analogy, the three-port network circuit can be represented as shown in Figures 13.28 and 13.29. Figure 13.28 is the representation of one pair of IDTs and Figure 13.29 for that of a pair of reflectors. The acoustic forces F are transformed to electrical equivalent voltages V and particle velocities at the surface v are transformed to equivalent currents I. These transformations can be written in terms of a proportionality constant ϕ as

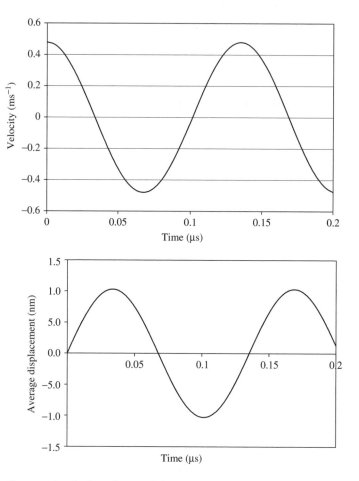

Figure 13.25 Computed velocity of a particle at the middle of the resonator at the resonant frequency. The negative sign indicates the up and down movement of the particle at the antinode with reference to the substrate plane

$$V = \frac{F}{\phi} \tag{13.88}$$

$$I = v\phi \tag{13.89}$$

where ϕ is related to the turns ratio of an equivalent acoustic-to-electric transformer that can be written in terms of its electromechanical coupling coefficient k, frequency f_0, and the total capacitance C of the IDT as

$$\phi = \sqrt{2 f_0 C k^2 A \rho v} \tag{13.90}$$

Hence, the mechanical characteristic impedance $Z_m = F/v$ of the substrate can be expressed as an equivalent transmission line characteristic impedance Z_0. For a substrate of density ρ and effective cross-sectional area A, the mechanical impedance for a propagating

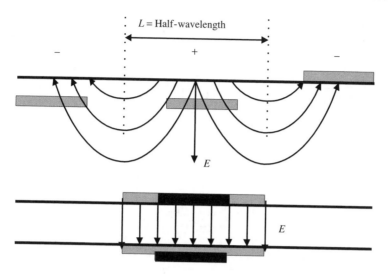

Figure 13.26 Computed displacement of a particle at the middle of the resonator at the resonant frequency

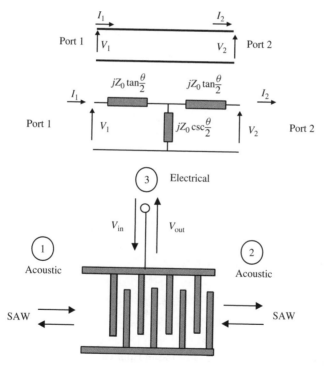

Figure 13.27 (a) Two-wire transmission line and its equivalent circuit representation and (b) schematic representation of the electrical and acoustic ports

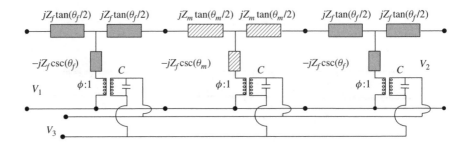

Figure 13.28 Cross-field model of IDT pair and its equivalent circuit representation

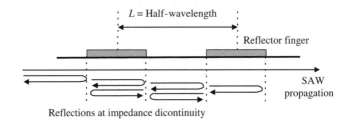

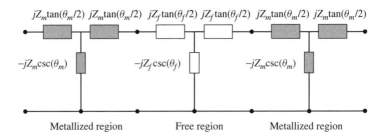

Figure 13.29 Representation of reflectors using equivalent circuit model

acoustic wave can be written as

$$Z_m = \rho v A \tag{13.91}$$

and the equivalent electric impedance is

$$Z_0 = \frac{Z_m}{\phi^2} \tag{13.91}$$

Once the unit cell of IDT pair is modeled, each model can be cascaded, depending on the number of pairs for the sensor IDTs. For the present case, we used 30 pairs. The Coriolis force is the input force, that is, the input voltage at one end of the constructed equivalent circuit, as shown in Figure 13.28.

The other end of the model is shorted with a substrate impedance, assuming no incoming wave. Simulation of this model has been achieved by using commercial RF circuit simulator, HP EESOF. Figure 13.30 shows a typical half-wavelength circuit

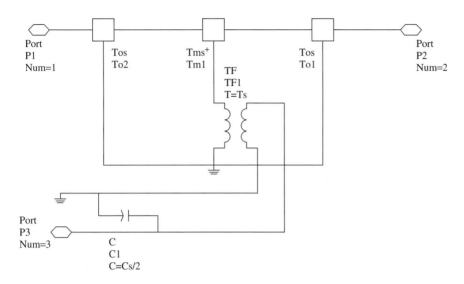

Figure 13.30   Typical half-wavelength equivalent circuit using HP EESOF

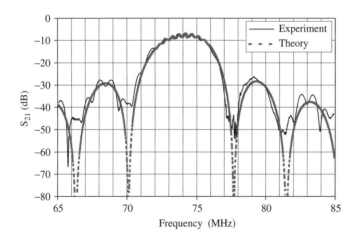

Figure 13.31   Computed and measured response of SAW sensor

constructed using EESOF. In order to calculate the overall response of the sensor, this three-port network has been cascaded electrically in parallel and acoustically in series. The S_{21} response is calculated for different frequencies and is presented in Figure 13.31. For comparison with the fabricated sensors, the sensors were connected to the HP 8510 C Network analyzer and the S_{21} response was measured and is also shown in Figure 13.31.

The SAW values generated by a Coriolis force are associated with a 3-D problem, which contains three vectors (reference vibration vector, input rotation vector, and resulting Coriolis vector) in three mutually perpendicular directions. Because the Coriolis force has been approximated by a one-dimensional analogy, the secondary SAW, generated by the Coriolis force, has also to be approximated using a one-dimensional analogy. If this

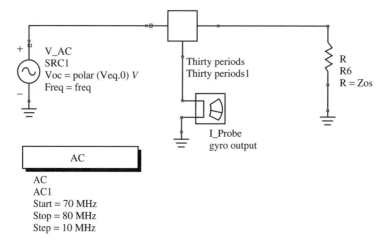

Figure 13.32   Equivalent circuit model for the gyroscope. Coriolis force is the input and the output is terminated to load impedance

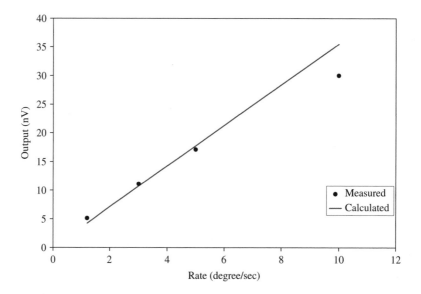

Figure 13.33   Measured and calculated response of the gyroscope using rate table

approximation is not used, a full 3-D finite element model (FEM) is necessary, which is computationally very demanding for SAW. The generated SAW, due to the Coriolis force, is over its effective cross-sectional area by a homogeneous bulk wave approximation. The effective area is where the SAW power is concentrated on the surface, which is a fraction of a wavelength times the beam width. The Coriolis force has been represented as an equivalent voltage input to one end of the sensor, which consists of 30 pairs of IDTs. As shown in Figure 13.32, the other end of the sensor is terminated with a load impedance (assuming that there are no reflections).

The output voltage can be calculated for a given Coriolis force, which is determined by the rotation rate (see Figure 13.33). The gyroscopes were fabricated by both the lift-off[4] and reactive ion-etching (RIE) techniques using the Pennsylvania State University Nanofabrication facility. The lift-off technique was adequate for 75 MHz gyroscopes, which have a minimum feature size of about 6 μm.

The response of the SAW gyroscope was evaluated using a rate table. The output response was measured using an HP dynamic signal analyzer fed by an RF lock-in amplifier. The gyroscope is fixed to a rate table, which, when excited at a constant frequency, causes the rate table to oscillate, and the gyroscope is measured for each setting. Both the frequency and the amplitude of the driving signal controls the amplitude of oscillations of the table, which in turn varies the rotation rate. Figure 13.33 presents the measured and computed output voltages from the SAW gyroscope for different rotation rate.

13.5 CONCLUDING REMARKS

In this chapter, we have described some of the many sensing applications to which a SAW-IDT microsensor may be applied. The principles that govern these IDT sensors have been presented along with the input–output equations. The versatility of this technology is very impressive and permits a wide variety of different sensing applications, such as strain, pressure, temperature, conductivity, and dielectric constant.

The next chapter is dedicated to the topic of IDT-MEMS sensors that represent an exciting development combining the fields of MEMS and SAW-IDT devices. The final example of a SAW-IDT device is described in Section 15.2 and is associated with a type of smart sensor called an *electronic tongue*. It complements the description of another smart sensor also given in the final chapter called the *electronic nose*.

REFERENCES

Achenbach, J. D. (1973). *Wave Propagation in Elastic Solids*, Elsevier North-Holland, Amsterdam, pp. 187–194.

Ballantine, D. S. *et al.* (1997). *Acoustic Wave Sensors: Theory, Design and Physico-Chemical Applications*, Academic Press, New York.

Bao, X. Q., Varadan, V. V. and Varadan, V. K. (1987). "SAW temperature sensor and remote reading system," *IEEE Ultrasonics Symp.*, **2**, 583–585.

Bao, X. Q., Varadan, V. V. and Varadan, V. K. (1994). "Wireless surface acoustic wave temperature sensor," *J. Wave Material Interact.*, **9**, 19–27.

Bell, D. T. and Li, R. C. (1976). "Surface acoustic wave resonators," *Proc. IEEE*, **64**, 711–721.

Campbell, C. K. (1998). *Surface Acoustic Wave Devices for Mobile and Wireless Communications*, Academic Press, New York.

Campbell, J. J. and Jones, W. R. (1968). "A method for estimating optimal crystal cuts and propagation directions for excitation of piezoelectric surface waves," *IEEE Trans. Sonics Ultrasonics*, **15**, 209–217.

Cross, P. S. and Schmidt, R. V. (1977). "Coupled surface acoustic wave resonator," *Bell Syst. Tech. J.*, **56**, 1447–1481.

Datta, S. (1986). *Surface Acoustic Wave Devices*, Ch. 3, Prentice-Hall, New Jersey.

[4] The lift-off process was described in Chapter 12.

Eddy, D. S. and Soarks, D. R. (1998). "Application of MEMS technology in automotive sensors and actuators," *Proc. IEEE*, **86**, 1747–1755.

Gere, J. M. and Timoshenko, S. P. (1990). *Mechanics of Materials*, PWS-Kent, Kent, UK.

Hauden, D., Jaillet, G. and Coquerel, R. (1981). Temperature sensor using SAW delay line, *Proc. IEEE Ultrasonic Symp.*, Chicago, IL, USA, pp. 148–151.

Haus, H. and Wright, P. V. (1980). "The analyses of grating structures by coupling-of-modes theory," *IEEE Ultrasonics Symp.*, **1**, 277–281.

Hollinger, R. D. *et al.* (1999). "Wireless surface acoustic wave-based humidity sensor," *Proc. SPIE*, **3876**, pp. 54–62.

Hoummady, M., Campitelli, A. and Wlodarski, W. (1997). "Acoustic wave sensors: design, sensing mechanisms and applications," *Smart Materials Struct.*, **6**, 647–657.

Kurosawa, M., Fukuda, Y., Takasaki, M. and Higuchi, T. (1998). "A surface-acoustic wave gyroscope sensor," *Sensors and Actuators A*, **66**, 33–39.

Lao, B. Y. (1980). "Gyroscopic effect in surface acoustic waves," *IEEE Ultrasonics Symp.*, **2**, 687–690.

Pierce, J. R. (1954). "Coupling-of-modes of propagation," *J. Appl. Phys.*, **25**, 179–183.

Reeder, T. M. and Cullen, D. E. (1976). "Surface acoustic wave pressure and temperature sensors," *Proc. IEEE*, **64**(5), 754–756.

Slobodnik, A. J. (1976). "Surface acoustic waves and SAW materials," *Proc. IEEE*, **64**, 581–595.

Soderkvist, J. (1994). "Micromachined gyroscopes," *Sensors and Actuators A*, **43**, 65–71.

Staples, E. J. (1974). "UHF surface acoustic wave resonators," *Proc. IEEE Ultrasonics Symp.*, **1**, 245–252.

Subramanian, H., Varadan, V. K., Varadan, V. V. and Vellekoop, M. J. (1997). "Design and fabrication of wireless remotely readable MEMS based microaccelerometers," *Smart Materials Struct.*, **6**, 730–738.

Suh, W. D. *et al.* (2000). "Design optimisation and experimental verification of wireless IDT based micro temperature sensor," *Smart Materials Struct.*, **9**, 890–897.

Varadan, V. K. *et al.* (2000a). "Conformal MEMS-IDT gyroscopes and their performance comparison with fiber optic gyroscope," *Proc. SPIE Smart Electronics MEMS*, **3990**, 335–344.

Varadan, V. K. *et al.* (2000b). "Design and development of a MEMS-IDT gyroscope," *J. Smart Materials Struct.*, **9**, 898–905.

Varadan, V. V. *et al.* (1997). "Wireless passive IDT strain microsensor," *Smart Materials Struct.*, **6**, 745–751.

Vlassov, Y. N. A., Kozlov, S., Pashchin, N. S. and Yakovkin, I. B. (1993). "Precision SAW pressure sensors," *IEEE Int. Frequency Control Symp.*, p. 635.

White, R. M. (1985). "Surface acoustic wave sensors," *Proc. IEEE Ultrasonics Symp.*, **1**, 490–494.

White, R. M. and Voltmer, F. W. (1965). "Direct piezoelectric coupling to surface acoustic waves," *Appl. Phys. Lett.*, **7**, 314–316.

Yazdi, N., Ayazi, F. and Najafi, K. (1998). "Micromachined inertial sensors," *Proc. IEEE*, **86**, 1640–1658.

14

MEMS-IDT Microsensors

14.1 INTRODUCTION

The combination of a microelectromechanical system (MEMS) device with an interdigital transducer (IDT) (surface acoustic wave (SAW)) microsensor is a relatively new concept. MEMS-IDT-based microsensors can offer some significant advantages over other MEMS devices, including the benefits of excellent sensitivity, surface conformability, greater robustness, and durability. These new MEMS-IDT combinations have fewer moving mechanical parts that ultimately gives rise to these valuable benefits. Indeed, the superior performance removes not only the need for associated electronic circuitry to balance (or measure) the movement of moving structures but also leads to even smaller devices.

In this chapter, we give a detailed description of the fabrication, characterisation, and testing of MEMS-IDT microsensors for inertial navigation systems. Inertial accelerometers generally consist of a suspended seismic mass that is displaced from its rest position under the influence of an external load. Thus, the position of the seismic mass depends on the momentary acceleration applied to the device. This displacement is usually transduced into either a change in resistance (in piezoresistive accelerometers) or a change in capacitance (in capacitive accelerometers). Each of these two basic electrical parameters has its own relative advantages and disadvantages: piezoresistive accelerometers tend to have a wide frequency range, low cost, and low precision, whereas capacitive accelerometers have high sensitivity but a small dynamic and frequency range. The topic of silicon microaccelerometers has already been discussed in Chapter 8 and further details on inertial devices may be found in the following references: Esashi (1994), Roylance and Angell (1979), Rudolf *et al.* (1987), Suzuki *et al.* (1990), Seidel *et al.* (1990), and Matsumoto and Esashi (1992).

Conventional accelerometers incorporate a large amount of electronics within the sensor package. They also require an internal battery to drive the associated electronics. These requirements result in sensors that are relatively bulky and require regular maintenance. The size of the sensors is a major drawback in applications in which the sensors need to be conformal to the structure and should not add excessive amounts of weight.

We now provide a detailed description of two worked examples of MEMS-IDT microsensors, namely, an accelerometer and a miniature inertial navigation system. The inertial navigation system uses both gyroscopes and accelerometers to measure the state of motion of an object (i.e. its kinematics) by sensing the changes to that state caused by the accelerations. The characteristic features required for many applications are a high level of precision and wide dynamic and frequency ranges. We also show that it is possible to create a wireless MEMS-IDT microsensor in a manner similar to that was shown as possible to make a wireless IDT microsensor in Chapter 13 (Subramanian *et al.* 1997).

14.2 PRINCIPLES OF A MEMS-IDT ACCELEROMETER

The Rayleigh wave is a surface wave in which the wave energy is almost completely confined within a distance of one wavelength above the substrate (Ballantine *et al.* 1997). There are basically two types of boundary conditions applicable with the surface of the piezoelectric being mechanically free. In one form, the surface has a thin conductive layer. This layer does not alter the mechanical boundary conditions but causes the surface to be equipotential and the propagating potential to be zero at the surface of the substrate. The other case is the one in which the surface is electrically free and the potential above the surface follows Laplace's equation. The potential vanishes as the distance above the substrate approaches infinity[1].

In a MEMS-IDT accelerometer, a conductive seismic mass is placed close to the substrate (at a distance of less than one acoustic wavelength). This serves to alter the electrical boundary condition discussed earlier. The seismic mass can be fabricated from polysilicon and incorporates reflectors and flexible beams (Figure 14.1).

The MEMS device functions as follows: An incoming electromagnetic wave at the IDT creates a radio frequency electrical field between the transducer fingers. Owing to the piezoelectric effect, mechanical deformations follow the signal and propagate along the surface of the piezoelectric substrate. The array of reflectors sends back this wave to the IDT. The phase of the reflected wave is dependent on the position of the reflectors. If the position of the reflectors is altered, the phase of the reflected wave also changes. As can be seen from the figure, the reflectors are part of the seismic mass. In response to an acceleration, the beam flexes, causing the reflectors to move. This can be measured as a phase shift of the reflected wave. By calibrating the phase shift measured with respect to

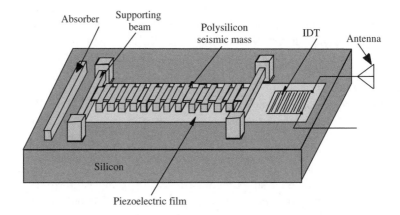

Figure 14.1   Basic arrangement of a MEMS-IDT accelerometer

---

[1] Readers are referred to Chapters 9 and 10 for a basic introduction to surface acoustic waves and SAW devices.

the acceleration, the MEMS device can be used as an acceleration sensor. Alternatively, the measurement can be done in the time domain, in which case the delay time of the reflection from the reflectors is used to sense the acceleration.

14.3 FABRICATION OF A MEMS-IDT ACCELEROMETER

We now illustrate the principles from the Worked Example 14.1 of a MEMS-IDT accelerometer.

Worked Example E14.1: MEMS-IOT Accelerometer Objective:

There are three steps involved: First, a Rayleigh wave device consisting of an IDT on a piezoelectric film is easily fabricated using conventional metal deposition or etching techniques. These techniques are standard in integrated circuit (IC) processing and details of SAW-IDT fabrication are provided in Chapter 12. Next, the reflector arrays can be fabricated using silicon micromachining techniques. Finally, flip-chip bonding can be used in order to reduce the handling of the substrate and hence maintain the performance of the SAW substrate. The specific details are as follows (see Varadan *et al.*):

The substrate chosen was silicon with a ZnO coating. The IDTs were then sputtered on the substrate. The fabrication steps involve mask preparation, lithography, and etching. The thickness of the metal for the IDTs should be at least 200 nm to make an adequate electrical contact. However, a very thick layer of metal can cause significant mass loading effects and so is detrimental to the device performance. The metallisation ratio used for the IDTs, that is, the ratio of finger width to repetition distance is 0.5. The number of fingers of the IDT and their aperture is chosen such that the IDTs have an impedance of nominally 50 Ω. Figure 14.2 shows a cross section of the basic design with the critical dimensions (in μm).

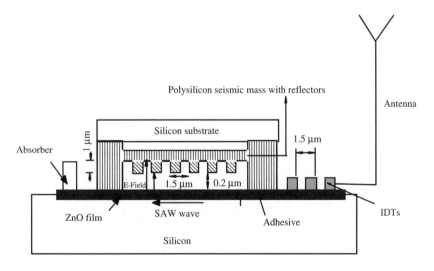

Figure 14.2   Cross section of MEMS-IDT accelerometer with dimensions

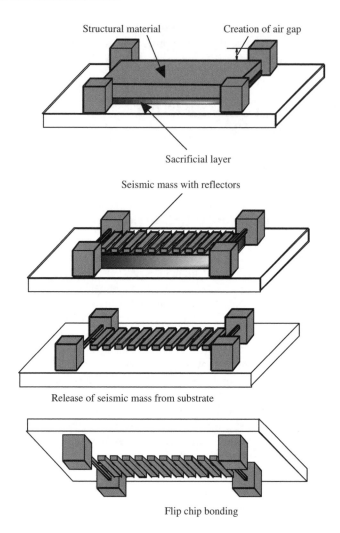

Structural material   Creation of air gap

Sacrificial layer

Seismic mass with reflectors

Release of seismic mass from substrate

Flip chip bonding

Figure 14.3   Basic steps in the fabrication of the seismic mass

Process Flow of Scismic Mass:

The steps involved in the fabrication of the seismic mass are outlined here and involve the sacrificial etching of silicon dioxide (see Figure 14.3)[2].

1. A sacrificial oxide is thermally grown on the silicon wafer.

2. A polysilicon (structural layer) is then deposited by low-pressure chemical vapour deposition (LPCVD) on the sacrificial layer. Polysilicon has good structural properties and is commonly used. This structural layer is thick enough (1 μm) to support itself. The polysilicon is patterned and etched using ethylenediamine pyrocatechol (EDP).

---

[2] Full details of lithography and etching techniques are described in earlier chapters.

3. The sacrificial layer is then etched with hydrofluoric acid (HF) to release the seismic mass.

4. The seismic mass is now ready to be flip-chip bonded to the SAW substrate.

These steps are only one of the many possible methods of realising these structures.

After the sacrificial oxide is removed in HF, the wafers are rinsed in deionised (DI) water and dried. The surface tension of the water under the structures tends to pull them down to the surface of the wafer, and in some cases, causes them to stick permanently. This problem can be avoided by using thick structural and sacrificial layers and short structures. There are several other methods of adhesion prevention that rely on avoiding the problem of surface tension. These include the following:

• Freeze drying (sublimation) of the final rinsing solution (DI water, t-butyl alcohol)

• Using an integrated polymer support structure during release etching and then ashing in oxygen

• Gradually replacing acetone with photoresist, then spinning and ashing the resist

• Electrically fusing (vaporising) micromechanical support structures

14.3.1 Fabrication of the SAW Device

The SAW device consists of a piezoelectric substrate on which IDTs are deposited. The IDTs can be sputtered through a mask or they can be etched using the lift-off technique as described in Chapter 12. The choice of the method is dependent on the minimum feature size to be fabricated and adhesion of the electrode metal to the piezoelectric substrate. The fabrication of the IDTs is essentially a single mask process.

For example, in the lift-off technique, fabrication begins with standard photolithography using standard equipment such as a wet bench, resist spinner, hot plates, and an evaporator. The lithium niobate wafers are cleaned using acetone, isopropanol, and trichloroethylene (in turn) at about 60 °C for about 10 minutes. The wafers are then thoroughly rinsed in DI water for about 5 minutes and subsequently heated at 125 °C (on a hot plate) for about 10 minutes to remove surface moisture. Upon cooling the wafer on a heat sinking plate, Shipley 1813 photoresist is spin-coated (at 4000 rpm for 55 seconds) on the wafer after soaking the top face with an adhesion agent called hexamethyl disilazane (HMDS). The wafer is then heated at 125 °C for 2 minutes in what is commonly referred to as the *soft-bake*. The wafer is then exposed to ultraviolet (UV) light (of 15 mW/cm^2) for 1.2 seconds such that the regions of the resist that are exposed become soluble to the developer (DI water and MF312 in a 1:1 ratio). A negative mask, whereby the patterns are glass set against a background of chrome, is used for this purpose. The wafer is developed until the sections that have been exposed to UV light and therefore soluble are etched away. The wafer is then hard-baked at 125 °C for 1 minute 30 seconds. The patterned wafer is coated with 20 nm of chromium using electron gun evaporation, which is deposited to improve the adhesion of the subsequent thermally evaporated 120 nm gold layer. The wafer is then submerged into acetone to facilitate the lift-off process.

14.3.2 Integration of the SAW Device and Seismic Mass

14.3.2.1 Wafer bonding

If two wafers with surface oxide layers are immersed in sulfuric acid or nitric acid and then brought into contact, they will stick immediately. If these wafers are subsequently annealed at high temperature, the resulting bond can be essentially perfect. The quality of the bond depends on the types of oxides used, the temperature, and the cleanliness of the process. If the wafers can be annealed at 1000 °C, any common (thermal, phosphosilicate (PSG), etc.) oxide will do. Because the oxide, rather than the silicon, is involved in the bonding, it is also possible to bond silicon to quartz or other glass substrates. This technique is often used in what is known as a '*dissolved wafer process*.' Further details of silicon-bonding techniques may be found in Sections 5.6 and 5.7.

14.3.2.2 Incorporation of electronics into the accelerometer

When the electromagnetic signal is converted to an acoustic signal on the surface of a piezoelectric, the initial wavelength is reduced by a factor of about 10^5. This results in the dimensions of acoustic wave devices becoming compatible with IC technology. Use of silicon substrates sputtered with piezoelectric zinc oxide instead of other piezoelectric materials, such as lithium niobate and stable temperature (ST)-quartz, makes it feasible to integrate electronics on the sensor substrate. This combination has the following distinct advantages:

- Onboard signal processing and data manipulation is possible

- The need for signal amplification may be eliminated in certain cases

- Information loss by transmission is eliminated

- The overall system is miniaturised and a smart sensor is created.

However, there are some drawbacks to this approach and these are as follows:

- Production processes must be compatible with each other

- The number of production steps (mask levels) may increase, which may make the yield low

- The contrasting demand that the electronics must be protected from the environment, while the sensor must be exposed to the environment. This complicates the packaging requirements

- Optimisation of the sensor and the electronics may involve conflicting requirements

14.4 TESTING OF A MEMS-IDT ACCELEROMETER

The MEMS-IDT accelerometer needs to be tested so that its design can be optimised. We will first describe, discuss, and evaluate the measurement setup and calibration procedure of the microsensor (Section 14.4.1–14.4.4). Then we will discuss the incorporation of a seismic mass to produce an inertial accelerometer (Section 14.4.5).

14.4.1 Measurement Setup

The vector network analyser and associated calibration techniques make it possible to measure accurately the reflection and transmission parameters of devices under test[3]. The basic arrangement of such a measurement system is illustrated in Figure 14.4. The network analyser system consists of a synthesized sweeper (10 MHz–20 GHz), the test set (40 MHz–40 GHz), HP 8510B network analyzer, and a display processor. The sweeper provides the stimulus and the test set provides the signal separation. The front panel of the HP 8510B is used to define and conduct various measurements. The various other instruments are also controlled by the network analyser through the system bus. The device to be tested is connected between the test Port 1 and Port 2. The point at which

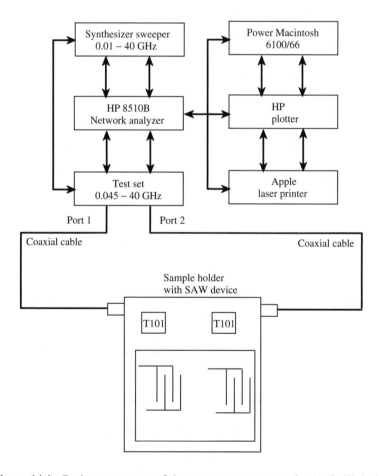

Figure 14.4   Basic arrangement of the measurement system for the SAW device

[3] A detailed explanation of SAW parameters and their measurement is given in Chapter 11.

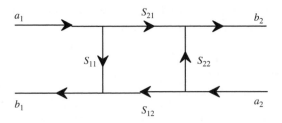

Figure 14.5 Signal flow in a two-port network

the device is connected to the test set is called the *reference plane*. All measurements are made with reference to this plane. The measurements are expressed in terms of scattering parameters referred to as *S parameters*. These describe the signal flow (Figure 14.5) within the network.

S parameters are defined as ratios and are represented by $S_{in/out}$, where the subscripts *in* and *out* represent the input and output signals, respectively. Figure 14.5 shows the energy flow in a two-port network. It can be shown that

$$b_1 = a_1 S_{11} + a_2 S_{12} \quad \text{and} \quad b_2 = a_1 S_{21} + a_2 S_{22} \tag{14.1}$$

and, therefore,

$$S_{11} = b_1/a_1, S_{21} = b_2/a_1 \quad \text{when } a_2 = 0; \quad S_{12} = b_1/a_2, S_{22} = b_2/a_2 \quad \text{when } a_1 = 0. \tag{14.2}$$

where S_{11} and S_{21} (S_{12} and S_{22}) are the reflection and transmission coefficients for Port 1(2), respectively.

14.4.2 Calibration Procedure

Calibration of any measurement system is essential in order to improve the accuracy of the system. However, accuracy is reduced because errors, which may be random or systematic, exist in all types of measurements. Systematic errors are the most significant source of measurement uncertainty. These errors are repeatable and can be measured by the network analyser. Correction terms can then be computed from these measurements. This process is known as *calibration*. Random errors are not repeatable and are caused by variations due to noise, temperature, and other environmental factors that surround the measurement system.

A series of known standards are connected to the system during calibration. The systematic effects are determined as the difference between the measured and the known response of the standards. These errors can be mathematically related by solving the signal-flow graph. The frequency response is the vector sum of all test setup variations in magnitude and phase with frequency. This is inclusive of signal-separation devices, test cables, and adapters. The mathematical process of removing systematic errors is called *error correction*. Ideally, with perfectly known standards, these errors should be completely characterised. The measurement system is calibrated using the full two-port calibration method. Four standard methods are used, namely, shielded open circuit, short circuit, load, and through. This method provides full correction for directivity, source match, reflection and transmission signal path frequency response, load match, and isolation for S_{11}, S_{12}, S_{21}, and S_{22}. The procedure involves taking reflection, transmission, and isolation measurements.

For the reflection measurements (S_{11}, S_{22}), the open, short, and load standards are connected to each port in turn and the frequency response is measured. These six measurements result in the calculation of the reflection error coefficients for both the ports. For the transmission measurements, the two ports are connected and the following measurements are conducted: forward transmission through (S_{21}-frequency response), forward match through (S_{21}-load), reverse transmission through (S_{12}-frequency response) and reverse match through (S_{12}-load). The transmission error coefficients are computed from these four measurements. Loads were connected to the two ports and S_{21} noise floor and S_{12} noise floor levels were measured. From these measurements, the forward and reverse isolation error coefficients are computed. The calibration is saved in the memory of the network analyser and the correction function is turned on to correct systematic errors that may occur.

14.4.3 Time Domain Measurement

The relationship between the frequency domain response and the time domain response is given by the Fourier transform, and the response may be completely specified in either domain. The network analyser performs measurements in the frequency domain and then computes the inverse Fourier transform to give the time domain response. This computation technique benefits from the wide dynamic range and the error correction of the frequency domain data.

In the time domain, the horizontal axis represents the propagation delay through the device. In transmission measurements, the plot displayed is the actual one-way travel time of the impulse, whereas for reflection measurements the horizontal axis shows the two-way travel time of the impulse. The acoustic propagation length is obtained by multiplying the time by the speed of the acoustic wave in the medium. The peak value of the time domain response represents an average reflection or transmission over the frequency range.

The *time band pass* mode of the network analyser is used for time domain analysis. It allows any frequency domain response to be transformed to the time domain. The Hewlett Packard (HP) 8510B network analyser has a time domain feature called *windowing*, which is designed to enhance time domain measurements. Because of the limited bandwidth of the measurement system, the transformation to the time domain is represented by a $\sin(x)/x$ stimulus rather than the ideal stimulus. For time band pass measurements, the frequency domain response has two cutoff points f_{start} and f_{stop}. Therefore, in the time band pass mode, the windowing function rolls off both the lower end and the higher end of the frequency domain response. The *minimum* window option should be used to minimise the filtering applied to the frequency domain data.

Because the measurements in the frequency domain are not continuous but, apart from Δf (in Hz), are taken at discrete frequency points, each time domain response is repeated every $1/\Delta f$ seconds. The amount of time defines the range of the measurement. Time domain response resolution is defined as the ability to resolve two close responses. The response resolution for the time band pass, using the minimum window, can be expressed as a parameter r:

$$r = \frac{1.2}{f_{span}} \tag{14.3}$$

where the frequency span f_{span} is expressed in Hz. Thus, if a frequency span of 10 MHz is used, the measurement system will not be able to distinguish between equal magnitude responses separated by less than 0.12 µs for transmission measurements.

Time domain range response is the ability to locate a single response in time. Range resolution is related to the digital resolution of the time domain display, which uses the same number of points as that of the frequency domain. The range resolution can be computed directly from the time span and the number of points selected. If a time span of 5 µs and 201 points are used, the marker can read the location of the response with a range resolution of 24.8 ns (5 µs/201). The resolution can be improved by using more points in the same time span.

14.4.4 Experimental

We now evaluate the suitability of using S_{11} measurement for the measurement of reflections in SAW delay line.

The operating principle of the device is based on the perturbation in the velocity of the acoustic wave due to the changes in the electrical boundary conditions. The two extreme electrical boundary conditions were applied in turn. An attempt was then made to detect these in various measurement options that are available in the vector network analyser. These electrical conditions represent the maximum change possible for the device designed and hence are useful in evaluating suitability of any measurement technique. The details of this device are given in the following paragraph. Split finger electrodes were used in order to reduce reflections from the electrodes:

- Number of finger pairs is 10

- Propagation path length is 6944 µm

- Operating frequency is 82.91 MHz

Two devices representing the two extreme electrical boundary conditions were used:

1. A SAW delay line with split finger IDTs and with an aluminum layer in the propagation path. This device represents the electrical boundary condition in which the electric field is shorted on the substrate surface.

2. A SAW delay line with split finger IDTs and without an aluminum layer in the propagation path. This device represents the electrical boundary condition in which the electric field decays at an infinite distance from the substrate surface.

Both these devices have a propagation path length of 6944 µm. The SAW propagation velocity on the substrate is 3980 m/s and the crystal size is (10×10) mm^2. The equipment

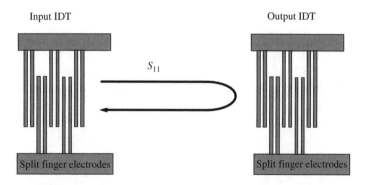

Figure 14.6   Measurement of the S_{11} parameter

used for the measurement was an HP Network Analyzer Model No.8753A operating in the range 300 kHz to 3 GHz.

The two ports are calibrated using test standards in the method described earlier. The devices are connected in turn and the reflection coefficient (S_{11}) was measured (see Figure 14.6). In the S_{11} measurement, the wave propagates from one set of IDTs to the other set of IDTs and the reflections due to the second set are measured at the first set.

It was also found that in the *linear* magnitude format, the reflection peak was more sharply defined than the one in the *log* magnitude format. The measurements were transformed into the time domain as the interpretation of the observations are much easier. The *gating* function of the network analyser was used to filter out the electromagnetic feed through. It also allows appropriate scaling of the desired signal.

In the case of the device with aluminum between the IDTs, the first reflection from the IDT occurred at 3.799 μs. The next peak beyond 3.799 μs is the reflection from the crystal edge. For the device without aluminum, a reflection was measured at 3.535 μs. It can be seen that for the same distance traveled, the wave velocity is greater in the case of the device without aluminum. The time difference between these two measurements (3.535 μs and 3.799 μs) is a measure of the coupling efficiency of the substrate as well as mass loading because of the aluminum layer between the IDTs. The theoretical calculations for this substrate leads us to expect the velocity of the wave to slow down by 136 m/s because of the change in the electrical boundary conditions. The observed slowing down of the wave was around 281 m/s. This difference is probably due to the mass loading effects of the aluminum.

The results of these experiments indicate that the effect of an aluminum conductor placed close to the surface should be seen in the region between 3.535 μs and 3.799 μs in the time domain measurement of S_{11}.

The experimental validation of the design and the concept was done in stages. The first step in this process was to conduct an experiment to qualitatively examine the effect of a conductor close to the surface and to devise a measurement method. The three samples used for the experiment are described here:

1. For the gross or qualitative evaluation of the effect, it is sufficient to place a conductor close to the surface. Three samples were prepared for this experiment. The sample consisted of a micromachined silicon trough in which aluminum was deposited. The

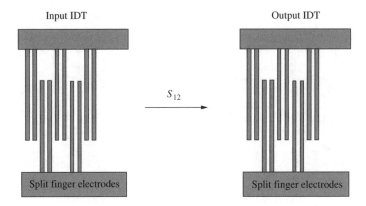

Input IDT   Output IDT

S_{12}

Split finger electrodes   Split finger electrodes

Figure 14.7 Measurement of the (S_{12}) parameter

trough is 1 µm deep. Within this trough, 600 nm of aluminum was deposited. This device allows the conductor to be placed 400 nm from the substrate.

2. This sample is the same as the one discussed earlier, except that there is a silicon dioxide layer 1 µm thick on the substrate. This sample allows the conductor to be placed 1.4 µm from the substrate.

3. The third sample is similar to the first sample. It consists of a micromachined trough that is 1 µm deep. Silicon is to serve as a conductor. This sample can be used to evaluate the suitability of silicon as a conductor for this application.

These samples are shown in Figure 14.8. These samples are flipped over and are placed on the substrate. They rest on spacers. The spacers lie outside the propagation path of the Rayleigh wave. The trough was big enough such that when it was placed on the substrate, it still left the substrate mechanically free. This can be easily tested by doing the S_{12} measurement (Figure 14.7). These observations were carried out in both the frequency and the time domain.

The following conclusions have been derived from the set of experiments mentioned previously. The arrangement of the spacer performs adequately in the placing of the conductor within one wavelength of the surface. Silicon instead of aluminum could be used for this device. For this application, it can almost be considered to be a conductor. The perturbation in the velocity of the wave is too small to be measured as a shift in the amplitude response in both frequency and time domain with the given resolution of the network analyser.

14.4.5 Fabrication of Seismic Mass

Following the aforementioned evaluation of the performance of the IDT microsensor, we will now discuss the addition of a seismic mass to the wafer to produce an accelerometer. The fabrication of a seismic mass is a two mask process. Here, the masks were designed for the process using the commercial software package of L-Edit (Tanner Tools Inc.). A

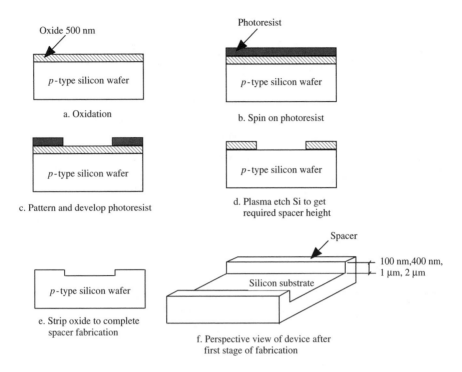

Figure 14.8   Basic steps in the fabrication of the spacers

4″ silicon wafer was chosen and four different wafers were processed, as each spacer height requires a separate wafer.

The first step in the fabrication process was the creation of the spacer of the desired height. The next step is the fabrication of the reflector arrays. These two stages are described with the help of Figures 14.8 and 14.9. The basic steps in the process involve the growth and patterning of an oxide mask, followed by dry etching of silicon by plasma.

The steps required to fabricate the spacers are as follows:

Four *p*-type wafers of silicon ⟨100⟩ of resistivity between 2 and 5 ohm.cm were used.

1. A 500 nm thick silicon dioxide is grown. This oxide layer will act as a mask for the dry etching (Figure 14.8(a)).

2. Photoresist is spun on the oxide layer (Figure 14.8(b)). The resist is baked to improve adhesion.

3. The first mask is aligned with respect to the flat of the wafer and the photoresist is patterned (Figure 14.8(c)). The oxide is then etched away in all areas except where it was protected (Figure 14.8(d)). The etching automatically stops when the etchant reaches silicon. The etchant is highly selective and etches only silicon dioxide. The above process of exposing and patterning the photoresist along with oxide etching is referred to as developing.

4. Silicon is dry-etched in plasma. The four wafers are etched to different depths, namely, 100 nm, 400 nm, 1 μm, and 2 μm (Figure 14.8(e)). This step results in the protected

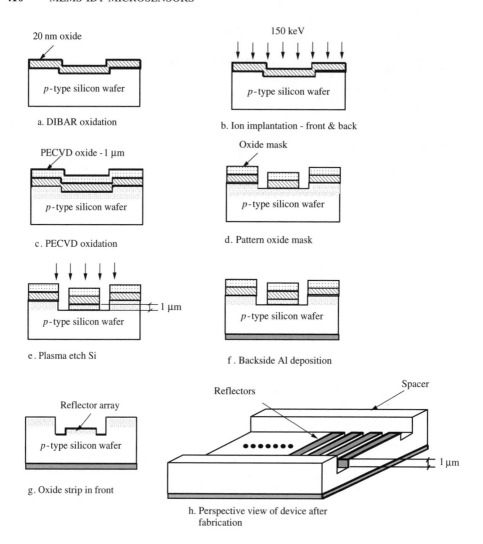

Figure 14.9   Basic steps in the fabrication of the reflectors

area being raised above the rest by the amounts indicated earlier. These raised regions are called *spacers*.

5. The wafers are cleaned and the oxide mask is then etched away. This completes the fabrication of spacers. The view of a single device after the aforementioned steps are completed is shown in Figure 14.8(f).

The process steps for the fabrication of reflectors are as follows:

1. A thin layer (20 nm) of silicon dioxide is grown in preparation for ion implantation (Figure 14.9(a)). Ion implantation on the wafer before the fabrication of the reflectors was done to make the reflectors more conductive with respect to the base of the wafer.

Ion implantation uses accelerated ions to implant the surface with the desired dopant. This high-energy process causes damage to the surface. The implantation was done using an LPCVD oxide in order to reduce the surface damage, as the surface planarity of the reflector is desired (Figure 14.9(b)).

2. Boron ions are implanted into the silicon wafer at 150 keV. The concentration of the dopant is 5×10^{15}/cm^2. Both the front and the back of the wafer are ion implanted. This dosage of ions will serve to make the doped region approximately ten times more conductive than the undoped region.

3. The wafers are then annealed to release any stress in the wafer. This is followed by plasma-enhanced chemical vapour deposition (PECVD) of a 1 μm thick oxide layer. This oxide layer will serve as a mask for the dry etching step to follow (Figure 14.9(c)).

4. The oxide is patterned and developed as described in the fabrication of the spacers (Figure 14.9(d)). The second mask is aligned to alignment marks that were put down during the fabrication of the spacers. This will ensure that the spacers and the reflectors are properly aligned with respect to each other.

5. The silicon is dry-etched in a plasma. The depth of the etch is 1 μm. This results in the formation of 1 μm thick reflectors (Figure 14.9(e)).

6. The backside of the wafer is sputtered with aluminum (0.6 μm) to allow grounding of the wafer (Figure 14.9(f)).

7. The oxide is finally stripped from the front (Figure 14.9(g)). The completed device is shown in Figure 14.9(h).

An array consisting of 200 reflectors is placed between the two IDTs. These reflectors cover nearly the entire space between the two IDTs. The spacer height was 100 nm. This allows the reflectors to be placed 100 nm above the substrate on which the Rayleigh wave propagates.

A study of this device showed the following:

1. The reflections from this set of reflectors was clearly seen in the region between 0 and 3.5 μs (Figure 14.10). The reflection is about 5 dB above the reference signal. The reflection is broadband because of the large number of reflectors in the array.

2. The purely electrical reflections are due to a suspended array of reflectors that can be detected, validating the design concept.

3. The spacer is able to place the reflector array adequately close to the substrate, allowing the electric field to interact with the reflectors. This is achieved without perturbing the mechanical boundary condition.

4. The reflection from a reflector array can be easily measured using the reflection coefficient (S_{11}) measurement of the network analyser.

With this experiment, the effect of moving the reflector array has been clearly demonstrated. In an accelerometer, this effect is due to the instantaneous acceleration sensed at that moment. Thus, the same method can be used to measure acceleration. Now, we are ready to build the accelerometer.

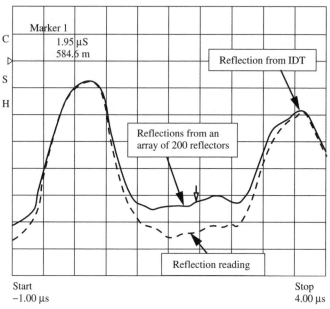

Figure 14.10   Reflections measured from an array of 200 reflectors

14.5 WIRELESS READOUT

The wireless accelerometer is finally created by the flip-chip bonding of the silicon seismic mass with 200 reflectors to that of the silicon substrate with 100 nm height above the SAW device. The IDTs are inductively connected to an onboard antenna, which is a dipole that communicates with the interrogating antenna, as shown in Figure 14.11. The inductive coupling permits an air gap between the SAW substrate and the antenna, which prevents stresses on the antenna from affecting the SAW velocity. Depending on the mounting and reader configuration, several techniques can be used to increase the gain of this antenna. For a planar configuration, a miniature Yagi-Uda antenna can be formed by adding a reflector and/or a director as in Figure 14.11. For a normal reader direction, a planar reflector behind the dipole can be used. In the case where the sensor is mounted on a metal structure, the structure itself is the reflector. By increasing the gain of the sensor antenna, the effective sensing range can be significantly increased. For example, doubling the gain will quadruple the signal strength sent back to the reader.

For the acceleration measurement, a simple geophone setup was used from Geospace. Figure 14.12 illustrates the layout of a geophone. The acceleration in the geophone causes relative motion between the coil and the magnet. This relative motion in a magnetic field causes a voltage that can be calibrated for the acceleration measurement. The geophone is attached to a plate on which the MEMS-IDT accelerometer is mounted. The plate is

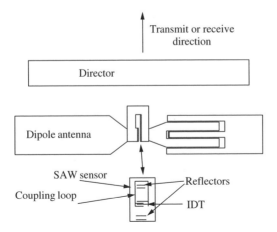

Figure 14.11 Remote antenna interface with SAW sensor. The loop on the SAW sensor is mounted in close proximity to the loop between the poles of the antenna

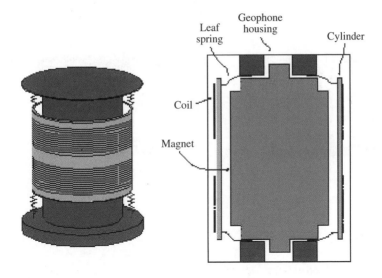

Figure 14.12 Basic arrangement of a geophone for acceleration measurement (Geospace, USA)

then subjected to acceleration. The acceleration is recorded from the output voltage of the geophone. Simultaneously, the phase shift of the SAW signal is also measured, as described earlier. The phase shift of the acoustic wave signal is then a measure of the acceleration of the device, and the results are plotted in Figure 14.13.

Programmable accelerometers can be achieved with split-finger IDTs as reflecting structures (Reindl and Ruile 1993). If IDTs are short-circuited or capacitively loaded, the wave propagates without any reflection, whereas in an open circuit configuration, the IDTs reflect the incoming SAW signal (see Figure 14.14). The programmable accelerometers can thus be achieved by using external circuitry on a semiconductor chip using hybrid technology.

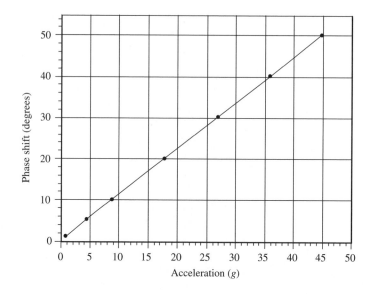

Figure 14.13 Effect of linear acceleration on the phase shift of MEMS microsensor

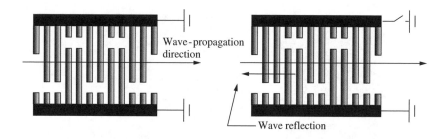

Figure 14.14 Design of programmable reflectors

14.6 HYBRID ACCELEROMETERS AND GYROSCOPES

The design of a MEMS device incorporating both an accelerometer and a gyroscope on a single silicon chip is shown in Figure 14.15. It consists of

1. IDTs for generating SAW waves

2. A floating seismic mass for sensing acceleration and a perturbation mass array for sensing the gyro motion

Again, silicon with a ZnO coating is chosen as the SAW substrate. The IDTs are sputtered on the substrate. The fabrication steps involve mask preparation, lithography, and etching. The thickness of the metal for the IDTs should again be at least 200 nm in order to make adequate electrical contact. The metallisation ratio for the IDTs is still 0.5.

The fabrication of the seismic mass is again realised by the sacrificial etching of silicon dioxide. The steps involved are as follows:

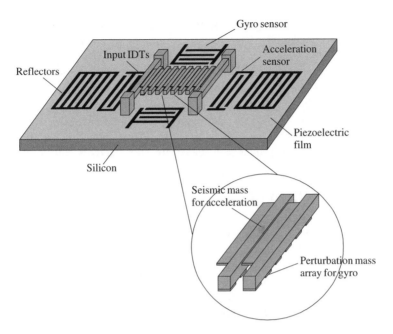

Figure 14.15 Basic design of a MEMS-IDT microsensor system that combines an accelerometer with a gyroscope on a signal chip

1. A sacrificial oxide is thermally grown on a second silicon wafer.

2. A polysilicon (structural layer) is then deposited by LPCVD on the sacrificial layer. The polysilicon is patterned to form the seismic mass and etched with EDP.

3. The perturbation mass array for gyro sensing is deposited on this seismic mass.

4. The sacrificial layer is then etched with HF to finally release the seismic mass and the perturbation mass array.

5. The seismic mass is then flip-chip bonded to the SAW silicon substrate.

The floating reflectors (seismic mass) can move relative to the substrate, and this displacement is proportional to the acceleration of the body to which the substrate is attached. This displacement is then measured as a phase difference of the reflected acoustic wave, which can be calibrated to measure the acceleration. This phase shift can be detected at the accelerometer sensor port of the device. It should also be noted that the strategically positioned metallic mass arrays on the underside of the seismic mass would change the coupling between the SAW at the Gyro sensor port because of the rotation and Coriolis force generation. This is sensed as the rate information for the gyroscope. When the electromagnetic signal is converted to an acoustic signal on the surface of a piezoelectric, the wavelength is reduced by a factor of 10^5. This allows the dimensions of acoustic wave devices to be compatible with IC technology.

The main advantages of a single device for the measurement of both angular rate and acceleration is the reduction in power requirements, signal-processing electronics, weight, and overall cost. These advantages are also important for its use in many commercial, military, and space applications. Thus, it has a number of advantages over the tuning fork and ring microgyroscopes, which were described in Chapter 8. Indeed, this type of MEMS-IDT device could revolutionise the MEMS industry with widespread application, for example, in geostationary positioning system (GPS), guidance systems, industrial platform stabilisation, tilt and shock sensing, motion-sensing in robotics, vibration monitoring, automotive vehicle navigation, automatic braking systems ABS, antiskid control, active suspension, integrated vehicle dynamics, three-dimensional mouse, head-mounted display, gaming, and medical products (wheel chairs, body movement monitoring).

14.7 CONCLUDING REMARKS

In this chapter, we have introduced the concept of combining a micromachined mechanical structure with an IDT microsensor to make a so-called MEMS-IDT microsensor. Accordingly, we have shown how to fabricate a MEMS-IDT accelerometer and gyroscope. This type of MEMS device is particularly attractive because it offers the possibility of a simple wireless and batteryless mode of operation. Such sensing devices will be needed in a wide variety of future applications from military through to the remote interrogation of surgical implants.

REFERENCES

Ballantine, D. S. *et al.* (1997). *Acoustic Wave Sensors – Theory, Design and Physico-Chemical Applications*, Academic Press, New York, pp. 72–73.

Esashi, M. (1994). "Sensors for measuring acceleration," in H. Bau, N. F. de Rooij, and B. Kloek, eds., *Mechanical Sensors*, Wiley-VCH, Verlag, p. 331.

Geospace, LP 7334 N. Gessner, Houston, Texas 77040 (*www.geospacelp.com*).

Matsumoto, Y. and Esashi, M. (1992). Technical Digest of the 11th Sensor Symposium, p. 47.

Reindl, L. and Ruile, W. (1993). "Programmable reflectors for SAW ID-tags," *Ultrasonics Symp. Proc.*, **1**, 125–130.

Roylance, L. M. and Angell, J. B. (1979). "A batch-fabricated silicon accelerometer," *IEEE Trans. Electron Devices*, **26**, 1911–1917.

Rudolf, F., Jornod, A. and Beneze, P. (1987). Digest of Technical Papers of Transducers '87, Institute of Electrical Engineers of Japan, Tokyo, p. 395.

Seidel, S. *et al.* (1990). "Capacitive silicon accelerometer with highly symmetrical design," *Sensors and Actuators A*, **21**, 312–315.

Subramanian, H. *et al.* (1997). "Design and fabrication of wireless remotely readable MEMS based accelerometers," *Smart Materials Struct.*, **6**, 730–738.

Suzuki, S. *et al.* (1990). "Semiconductor capacitance-type accelerometer with PWM electrostatic servo technique," *Sensors and Actuators A*, **21**, 316–319.

Varadan, V. K., Varadan, V. V., and Subramanian, H. (2001). "Fabrication, characterization and testing of wireless MEMS-IDT based microaccelerometers," *Sensors and Actuators A*, **90**, 7–19.

15

Smart Sensors and MEMS

15.1 INTRODUCTION

The adjective 'smart' is widely used in science and technology today to describe many different types of artefacts. Its meaning varies according to its particular use. For example, there is a widespread use of the term *smart material*, although functional material is also used and may be a more accurate description. A smart material may be regarded as an 'active' material in the sense that it is being used for more than just its structural properties. The latter is normally referred to as a *passive* material but could, perhaps, be called a *dumb* material. The classical example of a so-called smart material is a shape memory alloy (SMA), such as NiTi. This material undergoes a change from its martensitic to austenitic crystalline phase and back when thermally cycled. The associated volumetric change induces a stress and so this type of material can be used in various types of microactuator and microelectromechanical system (MEMS) devices (Tsuchiya and Davies 1998). Another example of a smart material is a magnetostrictive one, which is a material that changes its length under the influence of an external magnetic field. This type of smart material can be used to make, for example, a strain gauge as provided in the Worked Example 8.2 in Chapter 8 on Microsensors.

The term *smart* is also applied in the field of structures. However, a smart structure is, in general, neither a small structure nor one made of silicon. In this case, as we shall see later on, the term really implies a form of intelligence and is applied to civil buildings and bridges (Gandhi and Thompson 1992). A classic example of a smart structure is that of a building that contains a number of motion sensors together with an active damping system. Therefore, the building can respond to changes in its environment (e.g. wind loading) and modify its mechanical response appropriately (e.g. through its variable damping coefficient). Perhaps, a more familiar way that engineers would describe this type of structure is one with a closed-loop control system (Bissell 1994).

In this chapter, we are interested, specifically, in the topic of smart devices rather than either smart materials or smart structures. Readers interested in these other topics are referred to a book on '*Smart Materials and Structures*' (Culshaw 1996). The term *smart sensor* was first coined in the 1980s by electrical engineers and became associated with the integration of a silicon sensor with its associated microelectronic circuitry. Figure 15.1 shows the basic concept of a smart sensor in which a silicon sensor or microsensor (i.e. integrated sensor) is integrated with either a part or all of its associated processing elements (i.e. the preprocessor and/or the main processing unit). These devices are referred to here, for convenience, as smart sensor types I and II. For example, a silicon thermodiode could

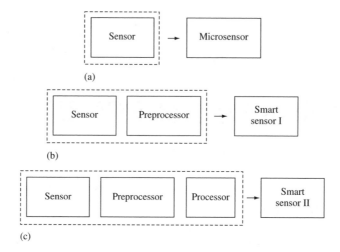

Figure 15.1 Basic concept of integrating the processing elements with an integrated sensor (microsensor) to make different types of smart sensor. The dotted lines show the integration process for one or more of the elements

be integrated with a constant current circuit to make a simple three-terminal voltage supply (+5 V DC), ground and output (0 to 5 V DC) smart device. Of course, this is a trivial example and barely deserves the title of smart; nowadays, the term tends to imply a higher degree of integration, such as the integration of an eight-bit microcontroller or microprocessor. This would be referred to here as a type II smart sensor. When the technologies and processes employed to make the microsensor are incompatible with the microprocessor, it is possible to make a hybrid rather than a true smart chip, as described earlier in Chapter 4. In this case, the term *smart* is sometimes used in a less formal sense, but hybrid would be a more accurate term.

The integration of part (type I) or all (type II) of the processing element with the microsensor in order to create a smart sensor is highly desirable when one or more of the following conditions are met:

- Integration reduces the unit manufacturing cost of the device.
- Integration substantially enhances the performance.
- The device would not work at all without integration.

These prerequisites make integration feasible when there is either a large potential market (i.e. millions or more units per year) demanding that the unit cost be kept low, or there is a specialised 'added value' market that can absorb the higher unit costs associated with smaller chip runs. Sometimes, these so-called market drivers are combined to define a performance–price (PP) indicator. This concept was introduced in Chapter 1 in which it was shown that there has been an enormous increase in the PP indicator during the past 20 years – first with silicon sensors (i.e. microsensors) and then with smart sensors.

The successful commercialisation of pressure and other smart sensors (see the following text) has led to a whole host of other types of smart devices, such as smart actuators, smart interfaces, and so on. Figure 15.2 gives a schematic representation of both a smart actuator and a smart microsystem. Of course, a MEMS device is one type of smart

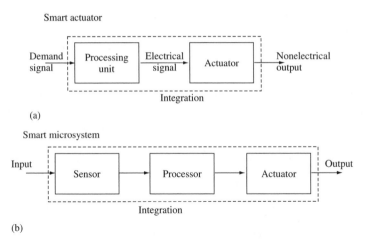

Smart actuator

Smart microsystem

Figure 15.2   Basic architecture of (a) a smart actuator and (b) a smart microsystem (or MEMS)

Table 15.1   Some different uses of the prefix 'smart' today

Description	Meaning	Example
Smart material	Material with a function other than passive mechanical support	Shape-memory alloy
Smart structure	Civil structure that adapts to changes in its environment	Building with an active damping system
Smart sensor	Microsensor with part, or all, of its processing unit integrated into one chip	Commercial (e.g. Motorola) automotive pressure sensor
Smart actuator	Actuator with part, or all, of its processing unit integrated into one chip	Micromotor
Smart controller	Microcontroller that automatically calibrates or compensates	Fuzzy controller
Smart electronics	Electronic systems have some embedded form of intelligence	Neuronal chip (analogue VLSI)
Smart microsystem	Sensor, processor, and actuator integrated in a single chip	MEMS or MOEMS devices

microsystem because a microsystem need not involve an electromechanical component. Some examples of these are given later.

Table 15.1 summarises the different uses of the term *smart* of today together with its meaning and an example of such a material, device, or other artefact.

Finally, we must draw a distinction between a smart device and an intelligent device. To the average person, the term *smartness* suggests a high level of intelligence rather than a high level of chip integration. Yet, there are a number of researchers who have used the term *intelligent instrument* to be one in which a microprocessor is used to control a piece of equipment (Barney 1985; Ohba 1992). For example, they would regard a large drilling machine controlled by a microprocessor as an intelligent instrument. This is quite different from the meaning of intelligence used by either cognitive scientists or, probably, the average person. Clearly, the term *intelligent* is a relative one, and so here we prefer to consider intelligence associated more with functionality than form, thus differentiating its usage from the term *smart*. This is consistent with an early definition of intelligent sensors proposed by Breckenbridge and Husson (1978):

'The sensor itself has a data processing function and automatic compensation function, in which the sensor detects and eliminates abnormal values or exceptional values. It incorporates an algorithm, which is capable of being altered, and has a certain degree of memory function. Further desirable characteristics are that the sensor is coupled to other sensors, adapts to changes in environmental conditions, and has a discrimination function.'

Nowadays, many of these so-called intelligent features are incorporated into smart sensors. So, this early definition of Breckenbridge and Husson (1978) can be updated by drawing upon more recent ideas published by others, such as Brignell and White (1994). The different possible forms (or classes) of an intelligent sensor are provided in Table 15.2, together with a working definition and an example of such a device.

The meaning of the term *intelligent* appears to be changing with time and it is generally used to describe a new device that is demonstrably superior in performance to those existing currently. Thus, the meaning of the word itself is subjective and evolving (or adapting) over the course of time. Consequently, the ability of a device simply to respond to its changing environment (e.g. temperature compensation) appears to be of relatively low level of intelligence today and hardly deserves the title. Instead, we tend to compare the intelligence of a device with the workings of a biological organism. Consequently, intelligent devices may now have embedded artificial intelligence algorithms, such as artificial neural networks (Fausett 1994) or expert systems (Sell 1986). These algorithms imbue the devices with humanlike features, such as fault-tolerance, adaptive learning, and complex decision making.

Table 15.2 Different types of device intelligence, starting with the lowest class

Class	Description	Example
1. Signal compensation	Device automatically compensates for changes in an external parameter, e.g. temperature	Temperature-compensated silicon microaccelerometer.
2. Structural compensation	Physical layout designed to reduce signal-to-noise ratio, enhances functionality.	Resistive gas sensor pair with differing geometry[1]
3. Self-testing	Device tests itself out and so has self-diagnostic capability.	ADC chips
4. Multisensing	Device combines together many identical or different sensors to improve performance.	Electronic nose[2]
5. Neuromorphic	Device shares characteristic with a biological structure, such as parallel architecture or neural network processor.	Cellular automata and neuromorphic VLSI chips

[1] See Gardner (1995); here we mean intelligent design.
[2] See Gardner and Bartlett (1999).

15.2 SMART SENSORS

There are many different types of smart sensor today, as defined by a microsensor that has had part, or all, of its processing unit integrated with it. We have already described in this book a number of different silicon micromachined sensors, and some of these possess on-chip (integrated) electronics, and so could merit the title of *smart sensor*. Here, we will describe some different examples of smart sensors and thus demonstrate some of the reasons for making smart sensors.

The most successful types of microsensors today are those that have been developed for the high-volume automotive industry. An important class is the silicon-based pressure sensors employed to measure manifold, barometric, exhaust gas, fuel, tyre, hydraulics, and climate control pressure (Section 8.4.5). The current market alone is estimated to be worth more than 750 million euros. The market for silicon pressure sensors is, perhaps, the most mature, and since it involves many different manufacturers, it has become very competitive. Consequently, there has been an enormous effort in recent years toward both cost reduction and added functionality of these devices through the integration of many processing functions onto a single chip. For example, Figure 15.3 shows the layout of a bulk micromachined pressure sensor with its analogue custom interface, eight-bit

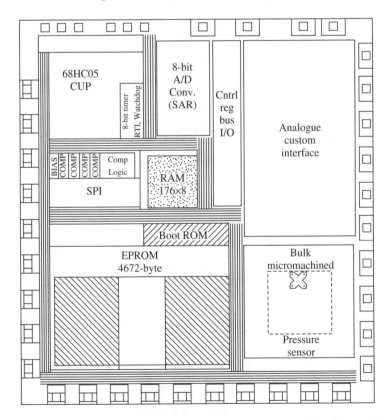

Figure 15.3 Layout of a smart, integrated silicon pressure sensor. The chip combines a bulk micromachined pressure, its analogue interface with a digital eight-bit microcontroller and local memory. See Appendix A for definition of abbreviations. Redrawn from Frank (1996)

analogue-to-digital converter, microprocessor unit (Motorola 68HC05), and the memory and serial port interface (SPI) in a single package (Frank 1996).

Front-side bulk silicon micromachining is now available with integrated complementary metal oxide semiconductor (CMOS) electronics. So, much of the CMOS circuitry can be integrated either before the CMOS process (pre-CMOS) or after the CMOS process (post-CMOS). In this way, it is possible to import the latest microprocessor dye and miniaturise the silicon chip still further.

The silicon pressure sensors available today not only cost a few euros but also have fewer connectors and so have enhanced their 'smartness.' Worked Example (6.8) has been presented in an earlier chapter and describes the integration of a capacitive pressure sensor with a local digital readout. The reduction of the pad count may seem trivial but is not since much of the cost to manufacture a chip is associated with its area (and so number of pads) and packaging requirements (wire/tab bonding).

The same incremental improvement in the other major type of smart automotive sensor – the microaccelerometer (Section 8.4.6) – has also been observed in recent years. The current US market is worth some 250 million euro and uses smart microaccelerometers in air bag, automatic braking, and suspension systems. For example, major manufacturers, such as Motorola, Analog Devices, Lucas NovaSensor and Bosch, make increasingly smart microaccelerometers with intelligent features.

- Damping and overload protection (fault-tolerance)

- Compensation for ambient temperature (e.g. -40 to $+80\,^\circ$C)

- Self-testing for fault-diagnostics

Figure 15.4 shows an interim two-chip solution to an accelerometer with the g-cell having a self-test facility and separate interface integrated circuits (ICs) (Frank 1996).

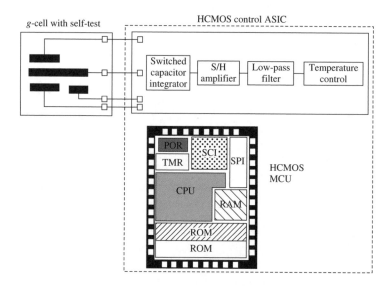

Figure 15.4   Schematic of a two-chip microaccelerometer featuring a g-cell with a self-test facility and an HCMOS sensor interface IC with an MCU (Redrawn from Frank (1996))

This solution provides both a shorter design cycle and lower initial cost for the sensor. Of course, larger price-driven markets favour a one-chip solution. The self-test feature provides a certain level of intelligence and is important in applications in which sensor failure is regarded as safety-critical.

Several books have been published on the topic of smart sensors, and automotive sensors feature strongly in many of them. Interested readers are referred to recent books by Chapman (1996), Frank (1996), Madou (1997) and van der Horn and Huijing (1998).

Another type of smart sensor is one that requires the integration of its associated electronics for functional rather than cost reasons. The most obvious case of this is the charge coupled device (CCD) array device. In a CCD, there is a large number of identical silicon elements (e.g. 1024 pixels[1]) that sense the level of light falling on them and because of their small size, produce a relatively low strength of signal. Consequently, on-chip electronics are needed to measure, first of all, the very small amounts of charge located on each silicon element and, secondly, this charge on a large number of identical elements in the array (e.g. 1 million).

Figure 15.5(a) shows a photograph of a commercial colour frame-transfer CCD image sensor (FXA1012 Philips) that is used in a CCD camera. The smart chip has two million active pixels. The integrated electronics use shift registers to output the light levels and then convert the signal to a standard format, for example, a data rate of up to 25 MHz and 5 frames per second. These chips are then used in various consumer electronic items, for example, digital cameras, videos, and so forth. Digital cameras are now manufactured in large quantities for a variety of applications, from security surveillance to robot vision[2]. The low-cost end of the market with low-resolution black and white chips (~100 euros) has now expanded enormously with the advent of cameras attached to the personal computer (PC) – the so-called *web camera* – that are rapidly becoming in common use in many offices and homes (Figure 15.5(b)).

Table 15.3 lists some commercially available optical CCD chips and some low-cost web cameras for PC mounting with an integrated digital serial (RS-232) or, increasingly, universal serial bus (USB) interface.

The latest and, perhaps, smartest type of optical array sensor being made today is the silicon retina. In this device, a large number of optical elements (pixels) are configured in an axisymmetric geometry to create a silicon eye (see Figure 15.5(c)). This geometry does not fit well with the rectilinear design rules used to layout most silicon chips but is an interesting concept and permits radially based pattern-recognition (PARC) algorithms.

Another smart sensor is the so-called *electronic nose* (Gardner and Bartlett 1999). An electronic nose has been defined by Gardner and Bartlett (1994) as follows:

'An electronic nose is an instrument, which comprises an array of electronic chemical sensors with partial specificity and an appropriate pattern-recognition system, capable of recognising simple or complex odours'

The electronic nose first aroused serious attention in the early 1980s with the first commercial versions beginning to appear in the mid-1990s. Figure 15.6 shows the basic architecture of an electronic nose in which an unknown complex odour j has a set of different odour concentrations c_j and these are detected by an array of n nonspecific

---

[1] A pixel is a 'picture element.'
[2] The topic of robot vision is well established and is described by Pugh (1986).

(a)   (b)

(c)

Figure 15.5   (a) Smart optical sensor: a colour frame transfer CCD image sensor with two million active pixels (1616 H by 1296 V) (Courtesy of Philips); (b) a web camera that contains a CCD device (Courtesy of Intel); and (c) a silicon retina (Courtesy of IMEC, Belgium)

Table 15.3   Some commercial smart optical microsensors: CCD chips and web cameras

Manufacturer	Model	Pixels (size)	Architecture	Max. frame rate (fps)	Dark current density (at 25 °C)/ typical unit price
CCD chips:					
Thomson	BV512AI	512 × 512 (19 μm)	Full frame	22	25/350 pA/cm^2
Philips	FXA 1012	1616 × 1296	Colour frame	5	–
Kodak	BV20CAC	2032 × 2044 (9 μm)	Full frame	2.2	10 pA/cm^2
Philips	BV40CAC	4096 × 4096 (9 μm)	Full frame	0.6	10 pA/cm^2
Web cameras:					
Logitech	Quick Cam Pro	540 × 480	Colour	30	140 euro
Philips	Vesta Pro	800 × 600 Super VGA	Colour	30	115 euro
Intel	Me2Cam™	640 × 480	16-bit colour	–	70 euro

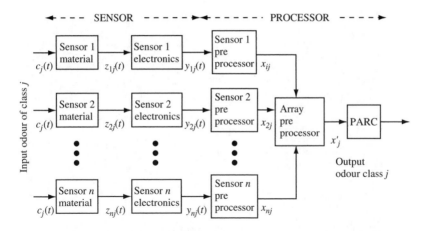

Figure 15.6 Basic architecture of an electronic nose. From Gardner and Bartlett (1999)

sensors (Gardner and Bartlett 1999). The signals from this sensor array are then processed through a number of stages via associated analogue and digital electronic circuitry before sophisticated algorithms classify the odour in a PARC system.

Figure 15.7 shows one of the first commercial electronic nose systems called the *Fox 2000*[3] and manufactured by Alpha MOS (France). It uses an array of the Taguchi-type resistive gas sensor described in Chapter 8. A modified unit has been applied recently to the problem of identifying an algae bloom called *cyanobacteria* from its odorous headspace. This bloom is found in lakes and reservoirs and can produce toxins that are a hazard to the health of both cattle and humans (Shin 1999). The electronic nose was used to analyse the headspace of both a toxin-producing strain and a nontoxin-producing strain of *cyanobacteria*. The output from the multidimensional array is shown as a principal components plot and a clear separation is observed between the two strains (i.e. PCC 7806 and 7941).

Today, there are more than 10 companies making electronic nose instruments that range from handheld units costing 8000 euros to large bench-top instruments costing more than 100 000 euros. Table 15.4 lists some of these commercial e-noses together with the type of sensors employed.

The most recent (1999) launch is a handheld electronic nose by Cyrano Sciences (USA) that incorporates an array of 32 carbon-polymer composite resistive sensors on a hybrid sensor substrate (Figure 15.8). The black polymer materials (dots on substrate) were first developed at Caltech and then the technology was transferred to Cyrano. A stand-alone unit costs around 8000 euros and is used to identify unknown odours or vapours.

Possibly, the most advanced smart e-nose is that reported by Baltes and Brand (2000) that uses CMOS technology to fabricate arrays of chemical microsensors and to integrate the associated electronics. Figure 15.9 shows two examples of CMOS chemical sensors. The first is an array of polymer-coated capacitors with integrated CMOS electronics, whereas the second shows two cantilever beams with a piezoresistive pickup element. The cantilever beams are coated with different vapour-absorbing polymer coatings, and the deflections are calibrated against known compounds such as *n*-octane, ethanol, and

---

[3] Based on work at the University of Warwick, UK.

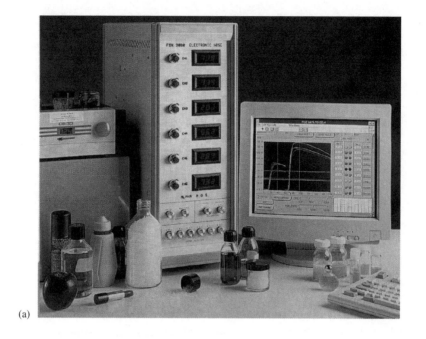

(a)

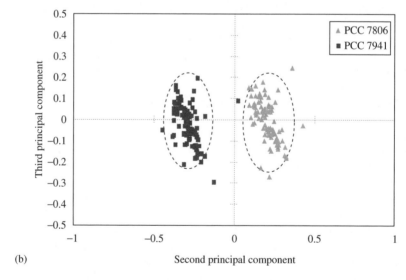

(b)   Second principal component

Figure 15.7   (a) Commercial electronic nose: Fox 2000 and (b) classification of a hazardous algal bloom (cyanobacteria) found in reservoirs and lakes by the characteristic smell of its headspace

toluene. The advantage of using either polymer or, indeed, polymer composite sensing materials is that the chip is both CMOS-compatible and requires low power.

For some applications, metal oxide materials are often appropriate and these require a high operating temperature (above 300 °C). Work on silicon micromachined hotplates has already been discussed in the section on Chemical Sensors in Chapter 8, and these have been used to fabricate low-power resistive gas sensors for electronic noses (Pike

Table 15.4   Some commercial electronic noses available today

Product name	Supplier	Started	Sensor no./type	Comments
Bloodhound	University of Leeds Innovations Ltd	1995	32 CP resistors	Small company. Instrument based on research at Leeds University.
Cyranose 320	Cyrano Sciences Inc, USA	1999	32 Polymer composite resistors	Small US company making handheld units based on materials from Caltech.
e-NOSE 4000	Neotronics Scientific Ltd, UK	1995	12 CP resistors	Medium-sized company. Available with autosampler and now part of Marconi plc (UK).
Fox 2000	Alpha MOS, France	1993	18 MOS (4000 unit)	Medium-sized company. Autosampler and air conditioning unit available.
Moses II	Lennartz Electronics, Germany	1996	MOS/QCM	Modular system based on research at the University of Tubingen, Germany.
Nordic sensors	Nordic sensors, Sweden	1995	4 MOSFET	Small company using devices developed by Linköping University. Now part of Applied Sensor Inc.
Olfactometer	HKR Sensorsysteme GmBH, Germany	1994	6 QCM	Small company. Based on research at the University of Munich, Germany.
Osmetech	Osmetech Plc, UK	1994	32 CP resistors	Medium-sized company. Market leader in 1997. Autosampler and air conditioning unit available.
Rhino	USA	1994	4 MOS	Early instrument may no longer be available.
ScanMaster II	Array Tec	1996	8 QCM	Small company. Launch November 1996.
Scentinel	Mastiff Electronic Systems Ltd	–	16 CP resistors	Small company. Instrument based on research at the Leeds University aimed at sniffing palms for personal identification.

1996; Pike and Gardner 1998). However, these cannot be readily integrated with the electronics to make a smart sensor. An alternative approach, proposed by Udrea and Gardner, is to use silicon-on-insulator (SOI) technology to make gas and odour sensors (Udrea and Gardner 1998). Figure 15.10 shows the basic principle of using a field-effect transistor (FET) microheater and SOI membrane to form a low-power platform with an integrated thermal management system. Multiple trench isolation can be used to reduce the heat lost by conduction through the SOI membrane. Simulations suggest that a p-type metal oxide semiconductor (p-MOS) FET heater permits higher operating temperatures (about 50 °C more) than an n-type metal oxide semiconductor (n-MOS) heater. Recent simulations have shown that temperatures of 350 °C are achievable with a FET heater

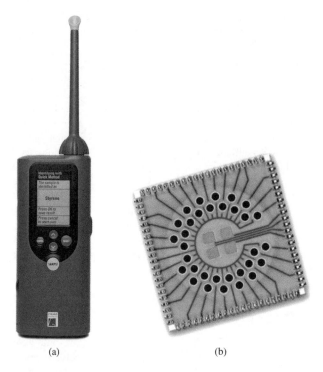

(a)   (b)

Figure 15.8   (a) A commercial handheld electronic nose from Cyrano Sciences and (b) its 32-element polymer composite (black dots) screen-printed chip. The four central squares are thermistors, and a heater pad has been printed on the reverse side

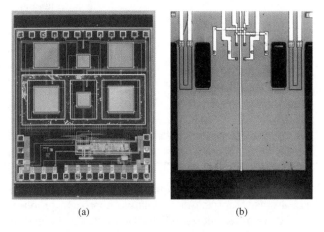

(a)   (b)

Figure 15.9   CMOS micronoses: (a) array of gas-sensitive polymer-coated capacitors and (b) a silicon cantilever beam with piezoresistive pickup. Courtesy of ETH, Zurich

with local oxide isolation of silicon (LOCOS) isolation on a thin SOI platform (Udrea *et al.* 2001). A power consumption of about 10 mW and a millisecond response should permit their use as low-power and, possibly, thermally modulated gas (or odour) sensors in handheld units.

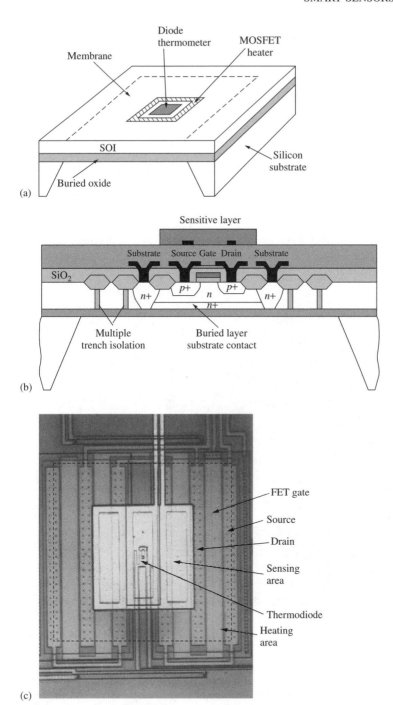

Figure 15.10 Smart gas sensor using SOI CMOS technology: (a) basic configuration of an SOI microhotplate; (b) cross section of the device layout with multiple trench isolation; and (c) SOI chip fabricated using a standard SOI CMOS process

Much of the research effort in the field of electronic noses today is directed toward making smaller, smarter, and more intelligent units. Some success has been achieved through the use of artificial neural networks (Gardner and Hines 1996) and, more recently, dynamical signal-processing techniques (Hines *et al.* 1999). These algorithms are required to overcome some of the undesirable problems that gas sensors currently tend to suffer from, such as

- Long-term drift in the baseline (in air) sensor signal

- Sensitivity to variations in ambient temperature and humidity

- Poisoning of the sensors by airborne contaminants (e.g. silicones, sulphur based compounds etc.)

These problems are not trivial and are a real issue in the practical application of electronic nose technology today. As mentioned earlier, researchers are increasingly applying methods employed by the human system to solve the problems encountered in sensors

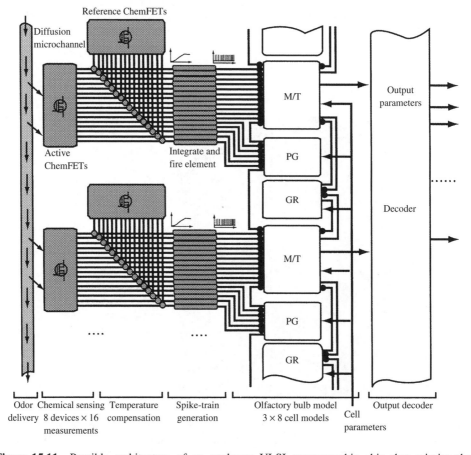

Figure 15.11   Possible architecture of an analogue VLSI neuromorphic chip that mimics the various parts of the human olfactory system

today – especially in the field of environmental monitoring. For example, Pearce and Gardner have proposed a neuromorphic analogue, a very large-scale integration (VLSI) chip (Figure 15.11) that mimics the functions of the olfactory epithelium and cortex. The silicon microchannel mimics the mucous coating of the olfactory epithelium, whereas ChemFETs (Chapter 8) are used to represent the olfactory cells. Integrate-and-fire elements are well known in the field of analogue VLSI and are used here to generate a spike train (cf. action potentials) and then the equivalent of the olfactory bulb. Although it is impractical today to emulate the 100 million olfactory neurones present in the entire human olfactory bulb, chips such as this, with added intelligence, help researchers to solve the complex odour classification problems faced.

There have already been reports of some silicon implementations of biological neurones in the field of analogue VLSI. Moreover, there are also some commercial products such as the Motorola MC14315 Neuron™ chip (Frank 1996). This chip is a communications and control processor with embedded LonTalk protocol for multimedia networking environments (Figure 15.12). To date, neither analogue VLSI chips nor neuronal chips have yielded the increase in processing speed envisaged originally. Moreover, because a single neurone comprises several transistors and capacitors, the need for relatively large areas of silicon has so far limited their exploitation because of the high production cost.

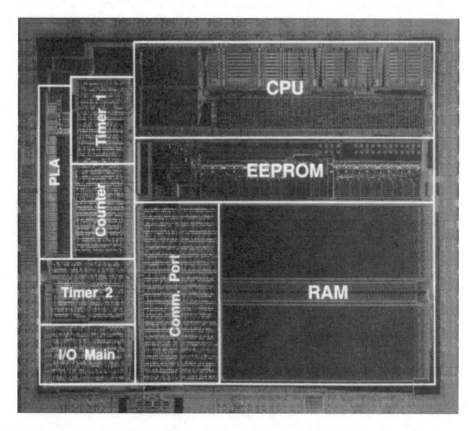

Figure 15.12   A commercial silicon neuronal chip: the Motorola MC143150 Neuron™ chip. From Frank (1996)

Nevertheless, there are many areas in which the process speed can be sacrificed for functionality and thus provide a role for VLSI chips.

Communication is an important aspect of smart sensors, and different types of interfaces have been integrated, ranging from a simple single port to J2058 (Chrysler, US), CAN (Bosch, Germany), A-Bus (Volkswagen AG, Germany), D^{2}B (Philips, Netherlands) and MI-Bus (Motorola, US) in the automotive industries. Interested readers are referred to Frank (1996) for a discussion of this topic.

For smart sensors, the most important development is likely to be wireless communications in which a radio frequency (RF) signal is used to both sense and communicate with other devices. Figure 15.13 shows the wide variety of applications of RF in the automotive industry covering the frequency range from 20 MHz to 100 GHz. The single most important one, probably, is the vehicle navigation and warning system (VNAW) based on geostationary positioning system (GPS) from satellite links and miniature gyroscopes and microaccelerators. Readers interested in silicon micromachined inertial devices are referred to Chapter 8.

One other example of a smart sensor employing wireless communication is the wireless IDT SAW device described in an earlier chapter. Figure 15.14(a) shows the basic arrangement of a passive interdigital transducer-surface acoustic waves (IDT-SAWs) sensor with an integrated antenna. The acoustic waves are excited by an RF pulse, which is then reflected back and the antenna transmits a return signal. This basic idea is being put to use to create a 'smart tongue' – that is, an electronic device that mimics the human sense of taste (Varadan and Gardner 1999). Figure 15.14(b) shows the classification of different liquids in terms of the relative amplitude and phase of the acoustic waves in a Love wave

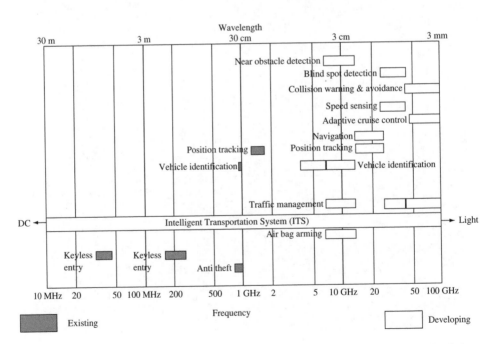

Figure 15.13 Various applications of radio frequency in smart devices for the automotive industry (Frank 1996)

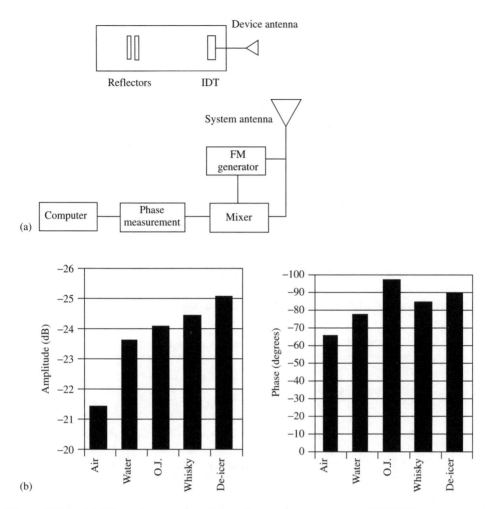

Figure 15.14 (a) Schematic representation of a wireless passive IDT-SAW sensor and (b) discrimination of different liquids through acoustic attenuation and phase parameters. The measurements were made at a peak frequency of 108.7 MHz. From Varadan and Gardner (1999)

IDT sensor.[4] The device may be regarded as an 'electronic tongue' because the sweetness (sugar level) and saltiness (ion concentration) will affect the electrical conductivity and dielectric permittivity of the liquid, whereas the mouth feel and, possibly, the freshness will relate to the viscoelasticity of the liquid. The use of a second Love wave can help to improve the discrimination process by removing (screening out) the effects of electrical coupling from the measurement system. The advantage of this type of liquid sensor is that it does not require a biological coating but relies upon a physical principle. This greatly enhances the lifetime of the sensor, although its specificity is clearly reduced. However, we think that the low cost and rapid response makes the development of a passive wireless smart tongue extremely attractive.

[4] For details of the principles of IDT-SAW microsensors, see Chapter 10

15.3 MEMS DEVICES

Silicon micromachining techniques are used to fabricate various micromechanical structures and many of these form micromechanical actuators. Throughout this book, we have
described a large number of different types of microactuators and some of these are,
perhaps, rather loosely referred to as MEMS devices. Table 15.5 lists some of the different
types of microactuators and MEMS described in the worked examples (WE) in this book.

A more comprehensive list of micromechanical structures is provided in the following
Table (15.6) and has been taken from Frank (1996). These include microvalves, micropumps, microgears, and so on.

MEMS devices can be considered as smart devices because they integrate sensors with
actuators (i.e. a smart microsystem), but the degree of integration can vary significantly.

There are many different types of MEMS devices being made today, but two of the most
exciting ones are used in optical and chemical instrumentation. The former is sometimes
referred to as *microoptoelectromechanical system* (MOEMS), and this is being driven
by the optical telecommunications and biotechnology industries. One example shown
in Figure 15.15 is a picture of an adaptive mirror IC. This array device permits the
electronic correction of an optical system in an adaptive manner. Clearly, the ability to

Table 15.5   Some examples of microactuators and MEMS

Type	Description	Reference
Microactuator	Linear motion actuator	WE 6.2
Microactuator	Rotor on a centre-pin bearing	WE 6.3
Microactuator	Rotor on a flange bearing	WE 6.4
MEMS	Centre-pin bearing side-drive micromotor	WE 6.6
MEMS	Gap comb-drive resonant actuator	WE 6.7
MEMS	Overhanging microgripper	WE 6.10

Table 15.6   Various micromechanical and MEMS devices made from silicon. (This table
is based on Frank (1996))

Cryogenic microconnectors	Microgears	Microprobes
Fibre-optic couplers	Micromoulds	Micropumps
Film stress measurement	Micromotors	Microswitches
Fluidic components	Micropositioners	Microvacuum tubes
IC heat-sinks	Microinterconnects	Nerve regenerators
Ink-jet nozzles	Microchannels	Photolithographic masks
Laser beam deflectors	Microrobots	Pressure switches
Laser resonators	Micromachines	Pressure regulators
Light modulators	Micromanipulators	RMS converters
Membranes	Micromechanical memory	Thermal printheads
Microaccelerometers	Microgyrometers	Thermopiles
Microairplanes	Microchromatographs	Torsion mirrors
Microaligners	Microinterferometers	Vibrating microstructures
Microbalances	Microspectrometers	
Microfuses	Micro-SEM	

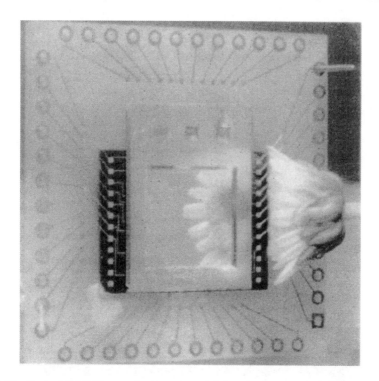

Figure 15.15   Smart MOEMS: an adaptive mirror IC for electronic correction of optical system (Courtesy of Delft University)

modify electronically the characteristics of a mirror offers the opportunity of automated focusing, beam direction and so forth.

A second example of a smart optical MEMS chip is a visible light microspectrometer based on a monolithic dielectric slab waveguide with monolithic integrated focusing echelle grating produced by a micromoulding process, as in Figure 15.16. The spectrum is generated by coupling the light through a silica fibre into a three-layer waveguide. The light is split by a self focussing reflection grating ($d = 0.2$ µm, $g = 2$ µm) with 625 lines per mm, and the different wavelengths are read out by a 256-element photodiode array. The device is small and has a wavelength reproducibility better than 0.1 nm. Measuring wavelength content is important in many sensing applications, from security labeling through to biological assays.

In the chemical analytical instrumentation area, Stanford reported the first silicon micromachined gas chromatography system as early as 1974. Figure 15.17(a) shows the basic layout of a gas chromatography system (Madou 1997) in which a gaseous sample is made to travel along a long capillary column. The various components in the sample have different retention times within the column and so appear at the far end as individual compounds. The chemical components are detected using, for example, a flame ionisation detector and the peak is identified according to known retention times. Both the capillary column and detector cavity has been micromachined. Figure 15.17(b) shows a minigas system, manufactured by MTI, which is a chromatograph that is essentially portable and so suitable for remote fieldwork.

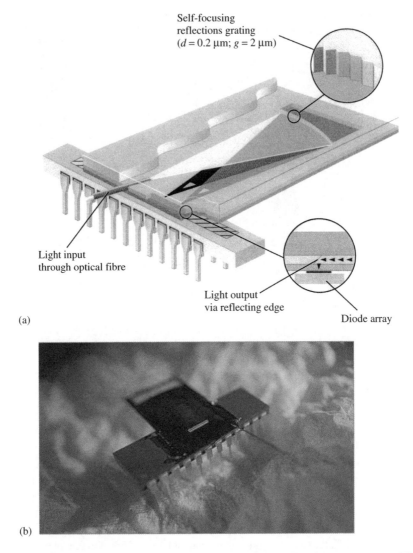

Self-focusing
reflections grating
($d = 0.2$ μm; $g = 2$ μm)

Light input
through optical fibre

Light output
via reflecting edge

Diode array

(a)

(b)

Figure 15.16 Smart MOEMS: (a) principle and (b) photograph of a microspectrometer IC. (Courtesy of MicroParts, Germany)

Efforts have also been made to make another miniature analytical chemical instrument, namely, the mass spectrometer. This instrument ionises the molecules after they have passed through an orifice and then separates the masses out in a mass filter by deflecting their motion with a quadrupole electrostatic or any other type of lens. Finally, the abundance of mass ions are measured using an ion detector (Figure 15.18(a) from Friedhoff *et al.* (1999)). Figure 15.18(b) shows a plot of the mass content of the headspace above a bacteria hazardous to human health, *Escherichia coli*, in two of its growth phases (Esteves de Matos *et al.* 2000). The measurements were taken with a conventional Agilent Technologies quadrupole mass spectrometer 4440. The difference between these mass spectrograms shows up when a pattern analysis technique, such as principal components

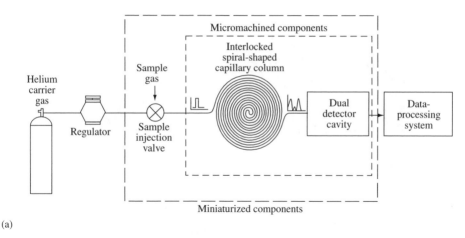

(a)

(b)

Figure 15.17 Microgas analysis: (a) basic arrangement of a gas chromatograph (Madou 1997) and (b) a portable gas chromatograph manufactured by MTI (Fremont, CA)

analysis or a neural network, is applied to the data. Thus, the growth phase of the bacteria can be identified and hence its response to antibiotics can be predicted. In fact, the Agilent 4440 system may be regarded as an 'intelligent instrument' because it couples the mass spectrometer together with sophisticated PARC software.

Conventional quadrupole mass spectrometers (e.g. Agilent Technologies Inc.) sell for 20 000 euros, are heavy (>50 kg), consume considerable power (>1 kW), and so there is a

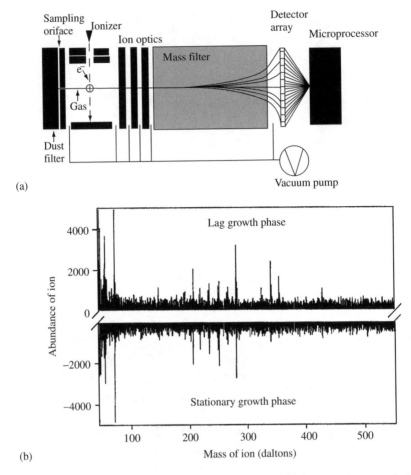

Figure 15.18   (a) Schematic layout of a mass spectrometer. The ions need to travel through a vacuum. Redrawn from Friedhoff *et al.* (1999) (b) Plot of the abundance of ions of differing mass in the headspace of a bacteria, *E. coli*, grown in a nutrient aqueous solution. The top plot shows the spectrum in the first growth phase of the bacteria (lag) and the second plot (inverted for the sake of clarity) shows the spectrum in the later stationary phase. From Esteves de Matos *et al.* (2000)

considerable demand for smaller, lighter, and portable units. Miniature mass spectrometers can be made using conventional precision machining techniques, but there have been recent efforts to make one using silicon micromachining techniques. Figure 15.19(a) shows an arrangement of a quadrupole lens employing silicon micromachined parts (Friedhoff *et al.* 1999). An optical microscopic view of the cross section is shown in Figure 15.19(b) (Syms *et al.* 1996). Making a complete silicon version is not straightforward because the device requires a good uniform field and vacuum. However, a unit that analyses a small mass range (up to 40 daltons) has been demonstrated successfully by Friedhoff *et al.* (1999) (Figure 15.19(c)), and further advances are expected shortly on ion sources and detectors. Biotechnology companies, such as Sequenome (USA), are developing MassArray™ chips to screen genetic defects called *single nucleotide*

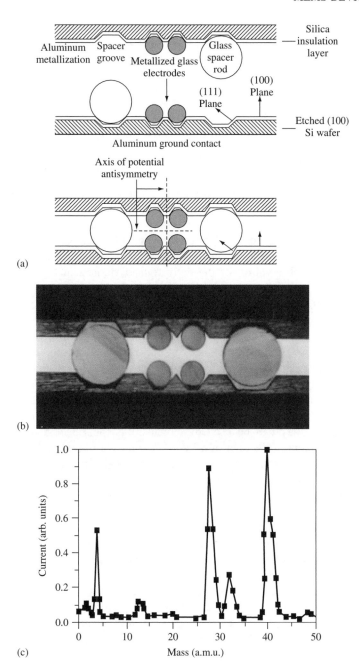

(a)

(b)

(c)

Figure 15.19 (a) Schematic of a micromachined quadrupole lens assembly for a MEMS mass spectrometer; (b) optical photograph of a cross section of the assembly quadrupole lens (Syms *et al*. 1996); (c) mass spectrum of a mixture of helium, argon, and air obtained with a 3 cm lens with 500 μm diameter electrodes driven at 6 MHz. (Friedhoff *et al*. 1999)

polymorphisms (SNPs). These require a mass spectrometer capable of detecting masses in the range of 5000 to 9000 daltons, and work is underway in DARPA (USA).

The advantages of carrying out chemical analysis and (bio)chemical reactions are primarily

- Small, portable, low-power units
- Small amounts of chemical reagents are required (microlitres or less)
- Rapid screening of multiple experiments (assay microarray chips)

The last reason is, probably, the most compelling, namely, that it is possible to perform hundreds, thousands, and, possibly, millions of (bio)chemical reactions on a single wafer, and thus the cost can be low for a large number of trials. This is essential in the rapidly emerging fields of genomics, proteomics, and pharmacogenomics. For example, Figure 15.20(a) shows the basic arrangement of a DNA probe. Here, light is shone on the elements and the observation of the activation of fluorescent markers permits the identification of the presence of certain genes. Figure 15.20(b) shows the biochip made using a 0.8 μm CMOS process with 128 DNA probes. The chip can be used in clinical diagnostics, for example, in screening for cancer.

There are now a plethora of emerging technologies and products in the Bio-MEMS area, and most of these do not involve silicon but use glass and polymers. These must tackle the problems of sample preparation, sample movement, and readout when potential is required to detect hundreds of thousands of genetic defects or SNPs. For example, disposable microtitration plates with a capillary fill of 96 reaction wells are now being manufactured by MicroParts in Germany (Figure 15.21), and when coupled with off-chip optical readout, they can be used to screen drugs (pharmacogenomics). There is also a rapid development in the area of a Lab-on-a-chip – a microfluidic chip to perform microbiology – that is to grow bacteria in microcells, challenge antibiotics, and analyse on-chip and various other Bio-MEMS chips such as Nanogen™, which is an APEX chip system, Spectrochip™, which is a microarray assay chip, and e-Sensor Systems (Motorola), which is a clinical microsensor. Such diagnostic Bio-MEMS chips permit rapid bacterial identification and antibody susceptibility.

In a related method, the use of combinatorial analysis makes it possible to try out, literally, millions of different chemical reactions on a single chip, and so help in synthesizing new compounds in a type of rapid prototyping.

Therefore, the rapid screening of biochemical reactions, identification of genetic defects, and testing of new drugs with a biochip are exciting prospects for smart MEMS devices. Such developments could revolutionise the field of medicine because it may be possible to individualise drug therapy; in other words, the rapid screening of people in clinics (and, perhaps, eventually at home) will result in the tailoring of both the choice of drug and its quantity delivered to them. It is essential when you consider the natural diversity in the response of individuals to both pathogenic agents and drug therapies. A medical advance like this should improve the targeting of drugs and, thereby, reduce the time for patient recovery. In addition, there are other medical applications of MEMS in surgery to enhance manipulators and catheters for better minimum access intervention (i.e. keyhole surgery) in smart drug delivery systems to control dosage, and in human implants to replace and augment body parts (e.g. cochlea, heart valves, pacemakers).

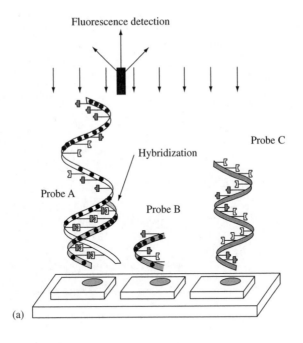

Figure 15.20 Rapid screening of biological material: (a) optical principle, and (b) CMOS array biochip for DNA analysis. From Caillat *et al.* (1999)

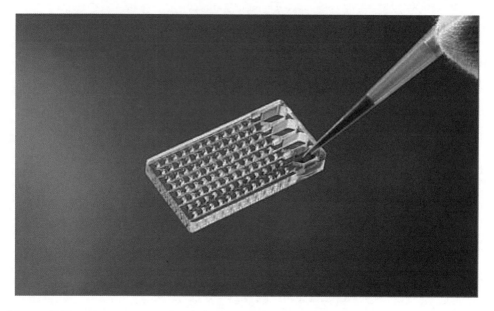

Figure 15.21 Microtiterplate for clinical microbiological applications: Lilliput chip from MicroParts (Germany)

15.4 CONCLUDING REMARKS

This chapter has described some of the recent advances in silicon micromachining techniques that seek to integrate the processor electronics with the sensor and actuators. These so-called smart sensors and MEMS are becoming increasingly sophisticated as the power of the processing unit increases from that provided by a simple eight-bit microcontroller (e.g. a 6805) through CISC technology (68 000-based) to RISC and digital signal-processing (DSP) technologies. It is now possible to perform on-chip a lot of intelligent features such as self-testing, fault-diagnostics, and adaptive control. Furthermore, there has been an increased use of artificial intelligence on sensors with the incorporation of artificial neural networks, expert systems, and fuzzy logic. Fuzzy controller ICs are already used today, for example, in buildings to operate lifts – we are just not aware of them!

Looking to the future, it seems that there are two areas that will develop further. The first is that we will be able to make increasingly sophisticated micromachines that parallel the human senses, such as the microgrippers, micronoses, microtongues etc. presented in this book. The second area is the way in which we communicate with these microdevices and especially micromachines, such as microrobots, microcars, and microplanes (Fujimasa 1996). The human–machine interface will probably become a limiting step and there will be a need to communicate remotely, perhaps via speech, with these intelligent micromachines. Perhaps, we will even see the day in which we implant these microdevices in our own body to *augment* our own limited senses, that is, we could have infrared (or night) vision, detect poisonous and toxic biological agents, and feel magnetic or electric fields. We could also have speech translators implanted in our ears, and so understand foreign languages, and, perhaps, attach devices that monitor our health and automatically warn us of imminent problems. It is hard to say how rapidly the development of technology will take us in the next few years. Let us hope that it is beneficial to our quality of life and is not misused.

REFERENCES

Baltes, H. and Brand, O. (2000). CMOS-based microsensors, Proceedings of Eurosensors XIV, Copenhagen, Denmark, August 27–30th, ISBN 87-89935-50-0.

Barney, G. C. (1985). *Intelligent Instrumentation*, Prentice-Hall, Englewood Cliffs, New Jersey, p. 532.

Bissell, C. C. (1994). *Control Engineering*, Chapman & Hall, London, p. 266.

Breckenbridge, R. A. and Husson, C. (1978). Smart sensors in spacecraft: the impact and trends, *Proc. of the AIAA/NASA Conf. on Smart Sensors*, Hampton, USA, pp. 1–5.

Brignell, J. E. and White, N. (1994). *Intelligent Sensor Systems*, IOP Publishing, Bristol, UK, p. 256.

Caillat, P. *et al.* (1999). Biochips on CMOS: an active matrix address array for DNA analysis.

Chapman, P. W. (1996). *Smart sensors*, Research Triangle Park, NC, 162 p.

Culshaw, B. (1996). *Smart Structures and Materials*, Artech House, Boston, p. 207.

Esteves de Matos, R., Mason, D. J., Dow, D. S. and Gardner, J. W. (2000). "Investigation of the growth characteristics of *E. coli* using headspace analysis," in J. W. Gardner and K. C. Persaud, eds., *Electronic Noses and Olfaction 2000*, IOP Publishing, Bristol, UK, p. 310.

Fausett, L. (1994). *Fundamentals of Neural Networks*, Prentice Hall, Englewood Cliffs, New Jersey, USA, p. 461.

Frank, R. (1996). *Understanding Smart Sensors*, Artech House, Boston, p. 269.

Friedhoff, C. B. *et al.* (1999). "Chemical sensing using non-optical microelectromechanical systems," *J. Vac. Sci. Technol.*, **17**, 2300–2307.

Fujimasa, I. (1996). *Micromachines: A New Era in Mechanical Engineering*, Oxford University Press, Oxford, UK, p. 156.

Gandhi, M. V. and Thompson, B. S. (1992). *Smart Materials and Structures*, Chapman Hall, London, p. 309.

Gardner, J. W. (1995). "Intelligent gas sensing using an integrated sensor pair," *Sensors and Actuators B*, **27**, 261–266.

Gardner, J. W. and Bartlett, P. N. (1994). "A brief history of electronic noses," *Sensors and Actuators B*, **18–19**, 211–220.

Gardner, J. W. and Bartlett, P. N. (1999). *Electronic Noses: Principles and Application*, Oxford University Press, Oxford, UK, p. 245.

Gardner, J. W. and Hines, E. L. (1996). Pattern analysis techniques, in E. Kress-Rogers ed. *Handbook of Biosensors: Medicine, Food and the Environment*, Ch.27, pp. 633–652.

Hines, E. L., Llobet, E. and Gardner, J. W. (1999). "Electronic noses: a review of signal processing techniques," *Proc. IEE Circuits, Devices Syst.*, **146**(6), 297–310.

Madou, M. J. (1997). *Fundamentals of Microfabrication*, CRC Press, Boca Raton, Florida, p. 589.

Ohba, R., ed. (1992). *Intelligent Sensor Technology*, Wiley and Sons, London.

Pike, A. (1996). "Design of chemoresistive silicon sensors for application in gas monitoring," PhD Thesis (Advisor J. W. Gardner), University of Warwick, Coventry, UK.

Pike, A. and Gardner, J. W. (1998). "Thermal modelling of micropower chemoresistive silicon sensors," *Sensors and Actuators B*, **45**, 19–26.

Pugh, A. (1986). *Robot sensors, vol. 1: vision*, IFS, UK also Springer-Verlag, Berlin.

Sell, P. S. (1986). *Expert Systems – A Practical Introduction*, Macmillan, London, p. 99.

Shin, H. W. (1999). "A hybrid electronic nose system for monitoring the quality of potable water," PhD Thesis (Advisor J. W. Gardner), University of Warwick, Coventry, UK.

Syms, R. R. A., Tate, T. J., Ahmad, M. M. and Taylor, S. (1996). "Fabrication of a microengineered quadrupole electrostatic lens," *Electronic Lett.*, **32**, 2094–2095.

Tsuchiya, K. and Davies, S. T. (1998). "Fabrication of TiNi shape memory alloy microactuators by ion beam sputter deposition," *Nanotechnology*, **9**, 67–71.

Udrea, F. and Gardner, J. W. (1998). Gas sensing semiconductor devices, International patent WO98/32009, July 1998.

Udrea, F. *et al.* (2001). "Design and simulations of a new class of SOI CMOS micro-hotplate gas sensors," *Sensors and Actuators B*, **78**, 180–190.

van der Horn, G. and Huijsng, J. H. (1998). *Integrated Smart Sensors: Design and Calibration*, Kluwer Academic Publishers, Dordrecht.

Varadan, V. K. and Gardner, J. W. (1999). Smart tongues and smart noses, *Proc. of SPIE on Smart Electronics and MEMS*, Newport Beach, USA, March.

Appendix A

List of Abbreviations

Abbrev.	Meaning
AC	Alternating current
ADC	Analogue to digital converter
AGC	Automatic gain control
aka	Also known as
APCVD	Atmospheric pressure chemical vapour deposition
APW	Acoustic plate mode
ASIC	Application-specific integrated circuit
BCC	Body-centred cubic
BGA	Ball grid array
BiCMOS	Bipolar complementary metal oxide semiconductor
BJT	Bipolar junction transistor
BSG	Borosilicate glass
BPSG	Borophosphosilicate glass
CAD	Computer-aided design
CB	Common base (configuration)
CC	Common collector (configuration)
CCD	Charge-coupled device
CE	Common emitter (configuration)
ChemFET	Chemically sensitive field effect transistor
CIU	Central interface unit
CMOS	Complementary metal oxide semiconductor
CNC	Computer numerically controlled
COM	Coupling of mode
COP	Poly(glycidyl methacrylate co-ethyl acrylate)
CP	Conducting polymer
CPU	Central processing unit
CTE	Coefficient of temperature expansion
C-V	Current voltage
CVD	Chemical vapour deposition
D.C.	Direct current

Abbrev.	Meaning
DES	Diethyl silane
DI	Deionised water
DIL	Dual in line (package)
DIP	Dual-in-line package
DMOS	Diffused-channel MOS (process)
DRAM	Dynamic random access memory
DSE	Doping selective etching
DSP	Digital signal processing
ECL	Emitter coupled logic
EDP	Ethylenediamine pyrocatechol
EEPROM	Electronically erasable programmable read-only memory
EFAB	Electrochemical fabrication
EGS	Electronic grade silicon
EM	Electromagnetic (radiation)
EPROM	Erasable programmable read only memory
FCC	Face-centred cubic
FEM	Finite element model
FET	Field-effect transistor
FFT	Fast Fourier transform
FM	Frequency modulated
FPGA	Field programmable gate array
FPW	Flexural plate wave
FSO	Full-scale operation
GaAs	Gallium arsenide
HCP	Hexagonal close packed
HMDS	Hexamethyl disilazane
HF	Hydrofluoric acid
HP	Hewlett Packard
HTCC	High-temperature co-fired ceramic
IC	Integrated circuit
IH	Integrated harden (resin)
IDT	Interdigitated transducer
IF	Intermediate frequency
IIL	Integrated injection logic
ISFET	Ion-selective field effect transistor
I/O	Input-output
I-V	Current voltage
JFET	Junction field-effect transistor
keV	Kilo electronvolt
LCD	Liquid crystal display
LOCOS	Local oxide isolation of silicon
LPCVD	Low-pressure chemical vapour deposition

Abbrev.	Meaning
LTCC	Low-temperature co-fired ceramic
LTO	Low-temperature oxide
MB	Megabyte
MBE	Molecular beam epitaxy
MCM	Multichip module
MCU	Main control unit
MEMS	Microelectromechanical system
MOEMS	Microoptoelectromechanical system
MFC	Mass-flow controller
MGS	Metallurgical-grade silicon
Mips	Million instructions per second
MISFET	Metal-insulator semiconductor field-effect transistor
MOS	Metal oxide semiconductor
MOSFET	Metal oxide semiconductor field-effect transistor
MS	Methyl silane
MSL	Microstereolithography
MST	Microsystems technology
MW	Molecular weight
N/A	Not available
nMOS	n-type metal oxide semiconductor
NRE	Nonreturning engineering (costs)
OCP	Open-circuit potential
PAA	Programmable analogue array
PARC	Pattern recognition
PCB	Printed circuit board aka PWB
PE	Piezoelectric
PECVD	Plasma-etched chemical vapour deposition
PGA	Programmable gate array
PHET	Photovoltaic electrochemical etchstop technique
PLA	Programmable logic array
PLD	Programmable logic device
pMOS	p-type metal oxide semiconductor
PMMA	Poly(methyl methacrylate)
p-n	Junction diode made from p-type and n-type materials
PP	Passivating potential
ppm	Parts per million
PSG	Phosphosilicate glass
PSI	Pounds per square inch
PTAT	Proportional to absolute temperature
PTFE	Poly(tetrafluoroethylene)
PVC	Poly(vinyl chloride)
PVD	Physical vapour deposition

Abbrev.	Meaning
PWB	Printed wiring board
PZT	Lead zinc titanate
QCM	Quartz crystal microbalance
RAM	Random access memory
RC	Resistance capacitance
REDOX	Reduction oxidation (reaction)
RIB	Reactive ion beam (milling)
RIE	Reactive ion etching aka RIB
RMS	Root-mean-square
ROM	Read-only memory
RF	Radio frequency
SAW	Surface acoustic wave
SC	Simple cubic
SCREAM	Single-crystal reactive etching and metallisation (process)
SEM	Scanning electron microscopy
SFF	Solid freeform fabrication
S/H	Sample and hold
SH-SAW	Shear horizontal SAW
SIMOX	Separation by implanted oxygen
SL	Stereolithography
SMA	Shape memory alloy
SMT	Surface mount technology
SOI	Silicon on insulator
SPI	Serial port interface
SRAM	Static random access memory
TAB	Tape-automated bonding
TCO	Temperature coefficient of operation
TCR	Temperature coefficient of resistance
TEOS	Tetraethoxy silane
TMS	Tetramethyl silane
TFT	Thin film transistor
TI	Texas Instruments
TTL	Transistor-transistor logic
UHF	Ultra high frequency
UV	Ultraviolet (light)
VCO	Voltage-controlled oscillator
VHF	Very high frequency
VLSI	Very large-scale integration
VNAW	Vehicle navigation and warning system
VP	Vapour pressure
VPE	Vapour-phase epitaxy

Appendix B

List of Symbols and Prefixes

Symbol	Meaning
Normal	
A_V	Voltage gain
Å	Angstrom (10^{-10} m)
b_m	Damping constant
c	Crystalline phase of material
c	Speed of light, phase velocity
$\boldsymbol{c}$	Elastic stiffness constant
c_m	Specific heat capacity
C	Electrical capacitance, concentration
C'	Electrical capacitance per unit area
C_d	Capacitance of dipole layer
C_s	Surface concentration
d	Distance
d_R	Thickness of resist
D	Diffusion coefficient
$\boldsymbol{D}$	Displacement field
$\mathbf{D}$	Spacing transfer matrix
e	Charge on an electron
E	Energy
$\boldsymbol{E}$	Electric (vector) field
E	Electric field strength
E_a	Activation energy
E_d	Ionisation energy of a donor atom
E_m	Young's modulus
f	Frequency (also Greek symbol υ)
F	Force
F	Flux (of photons)
g	Gaseous phase of material
g	Gravitational acceleration
g_{fs}	Forward transconductance with common-source configuration

Symbol	Meaning
G	Lame constant
$\mathbf{G}$	Reflector transfer matrix
h_{ie}	Input impedance
h_{fe}	Current gain
h_{oe}	Output conductance
h_{re}	Voltage gain
h	Planck's constant
$H(s)$	Transfer function (Laplace)
i	Alternating electrical current
I	Direct electrical current
I_B, I_C, I_E	Base, collector and emitter currents
I_{CO}	Reverse saturation current
I_m	Second moment of area
j	$\sqrt{-1}$
J	Electrical current density
k	Boltzmann constant
k	Wave number
k_m	Spring constant
K	Electromechanical coupling constant in SAW devices
K	Device constant
K_{gf}	Gauge factor
l	Liquid phase of material
l	Orbital angular momentum number
l_n, l_p	Length of space and charge regions in bipolar junction
L	Electrical inductance
L	Length of device channel
L_{min}	Minimum line width
L_n, L_p	Diffusion length in n- and p-type semiconductor materials
$\dot{L}$	Deposition rate of material
m	Mass, diode ideality factor
m_e	Mass of an electron
m^*	Effective mass of a charge carrier
M	Bending moment
n	Electron carrier concentration
n^+	Highly doped n-region
n	Refractive index
n_n	Donor concentration in n-type semiconductor material
n_p	Acceptor concentration in n-type semiconductor material
n_r	Refractive index
N_a, N_d	Acceptor and donor densities
p	Indicates type of silicon doping-electron acceptors
p^-	Highly-doped p-region
p_n	Donor concentration in p-type semiconductor material
p_p	Acceptor concentration in p-type semiconductor material
P	Pressure

Symbol	Meaning
q	Charge on an electron
Q	Flow rate
$r_{ds(on)}$	On-resistance of a transistor
R	Electrical resistance
$R/\square$	Electrical resistance per unit square, aka sheet resistance
s	Solid phase of material
s	Laplace parameter (complex)
S	Spin angular momentum number
S	Strain tensor
t	Time
T	Temperature in K or °C
T	Stress tensor
$\mathbf{T}$	IDT transfer matrix
T_{MP}	Melting point temperature
u	Particle displacement
v	AC Voltage
v	Linear velocity
v_R	Linear velocity of Rayleigh wave
V	DC voltage or potential
V_d	Potential barrier of dipole layer
V_T	Threshold voltage
w	Width
W	Width of device channel
x	Distance
y	Distance
Y	Yield strength of a material
z	Distance
Greek	
α_F	Common-base current gain
β_F	Common-base current gain
ε_O	Electrical permittivity of free space
ε_r	Relative dielectric constant
ε_m	Mechanical strain
ϕ	Electric potential
η	Viscosity of a liquid
φ	Phase angle
φ_d	Phase difference
κ	Thermal conductivity
λ	Wavelength
λ	Scaling factor
μ	Mobility of a charge carrier
μ	Rotation rate (gyroscope)
μ_n	Donor mobility
μ_p	Acceptor mobility

Symbol	Meaning
ρ	Electrical resistivity
ρ_m	Density
σ	Electrical conductivity
σ	Mechanical stress
τ	Time constant
τ	Time delay
τ	Mechanical shear stress
υ	Frequency (also symbol f)
Π	Piezoelectric constant
ω	Angular velocity ($2\pi\upsilon$)
ω	Angular velocity ($2\pi\upsilon$)
Ω	Angular velocity
ξ	Damping factor

Mathematical

$\dot{a}$	Derivative of a with respect to time t
$=$	Is equal to
$\neq$	Is not equal to
$\approx$	Is approximately equal to
$\sim$	Is of the order of
$\equiv$	Is identical to
$\propto$	Is proportional to
$\pm$	Plus or minus
$>$	Is greater than
$\geq$	Is greater than or equal to
$<$	Is less than
$\leq$	Is less than or equal to
$^\circ$	Degrees in angle or temperature

Prefixes

y	Yacto- (10^{-24})
z	Zepto- (10^{-21})
a	Atto- (10^{-18})
f	Femto- (10^{-15})
p	Pico- (10^{-12})
n	Nano- (10^{9})
μ	Micro- (10^{-6})
m	Milli- (10^{-3})
c	Centi- (10^{-2})
d	Deca- (10^{-1})
da	Deca- (10^{+1})
h	Hecto- (10^{+2})
k	Kilo- (10^{+3})
M	Mega- (10^{+6})
G	Giga- (10^{+9})

Symbol	Meaning
T	Tera- (10^{+12})
P	Peta- (10^{+15})
E	Exa- (10^{+18})
Z	Zetta- (10^{+21})
Y	Yotta- (10^{+24})

Appendix C

List of Some Important Terms

Term	Meaning
Actuand	The quantity being actuated by an actuator.
Actuator	A device that converts an electrical signal into a nonelectrical quantity (aka output transducer).
Device	An electronic part that contains one or more active elements, such as a diode, transistor, or integrated circuit.
Functional material	A material that is active or functional rather than passive, i.e. it does more than simply act as an inert substrate (aka smart material). The most common example is a shape memory alloy.
Intelligent sensor	A sensor that shows a high level of function. Early definitions argued for a sensor connected to a microprocessor was 'intelligent.' More recently, the terms require some superior functionality through tailored, structured, or processing design. Biomimetic features, such as adaptation, self-learning, and fault-tolerance, can be found in intelligent sensors. The term *smart* is reserved for device integration (see below).
Measurand	The quantity being measured by a sensor.
MEMS	A microelectromechanical system is a device made from extremely small parts. The term MST is widely used in Europe and is used to make MEMS.
MEMS device	This term is often used to describe the embodiment of a MEMS but, strictly (see preceding term), does not make sense.
Microactuator	A miniature device that converts an electrical signal into a nonelectrical quantity.
Microdevice	A miniature device that has been fabricated using microtechnology.
Microcontroller	A processing device used to control certain functions or parts. For instance, a simple microprocessor can be used as a microcontroller.

Term	Meaning
Microelectronics	The branch of electronics concerned with or applied to the realisation of electronic circuits or systems from extremely small (micron level) electronic parts.
Microprocessor	A miniature processor that has been fabricated using microtechnology, e.g. an 8-bit, 16-bit or 32-bit processing chip.
Microsensor	A sensor that has at least one physical dimension at the submillimetre level.
Microtechnology	The science and history of the mechanical and industrial arts used to make extremely small (micron level) parts. More commonly thought of as the methodologies and processes required to make a microstructure.
Multifunctional material	A miniature device that converts a nonelectrical quantity into an electrical signal.
Nanotechnology	The science and history of the mechanical and industrial arts used to make ultrasmall (nanometre level) parts. More commonly thought of as the methodologies and processes required to make a nanostructure.
Processor	A device that performs a set of logical or mathematical operations.
Sensor	A device that converts a nonelectrical quantity into an electrical signal (aka input transducer).
Smart actuator	An actuator with part or its entire associated processing element integrated in a single chip.
Smart electronics	Electronic circuitry that can take account of its environment, for example, self-compensating, self-diagnosing, self-repair, and so forth.
Smart material	A material that is active or functional rather than passive, that is it does more that simply act as an inert substrate (*aka* functional material). The most common example is a shape memory alloy.
Smart sensor	A sensor that has part (type I) or its entire (type II) processing element integrated in a single chip.
Smart structure	A structure that can take account of its environment, for example an instrumented bridge that has closed loop control of its motion. The term is usually applied to large objects, such as bridges, buildings and so forth.
System	An aggregation of parts between which there exists a relationship and together form a unity.
Transducer	A device that converts a nonelectrical quantity into an electrical signal and vice versa.

Appendix D

Fundamental Constants

Here is a list of the fundamental constants together with their value in SI units.

Constant	Symbol	Value in SI units
Avogadro's number	N_A	6.0225×10^{23} per mole
Acceleration due to gravity	g	9.8067 m s^{-2}
Bohr magneton	β	1.165×10^{-29} Wb m
Boltzmann's constant	k	1.3805×10^{-23} J K^{-1}
Compton wavelength of electron	λ_c	2.4263×10^{-12} m
Electronic charge	e	1.6021×10^{-19} C
Electron charge-to-mass ratio	e/m_e	1.7588×10^{11} C kg^{-1}
Electronic rest mass	m_e	9.1096×10^{-31} kg
Electronic radius (classical)	r_e	2.8179×10^{-15} m
Electron volt	eV	1.6021×10^{-19} J
Faraday's constant	F	9.6487×10^4 C mol^{-1}
Fine structure constant	$\dfrac{e^2}{2h\varepsilon_0 v_c}$	7.2974×10^{-3}
Gas constant	R	8.3143 J K^{-1} mol^{-1}
Gravitational constant	G	6.670×10^{-11} N m^2 kg^{-2}
Impedance of free space	Z_0	376.73 Ω
Loschmidt's constant	n_L	2.6872×10^{-25} m^{-3}
Neutron rest mass	m_N	1.6748×10^{-27} kg
Permeability of free space	μ_0^B	$4\pi \times 10^{-7}$ H m^{-1}
Permittivity of free space	ε_0	8.8542×10^{-12} F m^{-1}
Planck's constant	h	6.6256×10^{-34} J s
Proton rest mass	m_P	1.6725×10^{-27} kg
Quantum charge ratio	h/e	4.1357×10^{-15} J s C^{-1}
Speed of light in vacuum	v_c	2.9979×10^8 m s^{-1}
Stefan-Boltzmann constant	σ	5.6697×10^{-8} W m^{-2} K^{-4}
Volume of 1 mole of ideal gas at STP	–	22.4 litres

Appendix E

Unit Conversion Factors

Unit	SI or cgs	fps
Length	1 cm = 0.39370 in 1 km = 0.62137 mi	1 in = 2.5400 cm 1 mi = 1.6093 km
Area	$1\ cm^2 = 0.1550\ in^2$ 1 hectare = 2.4711 acre	$1\ in^2 = 6.4516\ cm^2$ $1\ acre = 4046.9\ m^2$
Volume	$1\ cm^3 = 0.061024\ in^3$ 1 litre = 0.21997 UK gallon	$1\ in^3 = 16.387\ cm^3$ 1 UK gallon = 4.5461 litre 1 US gallon = 0.8327 UK gallon
Velocity	1 m/s = 2.2369 mile/hr 1 km/hr = 0.62137 mile/hr	1 mile/hr = 0.44704 m/s 1 mile/hr = 1.6093 km/hr
Mass	1 kg = 2.20462 lb 1 tonne = 0.98421 ton	1 lb = 0.45359 kg 1 ton = 1.0160 tonne
Density	$1\ g/m^3 = 0.036127\ lb/in^3$	$1\ lb/in^3 = 27.680\ g/cm^3$
Force	1 N = 0.22481 lb force $1\ N = 10^5\ dyne$ 1 N = 7.2330 poundal 1 N = 0.10197 kgf	1 lbf = 4.4482 N $1\ dyne = 10^{-5}\ N$ 1 poundal = 0.13826 N –
Torque	1 N m = 0.7375 lbf ft	1 lbf ft = 1.356 N m
Pressure	$1\ N/m^2 = 1.4504 \times 10^{-4}\ lb/in^2$ $1\ kg/cm^2 = 14.223\ lb/in^2$ 1 Pa = 0.0075006 torr $1\ N/m^2 = 10\ dynes/cm^2 = 1\ Pa$ $1\ Pa = 10^{-5}\ bar =$ $9.8692 \times 10^{-6}\ atmos$	$1\ lb/in^2 = 6,894.8\ N/m^2$ $1\ lb/in^2 = 0.070307\ kg/cm^2$ 1 torr = 133.322 Pa – –
Energy	1 J = 0.23885 calorie $1\ J = 9.4781 \times 10^{-4}\ btu$ $1\ J = 10^7\ erg$ 1 kW h = 3.6 MJ	1 calorie = 4.1868 J 1 btu = 1,055.1 J $1\ erg = 10^{-7}\ J$ –

Unit	SI or cgs	fps
	$1 \text{ J} = 1 \text{ N m} = 1 \text{ W s}$	–
	$1 \text{ eV} = 1.6021 \times 10^{-19} \text{ J}$	–
Viscosity	$1 \text{ Pa s} = 10 \text{ g/cm/s} = 10 \text{ poise}$	$1 \text{ lbf s/in}^2 = 6,895 \text{ Pa s}$
Photometric	$1 \text{ cd} = 0.982 \text{ int. candles}$	$1 \text{ candle} = 1.018 \text{ cd}$
	$1 \text{ lx} = 0.09290 \text{ ln/ft}^2 \text{ or fc}$	$1 \text{ fc} = 10.764 \text{ lx}$
	$1 \text{ cd} = 1 \text{ lm/sr}$	–
	$1 \text{ lx} = 1 \text{ lm/m}^2$	–

Appendix F

Properties of Electronic & MEMS Metallic Materials

Table F.1 Physical properties of some common metallic passive materials used in microsensor and MEMS technology. Measurements taken at 20 °C where appropriate unless stated otherwise. Source: most of the values were taken from either the *Handbook of Chemistry and Physics* (CRC Press, Inc.) or *MacMillans Chemical and Physical Data* (James and Lord, MacMillans Press Ltd, 1992). These values are intended only as a guide and we recommend, wherever possible, validation against other sources

Property:	Al	Ag	Au	Cr	Cu	In	Ti
Density, ρ_m (kg/m^3)	2702	10,50	19,32	7194	8920	7290	4,508
Melting point, T_{mp}(°C)	660	962	1064	1857	1083	157	1,660
Boiling point, T_{bp}(°C)	2467	2212	2807	2672	2567	2080	3,287
Electrical conductivity, σ (10^3 S/cm)	377	630	488	79	607	340	26
Temperature coefficient of resistance, α_r(10^{-4}/K)	39	38	34	30	39	39	38
Work function, ϕ (eV)	4.3	4.3	5.1	4.5	4.5	4.1	4.3
Thermal conductivity, κ(W/m/K)[1]	236	428	319	97	403	84	22
Specific heat capacity, c_p(J/K/kg)	904	236	129	448	385	243	522
Linear expansivity, α_l(10^{-6}K^{-1})	23.1	18.9	14.2	4.9	16.5	32.1	8.6
Young's modulus, E_m (GPa)	70	83	78	279	130	–	~40
Yield strength[2], Y(MPa)	50	–	200	–	150	–	480
Poisson's ratio, ν	0.35	0.37	0.44	0.21	0.34	–	0.36

[1] Value taken at 0 °C.
[2] The values stated here are typical but can vary by as much as 30%.

Table F.2 Physical properties of some common refractory metallic materials used in microsensor and MEMS technology. Measurements taken at 20 °C where appropriate unless stated otherwise. Source: most of the values were taken from either the *Handbook of Chemistry and Physics* (CRC Press, Inc.) or *MacMillans Chemical and Physical Data* (James and Lord, MacMillans Press Ltd, 1992). These values are intended only as a guide and we recommend, wherever possible, validation against other sources

Property:	Material:						
	Fe	Ni	Pb	Pd	Pt	Sb	W
Density, ρ_m (kg/m^3)	7860	8902	11,34	12,02	21,45	6684	19,35
Melting point, T_{mp}(°C)	1535	1453	328	1552	1772	631	3,410
Boiling point, T_{bp}(°C)	2750	2732	1740	3140	3827	1750	5,660
Electrical conductivity, σ (10^3 S/cm)	112	146	48	100	94	26	183
Temperature coefficient of resistance, α_r(10^{-4}/K)	50	60	39	33	30	36	45
Work function, ϕ (eV)	4.2	5.3	5.4	5.4	5.6	4.6	4.6
Thermal conductivity[1], κ(W/m/K)	84	94	36	72	72	26	177
Specific heat capacity, c_p(J/K/kg)	449	456	134	244	133	207	134
Linear expansivity, α_l(10^{-6}K^{-1})	11.8	13.4	28.9	11.8	8.8	11	4.5
Young's modulus, E_m (GPa)	152	219	16	–	168	–	411
Yield strength, Y(MPa)	160	148	15		<14	-	750
Poisson's ratio, ν	0.27	0.31	0.44	-	0.38	-	0.28

[1] Values taken at 0 °C.
[2] Young's modulus value is for cast iron.

Appendix G

Properties of Electronic & MEMS Semiconducting Materials

Table G.1 Physical properties of common semiconductor materials used in microsensor and MEMS technology. Values are taken at 20°C where appropriate unless stated otherwise. Source: most of the values were taken from either the *Handbook of Chemistry and Physics* (CRC Press, Inc.) or *MacMillans Chemical and Physical Data* (James and Lord, MacMillans Press Ltd, 1992). These values are intended only as a guide and we recommend, wherever possible, validation against other sources

Property	Material:				
	Si(c)	Poly-Si	Ge	C[1]	GaAs
Density, ρ_m (kg/m^3)	2330	2320	5350	3510	5316
Melting point, T_{mp}(°C)	1410	–	937	3827	1238
Boiling point, T_{bp} (°C)	2355	–	2830	4827	N/A
Electrical conductivity, σ (10^3 S/cm)	4×10^{-3}	–	3×10^{-5}	$\sim 10^{-17}$	$\sim 10^{-5}$
Energy band gap, E(eV)	1.1	1.1	0.67	5.4	1.35
Thermal conductivity, κ(W/m/K)	168	34	60	1000–2600	370
Specific heat capacity, c_p(J/K/kg)	678	678	310	523	–
Temperature expansivity, α_l(10^{-6}/K)	2.6	2–2.8	5.7	1	5.7
Dielectric constant, ε_r	11.7	–	16.3	5.1	12
Young's modulus, E_m (GPa)	190[1]	161	–	542	–
Yield strength, Y (GPa)	6.9	–	–	–	–
Breakdown field, (MV/cm)	0.3	–	0.1	–	0.5

[1]In diamond form and [111] Miller index.

Appendix H

Properties of Electronic & MEMS Ceramic and Polymer Materials

Table H.1 Physical properties of common ceramic and insulating materials used in microsensor and MEMS technology. Values are taken at 20 °C where appropriate unless stated otherwise. Source: most of the values were taken from either the *Handbook of Chemistry and Physics* (CRC Press, Inc.) or *MacMillans Chemical and Physical Data* (James and Lord, MacMillans Press Ltd, 1992). These values are intended only as a guide and we recommend, where possible, validation against other sources

Property	Material:						
	Al_2O_3	SiO_2	Quartz ‖c	Quartz ⊥c	SiC	Si_3N_4	ZnO
Density, ρ_m (kg/m^3)	3965	2200	2650	2650	3216	3100	5606
Melting point, T_{mp}(°C)	2045	1713	–	–	3070	1900	1975
Boiling point, T_{bp} (°C)	2980	2230	–	–	–	–	–
Thermal conductivity, κ(W/m/K)	38	1.4	12	6.7	110	20	6
Specific heat capacity[1], c_p(J/K/kg)	730	730	730	730	710	600–800	–
Temperature expansivity, α_l(10^{-6}/K)	–	–	6.8	12.2	3.3	–	–
Dielectric constant, ε_r	85	3.8	4.4	4.4	–	–	–
Energy band gap, $E_{g(eV)}$	18–23	9	–	–	3	5	3.35
Young's modulus, E_m (GPa)	–	57–85	72	72	440	304	–
Breakdown strength (MV/cm)	39	10	–	–	–	2.3	–

[1]Values at a temperature of 0 °C.

Table H.2 Physical properties of common polymer and plastic materials used in microsensor and MEMS technology. Values are taken at 20 °C where appropriate unless stated otherwise. Source: most of the values were taken from either the *Handbook of Chemistry and Physics* (CRC Press, Inc.) or *MacMillans Chemical and Physical Data* (James and Lord, MacMillans Press Ltd, 1992). These values are intended only as a guide and we recommend, where possible, validation against other sources

| Property | Material: | | | | | |
	Nylon	Polyimide	Polythene[2]	PTFE[1]	PVC	PVDF
Density, ρ_m (kg/m^3)	1120–1170	1000–1600	926–941	2100–2300	1300–1400	1750–1780
Maximum working point, T_{max} (°C)	100	–	71–93	260	70–74	150
Thermal conductivity, κ (W/m/K)	0.25–0.27	0.15	0.33–0.42	0.24–0.25	0.16	0.1
Specific heat capacity, c_p (J/K/kg)	1600–1900	1100	1900	1050	840–1170	–
Temperature expansivity, α_l (10^{-5}/K)	28	–	14–16	10	5–18	8–14
Dielectric constant, ε_r	3.7–5.5	–	2.3	2	3.0–4.0	2.9
Young's modulus, E_m (GPa)	1–4	~3.1	0.4–1.3	0.4	2.9	2.1
Tensile strength, Y_m (MPa)	50–90	69–104	8–24	10–31	34–62	36–56

[1] Trade name is Teflon.
[2] Medium density.

Appendix I

Complex Reciprocity Relation and Perturbation Analysis

I1 THE COMPLEX RECIPROCITY RELATION

The propagation of waves in a waveguide structure, such as the SAW substrate, with a thin film overlay can be accomplished using the technique of modal analysis. Any waveguide can support the propagation of an infinite number of solutions, or waveguide modes. As long as the set of modes is mathematically complete, any function can then be expressed as an infinite sum of these waveguide modes. Modal analysis also requires that the individual modes must be orthogonal, just as the sine and cosine functions in Fourier analysis are orthogonal. The proof of orthogonality requires the derivation of the complex reciprocity relation. The quasi-static and electromagnetic field equations needed derive the reciprocity. If an acoustic force vector $\vec{F}$ is included, the tensor equation given in Chapter 10 may now be written as follows

$$T_{ij,j} = \rho \ddot{u}_i - F_i,$$

(I.1)

and

$$T_{ij} = c_{ijkl} u_{k,l} + e_{ijk} \phi_{,k}$$

(I.2)

Another necessary equation is the time derivative of the electrical displacement equation

$$\dot{D}_i = e_{ikl} \dot{u}_{k,l} - \varepsilon_{ij} \dot{\phi}_{,j}$$

(I.3)

A second solution to the equations can be denoted by primed field quantities $(T', u', f', \ldots)$ and the complex conjugate of the field equations can be written

$$T'^{*}_{ij,j} = \rho^{*} \ddot{u}'^{*}_i - F'^{*}_i$$

(I.4)

$$T'^{*}_{ij} = c^{*}_{ijkl} u'^{*}_{k,l} + e^{*}_{ijk} \phi'^{*}_{,k}$$

(I.5)

$$\dot{D}'^{*}_j = e^{*}_{ikl} \dot{u}'^{*}_{k,l} - \varepsilon^{*}_{ij} \dot{\phi}'^{*}_{,j}$$

(I.6)

If Equations (I.1),(I.2) and (I.3) are multiplied by $\dot{u}_i^{'*}$, $\dot{u}_{i,j}^{'*}$ and $-\phi_{,i}^{'*}$ added, then we have

$$T_{ij,j}\dot{u}_i^{'*} + T_{ij}\dot{u}_{i,j}^{'*} - \dot{D}_i\phi_{,i}^{'*}$$
$$= \left\{ \rho\ddot{u}_i\dot{u}_i^{'*} - F_i\dot{u}_i^{'*} + c_{ijkl}u_{k,l}\dot{u}_{i,j}^{'*} + e_{ijk}\phi_{,k}\dot{u}_{k,l}^{'*} - e_{ikl}\dot{u}_{k,l}\phi_{,i}^{'*} + \varepsilon_{ij}\dot{\phi}_{,j}\phi_{,i}^{'*} \right\} \qquad (I.7)$$

Similarly, Equations (I.4),(I.5) (I.6) can be multiplied by $\dot{u}_i$, $\dot{u}_{i,j}$ and $-\phi_{,i}$ added to give

$$T_{ij,j}^{'*}\dot{u}_i + T_{ij}^{'*}\dot{u}_{i,j} - \dot{D}_i^{'*}\phi_{,i}$$
$$= \left\{ \rho^*\ddot{u}_i^*\dot{u}_i - F_i^{'*}\dot{u}_i + c_{ijkl}^*u_{k,l}^{'*}\dot{u}_{i,j} + e_{ijk}^*\phi_{,k}^{'*}\dot{u}_{i,j} - e_{ikl}^*\dot{u}_{k,l}^{'*}\phi_{,i} + \varepsilon_{ij}^*\dot{\phi}_{,j}^{'*}\phi_{,i} \right\} \qquad (I.8)$$

In a lossless material, the material constant matrices are real and symmetric and therefore satisfy the relations,

$$\rho = \rho^*, \quad c_{ijkl} = c_{klij}^*, \quad \varepsilon_{ij} = \varepsilon_{ij}^* \quad \text{and} \quad e_{ijk} = e_{kij}^* \qquad (I.9)$$

Under these conditions, Equations (I.7) and (I.8) can be added to get

$$\left\{ \begin{array}{l} T_{ij,j}\dot{u}_i^{'*} + T_{ij}^{'*}\dot{u}_{i,j} + T_{ij,j}^{'*}\dot{u}_i \\ +T_{ij}^{'*}\dot{u}_{i,j} - \dot{D}_i\phi_{,i}^{'*} - \dot{D}_i^{'*}\phi_{,i} \end{array} \right\} = \left\{ \begin{array}{l} \rho(\ddot{u}_i\dot{u}_i^{'*} + \ddot{u}_i^{'*}\dot{u}_i) - F_i\dot{u}_i^{'*} - F_i^{'*}\dot{u}_i \\ +c_{ijkl}(u_{k,l}^{'*}\dot{u}_{i,j} + u_{i,j}\dot{u}_{k,l}^{'*}) \\ +\varepsilon_{ij}(\phi_{,i}\phi_{,j}^{'*} + \phi_{,i}\phi_{,j}^{'*}) \end{array} \right\} \qquad (I.10)$$

By combining various derivative terms, it is possible to rewrite the preceding equation as

$$\left\{ (T_{ij}\dot{u}_i^{'*} + T_{ij}^{'*}\dot{u}_i)_{,j} - D_i\dot{\phi}_{,i}^{'*} - \dot{D}_i^{'*}\phi_{,i} \right\}$$
$$= \left\{ \frac{\partial}{\partial t}(\rho\dot{u}_i\dot{u}_i^{'*} + c_{ijkl}u_{i,j}u_{k,l}^{'*} + \varepsilon_{ij}\phi_{,i}\phi_{,j}^{'*}) - F_i\dot{u}_i^{'*} - F_i^{'*}\dot{u}_i \right\} \qquad (I.11)$$

In order to take into account an electrical forcing function, it can be assumed that some free charge distribution (q) is present. The equation $D_{i,i} = q$ simply describes Gauss's Law. This can be brought into the equation preceding by presenting equation in the following form:

$$\left(D_i\dot{\phi}^{'*} + D_i^{'*}\dot{\phi} \right)_{,i} = D_i\dot{\phi}_{,i}^{'*} + \dot{D}_{i,i}\phi^{'*} + \dot{D}_i^{'*}\phi_{,i} + \dot{D}_{i,i}^{'*}\phi \qquad (I.12)$$

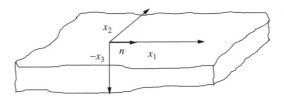

Figure A1   Coordinate system for SAW waves showing the propagation vector.

Rearranging this would give

$$\dot{D}_i\phi'^*_{,i} + \dot{D}'^*_i\phi_{,i} - \left(\dot{D}_i\phi'^* + \dot{D}'^*_i\phi\right)_{,i} = -\dot{q}\phi'^* - \dot{q}'^*\phi \qquad (I.13)$$

Adding Equations (I.12) and (I.13) and multiplying the result by -1 gives

$$\left\{ \begin{matrix} -T_{ij}\dot{u}'^*_i - T'^*_{ij}\dot{u}_i \\ +\dot{D}_j\phi'^* + \dot{D}'^*_j\phi \end{matrix} \right\}_{,j} = \left\{ \begin{matrix} -\dfrac{\partial}{\partial t}\left(\rho\dot{u}_i\dot{u}'^*_i + c_{ijkl}u_{i,j}u'^*_{k,l} + \varepsilon_{ij}\phi_{,i}\phi'^*_{,j}\right) \\ +F_i\dot{u}'^*_i + F'^*_i\dot{u}_i + \dot{q}\phi'^* + \dot{q}'^*\phi \end{matrix} \right\} \qquad (I.14)$$

The preceding equation is known as *complex reciprocity relation*.

On the right-hand side of the Equation (I.14), each of the term inside the parentheses contains the product of a conjugated function with a nonconjugated function. Because the functional form of $e^{i\omega t}$ is assumed in modal analysis, any term of the form $A * B$ will have no time dependence ($e^{-j\omega\tau}\varepsilon^{\varphi\omega\tau} = 1$); therefore, the derivative term will be zero. The complex reciprocity relation for $e^{i\omega t}$ variations is then reduced to

$$\left\{-T_{ij}\dot{u}'^*_i - T'^*_{ij}\dot{u}_i + \dot{D}_j\phi'^* + \dot{D}'^*_j\phi\right\}_{,j} = \left\{F_i\dot{u}'^*_i + F'^*_i\dot{u}_i + \dot{q}\phi'^* + \dot{q}'^*\phi\right\} \qquad (I.15)$$

I2 PERTURBATION THEORY

The perturbation analysis begins with the complex reciprocity relation with the added assumption that no sources (body force or charge density variations) are present within the substrate.

$$\left(-T^*_{ij}\dot{u}'_i - T'_{ij}\dot{u}^*_i + \dot{D}^*_j\phi' + \dot{D}'_j\phi^*\right)_{,j} = 0 \qquad (I.16)$$

The unperturbed structure that will be analysed is shown in Chapter 10, Figure 10.3. The semi-infinite piezoelectric slab extends to infinity in the $\pm x_1$ and $\pm x_2$ and $-x_3$ directions. The electromagnetic quantities extend out of the substrate into vacuum that extends to infinity in the $+x_3$ direction. The field quantities are independent of x_2 and propagation is in the $+x_1$ direction. Under these conditions, all $j = 2$ terms in Equation (I.16) are zero and integration with respect to x_3 can be performed across the substrate resulting in

$$\int_{-\infty}^{0} \frac{\partial}{\partial x_1}\left(\begin{matrix} -T^*_{i1}\dot{u}'_i - T'_{i1}\dot{u}^*_i \\ +\dot{D}^*_1\phi' + \dot{D}'^*_1\phi \end{matrix} \right) dx_3 = -\left[-T^*_{i3}\dot{u}'_i - T'_{i3}\dot{u}^*_i + \dot{D}^*_3\phi' + \dot{D}'^*_3\phi^*\right]_{-\infty}^{0} \qquad (I.17)$$

The unprimed quantities represent the unperturbed field quantities in the bare substrate, whereas the primed quantities will represent the perturbed field quantities in the layered structure. The unperturbed quantities have the functional form e^{jkx_l} because the substrate is lossless, and the perturbed quantities vary as $e^{jk'x_l}$. If Δk is defined as

$$\Delta k = k' - k,$$

then Equation (I.17) becomes

$$
j\Delta k \int_{-\infty}^{0} \frac{\partial}{\partial x_1} \left(\begin{array}{c} -T_{i1}^* \dot{u}_i' - T_{i1}' \dot{u}_i^* \\ +\dot{D}_1^* \phi' + \dot{D}_1' \phi^* \end{array} \right) dx_3 = - \left[-T_{i3}^* \dot{u}_i' - T_{i3}' \dot{u}_i^* + \dot{D}_3^* \phi' + \dot{D}_3' \phi^* \right]_{-\infty}^{0} \quad (\text{I.18})
$$

or

$$
\Delta k = \frac{j \left[-T_{i3}^* \dot{u}_i' - T_{i3}' \dot{u}_i^* + \dot{D}_3^* \phi' + \dot{D}_3' \phi^* \right]_{-\infty}^{0}}{\int_{-\infty}^{0} \frac{\partial}{\partial x_1} \left(-T_{i1}^* \dot{u}_i' - T_{i1}' \dot{u}_i^* + \dot{D}_1^* \phi' + \dot{D}_1' \phi^* \right) dx_3} \quad (\text{I.19})
$$

In order to calculate the change in the propagation constant Δk, using Equation (I.19), it is necessary to know the perturbed field distributions. Without this knowledge, it is necessary to make several approximations to simplify the analysis. The first approximation can be made by replacing the perturbed quantities in the denominator with unperturbed quantities – because the perturbation is small. The denominator can then be simplified using the power flow equation for piezoelectric media that is given by

$$
P = \frac{1}{2} \text{Re} \int_{-\infty}^{0} \left(-T_{i1} \dot{u}_i^* + \dot{D}_1^* \phi \right) dx_3 \quad (\text{I.20})
$$

where P is the average power flow in the x_1 direction per unit width. Because the real part of any complex variable x can be calculated using

$$
\text{Re}(x) = \frac{x + x^*}{2} \quad (\text{I.21})
$$

Equation (I.20) can be written as

$$
P = \frac{1}{4} \int_{-\infty}^{0} \left(-T_{i1} \dot{u}_i^* - T_{i1}^* \dot{u}_i + \dot{D}_1^* \phi + \dot{D}_1 \phi^* \right) dx_3 \quad (\text{I.22})
$$

The assumption that all quantities in the denominator are unperturbed quantities allows Equation (I.10) to be written as

$$
\Delta k = \frac{j \left[-T_{i3}^* \dot{u}_i' - T_{i3}' \dot{u}_i^* + \dot{D}_3^* \phi' + \dot{D}_3' \phi^* \right]_{-\infty}^{0}}{4P} \quad (\text{I.23})
$$

The next assumption that greatly simplifies the calculation of Equation (I.23) is that small mechanical and electrical perturbations are independent and can therefore be calculated separately. For mechanical boundary perturbations, the electrical boundary conditions are assumed to be unperturbed, and for electrical boundary perturbations, the mechanical boundary conditions are considered unperturbed. When both types of boundary perturbations are present, the perturbations are calculated separately and added.

When perturbation involves only mechanical effects, the last two terms in the numerator of Equation (I.23) are ignored. In addition, unperturbed situation includes a stress free surface at $x_3 = 0$ (i.e. $T_{i3} = T_{3i} = 0$) and all field quantities disappear at $x_3 = -\infty$;

therefore, only the second term evaluated at $x_3 = 0$ is nonzero. Because u^* has the time dependence $e^{j\omega t}$ the equation reduces to

$$\Delta k = -\frac{\omega T'_{i3} u_i^*|_0}{4P}$$

At this point, it is helpful to eliminate the stress terms by calculating T'_{i3} in terms of particle displacements.

If the perturbed SAW propagation is in the Love wave layer structure, the surface at $x = h$ is stress-free. If the film is isotropic, the Lamé constants λ and μ, can be used to define the elastic constants as,

$$c_{11} = c_{22} = c_{33} = \lambda + 2\mu$$

$$c_{12} = c_{21} = c_{13} = c_{31} = c_{23} = c_{32} = \lambda$$

$$c_{44} = c_{55} = c_{66} = \mu$$

and all other $c_{ij} = 0$

where the subscripts are Voigt index subscripts. Under these conditions and in the absence of external body forces, Equation (I.1) becomes

$$T'_{13,3} + jk'T'_{11} = -\omega^2 \rho u'_1$$
$$T'_{23,3} + jk'T'_{21} = -\omega^2 \rho u'_2 \qquad (I.24)$$
$$T'_{33,3} + jk'T'_{31} = -\omega^2 \rho u'_3$$

where ρ is the density of the film.

At this point, it is beneficial to describe the field quantities in the film as Taylor series about $x_3 = h$, the film surface. Any quantity f that is a function of x_3 is then approximated by

$$f(x_3) = f(h) + (x_3 - h) \left.\frac{\partial f}{\partial x_3}\right|_h + \frac{(x_3 - h)^2}{2!} \left.\frac{\partial^2 f}{\partial x_3^2}\right|_h + \cdots \qquad (I.25)$$

where only first two terms will be significant when h is small. Because the boundary is stress-free at $x_3 = h$ and the stress tensor is symmetric in the absence of external body torques, it is also true that

$$T'_{13}(h) = T'_{31}(h) = T'_{23}(h) = T'_{32}(h) = T'_{33}(h) = 0 \qquad (I.26)$$

If the Equations (I.25) and (I.26) are substituted for the field quantities in Equation (I.17) and all terms containing $x_3 = h$ are ignored ($x_3 - h$ is small), Equation (I.24) becomes

$$T'_{13,3}(h) + jk'T'_{11}(h) = -\omega^2 \rho u'_1(h)$$

$$T'_{23,3}(h) + jk'T'_{21}(h) = -\omega^2 \rho u'_2(h) \tag{I.27}$$

$$T'_{33,3}(h) = -\omega^2 \rho u'_3(h)$$

In order to simplify Equation (I.27), it is necessary to calculate T'_{11} and T'_{21} in terms of the particle displacements. Using Equations E.2, the elastic constants in terms of Lame constants and (I.26) the Voigt index notation described in [31], it can be shown that

$$T'_{21}(h) = -jk'\mu u'_2(h)$$

$$T'_{11}(h) = -jk'(\lambda + 2\mu)u'_1(h) + \lambda u'_{33}(h) \tag{I.28}$$

$$T'_{33}(h) = 0 = jk'\lambda u'_1(h) + (\lambda + 2\mu)u'_{33}(h)$$

If the last two equations are solved simultaneously to eliminate $u'_{33}(h)$, the result is

$$T'_{11}(h) = jk'4\lambda \frac{\lambda + \mu}{\lambda + 2\mu} u'_1(h) \tag{I.29}$$

The use of Equations (I.25) and (I.26) leads to the following derivation for the stress derivative terms

$$T'_{i3}(0) = T'_{i3}(h) - hT'_{i33}(h) \tag{I.30}$$

Therefore,

$$T'_{i3}(0) = -hT'_{i33}(h) \tag{I.31}$$

Substituting Equations (I.28) and (I.27) determines $T'_{i33}(h)$ and therefore $T'_{i3}(0)$. The resultant equations are

$$T'_{13}(0) = h\omega^2 \left(\rho - \frac{4\mu}{v^2} \frac{\lambda + \mu}{\lambda + 2\mu} \right) u'_1(h)$$

$$T'_{23}(0) = h\omega^2 \left(\rho - \frac{\mu}{v^2} \right) u'_2(h) \tag{I.32}$$

and

$$T'_{33}(0) = h\omega^2 \rho u'_3(h)$$

where $v = \omega/\kappa \cong \omega/k' = $ the velocity of the wave.

Equation (I.28) can be used to rewrite $T'_{33}(0) = h\omega^2 \rho u'_3(h)$ as

$$\frac{\Delta k}{k} = -\frac{vh\omega^2}{4p} \left[\begin{array}{l} \left(\rho - \frac{4\mu}{v^2} \frac{\lambda + \mu}{\lambda + 2\mu} \right) u'_1(h)u_1^*(0) \\ + \left(\rho - \frac{\mu}{v^2} \right) u'_2(h)u_2^*(0) + \rho u'_3(h)u_3^*(0) \end{array} \right] \tag{I.33}$$

Equation (I.33) can still not be used to calculate the perturbation to the film because it requires knowledge of the particle displacements within the perturbing film. The equation can be simplified by making two further assumptions. The first assumption is that the addition of the thin film does not change the particle displacements at the surface of the substrate. The second assumption is that the particle displacements in the film are constant with respect to x_3. The result of these two assumptions can be written,

$$u_i'(h) = u_i'(0) = u_i(0)$$

With these assumptions made, Equation (I.33) takes the form

$$\frac{\Delta k}{k} = -\frac{vh\omega^2}{4p} \left[\left(\rho - \frac{4\mu}{v^2} \frac{\lambda + \mu}{\lambda + 2\mu} \right) |u_1(0)|^2 + \left(\rho - \frac{\mu}{v^2} \right) |u_2(0)|^2 + \rho |u_3(0)|^2 \right] \quad (I.34)$$

when Equation (I.34) defines the changes in propagation constant, when a lossless SAW substrate is coated by a thin, lossless, isotropic film.

For a Love wave device, u_1 would typically be 0 (because of the absence of a longitudinal component of displacement), and rearranging the equation using the aforementioned fact gives

$$\frac{\Delta v}{v} = -\frac{v_h}{4P} \left[\left(\rho' - \frac{\mu'}{v^2} \right) v_2^2 + \rho' v_3^2 \right] \quad (I.35)$$

Equation (I.35) is the expression for the elastic mass-loading effect of the ice film on the Love wave device. The ice film overlay is described as an isotropic film with a density of ρ', Lamé constant of μ', and a thickness of h uniformly loading the propagation path.

Frequency shift because of acoustic-electric perturbation

To obtain the frequency shift because of an electrical perturbation, the first and second terms of Equation (I.23) (indicating the mechanical characteristics) are made zero and the wave is considered to be perturbed only by the third and fourth terms (indicating electrical characteristics). Hence, the perturbation solution the electrical perturbation due to as obtained from Equation (I.16) neglecting mechanical perturbation is as follows:

$$\Delta k = \frac{j \left[\dot{D}_3^* \phi' + \dot{D}_3'^* \phi^* \right]_{-\infty}^0}{4P} \quad (I.36)$$

The electrical perturbation in air has been derived by Auld from Equation (I.36) and is given as

$$\frac{\Delta \beta}{\beta} = \left(\frac{\Delta V}{V_l} \right)_{SC} \varepsilon_0 \frac{1 - j Z_e'(0)}{\varepsilon_0 + j \varepsilon_p^T Z_e'(0)} \quad (I.37)$$

where $((\Delta V / V_l)_{SC})$ is the velocity change rate when the substrate surface loaded by air is short and open, V_l is the propagation velocity of the Love wave when the propagation surface is open, and Z'_e is the surface impedance (defined in Auld, V.1, Section 4.5.2) after perturbation is normalised by the one before. Also,

$$\varepsilon_p^T = \sqrt{\varepsilon_{11}^T \varepsilon_{33}^T - (\varepsilon_{31}^T)^2}$$

is the effective permittivity of the substrate.

Now we could extend this to electrical perturbation in a liquid. The relevant parameters for the electrical characteristics of the liquid are the conductivity σ and the relative permittivity ε_r. These parameters are expressed by the complex permittivity as

$$\varepsilon_{\text{water}} = \varepsilon_{rl}\varepsilon_0 - \frac{j\sigma}{\omega}. \tag{I.38}$$

where $\varepsilon_{\text{water}}$ is equal to $\varepsilon_{\text{perturbed}}$ and ε_{rl} is equal to the relative permittivity of the liquid.

The unperturbed air is assumed to have $\varepsilon_{\text{air}} = \varepsilon_{ra}\varepsilon_0$ and the conductivity is zero, where ε_{ra} is the relative permittivity of air and ε_{air} is equal to $\varepsilon_{\text{unperturbed}}$.

In terms of these parameters, the normalised surface impedance after perturbation is

$$Z'_e(0) = -\frac{j\varepsilon_{\text{air}}}{\varepsilon'_l} = -j\frac{\varepsilon_{\text{air}}}{\varepsilon_{\text{water}}} \tag{I.39}$$

Substituting (I.39) in (I.37) we get

$$\frac{\Delta\beta}{\beta} = \left(\frac{\Delta V}{V_l}\right)_{\text{SC}} \varepsilon_0 \frac{1 - j\left(\dfrac{-j\varepsilon_{\text{air}}}{\varepsilon_{\text{water}}}\right)}{\varepsilon_0 + j\varepsilon_p^T\left(\dfrac{-j\varepsilon_{\text{air}}}{\varepsilon_{\text{water}}}\right)} \tag{I.40}$$

$$\frac{\Delta\beta}{\beta} = \left(\frac{\Delta V}{V_l}\right)_{\text{SC}} \varepsilon_0 \frac{\varepsilon_{\text{water}} - \varepsilon_{\text{air}}}{\varepsilon_0\,\varepsilon_{\text{water}} + \varepsilon_p^T\varepsilon_{\text{air}}} \tag{I.41}$$

$$\frac{\Delta\beta}{\beta} = \left(\frac{\Delta V}{V_l}\right)_{\text{SC}} \varepsilon_0 \frac{\varepsilon_{\text{water}} - \varepsilon_{\text{air}}}{\varepsilon_0\,\varepsilon_{\text{water}} + \varepsilon_p^T\varepsilon_0\varepsilon_{ra}} \tag{I.42}$$

$$\frac{\Delta\beta}{\beta} = \left(\frac{\Delta V}{V_l}\right)_{\text{SC}} \frac{\varepsilon_{\text{water}} - \varepsilon_{\text{air}}}{\varepsilon_{\text{water}} + \varepsilon_p^T} \tag{I.43}$$

as $\varepsilon_{ra} \to 1$.

Thus, Equation (I.43) could be rewritten in terms of perturbed and unperturbed quantities as

$$\Delta\beta' = \frac{\Delta\beta}{k} \cong \frac{\Delta\beta}{\beta} = -\left(\frac{\Delta V}{V_l}\right)_{\text{SC}} \frac{\varepsilon_{\text{perturbed(water)}} - \varepsilon_{\text{unperturbed(air)}}}{\varepsilon_{\text{perturbed(water)}} + \varepsilon_p^T} \tag{I.44}$$

Substituting E.38 and $\varepsilon_{\text{air}} = \varepsilon_{ra}\varepsilon_0$ for the perturbed and unperturbed quantities mentioned earlier, we obtain

$$\frac{\Delta\beta}{\beta} = \left(\frac{\Delta V}{V_l}\right)_{\text{SC}} \frac{\varepsilon_{rw}\varepsilon_0 - j(\sigma/\omega) - \varepsilon_{\text{air}}}{\varepsilon_0\varepsilon_{rw} - j(\sigma/\omega) + \varepsilon_p^T} \tag{I.45}$$

From the product of Equation (I.45) and the complex conjugate of the denominator, we obtain the variations in propagation velocity due to acoustic-electric interaction as equal to

$$\frac{\Delta v}{v_l} = -\left(\frac{\Delta V}{V_l}\right)_{SC}\left[\frac{(\sigma'/\omega)^2 + \varepsilon_0(\varepsilon_{rw} - \varepsilon_{air})(\varepsilon_{rw}\varepsilon_0 + \varepsilon_p)}{(\sigma'/\omega)^2 + (\varepsilon'_{rw}\varepsilon_0 + \varepsilon_p)^2}\right] \tag{I.46}$$

where

$$\left(\frac{\Delta V}{V_l}\right)_{SC} = -\omega\left(\varepsilon_{air} + \varepsilon_p{}^T\right)\frac{|\phi|^2}{4P} \tag{I.47}$$

and

$$\varepsilon_p{}^T = \sqrt{\varepsilon_{11}^T\varepsilon_{33}^T - (\varepsilon_{31}^T)^2} \tag{I.48}$$

Appendix J

Coupled-mode Modeling of a SAW Device

Material and geometry constants or parameters for coupled-mode modeling of a SAW device are listed here. Numerical values can be easily found from the references or in general literature for a given material.

v is the SAW velocity.

K^2 is the electromechanical coupling coefficient.

C_0 is the capacitance per finger pair per unit length on a given piezoelectric material.

Λ is the periodicity of IDT and reflector.

W is the aperture of IDT.

N_t is the number of IDT fingers.

N_p is the number of IDT finger pairs.

L is the reflector length.

Z_e is the load impedance.

R_s is the lead resistance for IDT, wires, etc.

h is the thickness of deposited metal.

f_0 is the centre operating frequency of IDT.

C_s is the capacitance per finger pair $= C_0 W$.

C_t is the total capacitance for an IDT.

α is the SAW attenuation constant.

k_{11} is the self-coupling coefficient responsible for velocity shift because of metallisation.

k_{12} is the mutual coupling coefficient between opposing SAW.

β_0 is the phase constant.

δ is the frequency detuning parameter showing centre frequency shift due to k_{11} and given by

$$\delta = 2\pi \frac{(f - f_0)}{v} + k_{11} \qquad (J.1)$$

σ is the reflector grating attenuation constant and given by

$$\sigma = \sqrt{k_{12}^2 - (\delta - \alpha i)^2} \qquad (J.2)$$

θ_t is the transit angle and equal to $N_t \Lambda \delta$.

θ_3 is the reference phase at the start of each array
G_0 is the characteristic radiation conductance and given by

$$G_0 = 8K^2 C_S f_0 \tag{J.3}$$

G_a is the radiation conductance and is defined by

$$G_a = G_0 (N_t - 1)^2 \quad \text{if } \omega = \omega_0; \text{ or}$$

$$G_a = G_0 (N_t - 1)^2 \frac{\sin^2 (\theta_t/2)}{(\theta_t/2)^2} \quad \text{if } \omega \neq \omega_0 \tag{J.4}$$

B_a is the radiation susceptance and defined as

$$B_a = 0 \quad \text{if } \omega = \omega_0; \text{ or } B_a = 2G_0 (N_t - 1)^2$$

$$\times \frac{\sin \theta_t - \theta_t}{\theta_t^2} \quad \text{if } \omega \neq \omega_0 \tag{J.5}$$

G is the reflector transfer function and is equal to

$$G = C \begin{bmatrix} a & b \\ c & d \end{bmatrix} \tag{J.6}$$

and the parameters defining G are

$$a = \exp(i\beta_0 L) \left[\frac{\sigma}{k_{12}} + i \tanh(\sigma L) \frac{\delta - \alpha i}{k_{12}} \right]$$

$$b = \exp(i\beta_0 L) i \tanh(\sigma L) \exp(-i\theta_3)$$

$$c = -\exp(-i\beta_0 L) i \tanh(\sigma L) \exp(-i\theta_3)$$

$$d = \exp(-i\beta_0 L) \left[\frac{\sigma}{k_{12}} - i \tanh(\sigma L) \frac{\delta - \alpha i}{k_{12}} \right]$$

$$C = \frac{k_{12} \cosh(\sigma L)}{\sigma}$$

D is the spacing transfer function and is given by

$$D = \begin{bmatrix} \exp(\beta di) & 0 \\ 0 & \exp(-\beta di) \end{bmatrix} \tag{J.7}$$

$[T]$ is the IDT transfer matrix and is given by

$$[T] = \begin{bmatrix} t & \tau \\ \tau_s & t_{33} \end{bmatrix} \tag{J.8}$$

where the various terms are

$$
t = \begin{bmatrix} s\exp(\theta_t i)\left[1 + \dfrac{G_a(R_s + Z_e)}{1 + \theta_e i}\right] & -s\dfrac{G_a(R_s + Z_e)}{1 + \theta_e i} \\[4mm] s\dfrac{G_a(R_s + Z_e)}{1 + \theta_e i} & s\exp(-\theta_t i)\left[1 - \dfrac{G_a(R_s + Z_e)}{1 + \theta_e i}\right] \end{bmatrix}
$$

$$(J.9)$$

with θ_e equal to $(\omega C_t + B_a)(R_s + Z_e)$

$$
\tau = \begin{bmatrix} \dfrac{\sqrt{2G_a Z_e}}{1 + \theta_e i}\exp(\theta_t i/2) \\[4mm] \dfrac{\sqrt{2G_a Z_e}}{1 + \theta_e i}\exp(\theta_t i/2)\exp(-i\theta_t) \end{bmatrix}
$$

$$(J.10)$$

$$
\tau_s = \begin{bmatrix} \dfrac{\sqrt{2G_a Z_e}}{1 + \theta_e i}s\exp(\theta_t i/2) - \dfrac{\sqrt{2G_a Z_e}}{1 + \theta_e i}s\exp(\theta_t i/2)\exp(-i\theta_t) \end{bmatrix} \quad (J.11)
$$

$$
t_{33} = 1 - \dfrac{2\theta_c i}{1 + \theta_e i}
$$

$$(J.12)$$

with θ_c equal to $\omega C_t(R_s + Z_e)$

REFERENCES

Cross, P. S. and Schmidt, R. V. (1977). "Coupled surface acoustic wave resonators," *Bell Sys. Tech. Journal*, **56**, 1447–1482.

Campbell, C. (1998). *SAW Devices for Mobile and Wireless Communications*, Academic Press, New York.

Appendix K
Suggested Further Reading

Readers may find the following list of books available of interest to them. Some of the books listed are rather introductory in nature and are intended as exemplars for background or supplementary information – these are labelled by the letter 'I.' Other books are more advanced in nature and provide further technical detail to the reader than that given in our book – these are labeled by the letter 'A.'

The list of books coming under the topics of microsensors, MEMS, and smart devices is intended to be comprehensive – that is all of the ones that we could find through library and database searches. A number of these are the published, edited proceedings of technical conferences and so may not be suitable to the less knowledgeable reader – these are labeled by the letter 'P.'

(1) Semiconductor microtechnology

- *Introduction to Semiconductor Microtechnology*, D. V. Morgan and K. Board, John Wiley & Sons, London, 1990. (I)

- *Introduction to Microelectronic Fabrication*, R. C. Jaeger, Addison-Wesley, Reading, p. 232, 1993. (I)

- *Microelectronics Processing and Device Design*, R. A. Colclaser, John Wiley & Sons, New York, p. 333, 1980.

- *Modern Vacuum Practice*, N. Harris, McGraw-Hill, New York, 1989.

- *Modern Semiconductor Fabrication Technology*, P. Gise and R. Blanchard, Prentice Hall, Englewood Cliffs, p. 265, 1986. (I)

(2) Microfabrication and Micromachining

- *Fundamentals of Microfabrication*, M. Mardou, CRC Press, Boca Raton, p. 589, 1997.

- *Handbook of Microlithography, Micromachining and Microfabrication*, 2 vols., P. Rai-Choudrey, ed., SPIE Press, Bellingham, Washington, USA, 1997. (A)

- *Laser Microfabrication: Thin Film Processes and Lithography*, D. J. Ehrlich and Y. Jeffrey, Academic Press, Boston, 1989.

- *Nanofabrication and Biosystems*, H. C. Hoch, L. Jelenski and H. G. Craighead, eds., Cambridge University Press, Cambridge, 1996.

(3) Electronic packaging and interconnection technologies

- *Electronic Equipment Packaging Technology*, G. L. Ginsberg, van Nostrand Reinhold, New York, p. 279, 1992. (I)

- *Electronic Packaging and Interconnection Handbook*, C. A. Harper, McGraw-Hill, New York, 1997.

- *Micromachining and Micropackaging of Transducers*, C. D. Fung *et al.* eds., Elsevier, Amsterdam, p. 244, 1985.

- *Multichip Module Technologies and Alternatives*, D. A. Doane and P. D. Franzon, eds., van Nostrand Reinhold, New York, p. 875, 1993.

- *Multichip Modules with Integrated Sensors*, W. K. Jones and G. Harsanyi, eds., Kluwer Academic Publishers, Dordrecht, p. 324, 1995. (P)

(4) Microelectronic devices

- *An Introduction to Application Specific Integrated Circuits*, M. R. Haskard, Prentice Hall, New York, p. 146, 1990. (I)

- *Analogue VLSI Devices: nMOS & CMOS*, M. R. Haskard and I. C. May, Prentice Hall, New York, 1988. (I)

- *Designing Field-Effect Transistors*, E. Oxner, ed., McGraw-Hill, New York, p. 296, 1990. (A)

- *Electrical and Thermal Characterisation of MESFETs, HEMTs, and HBTs*, R. Ahholt, Artech House, Boston, p. 310, 1995. (A)

- *Electronic Devices and Circuit: Discrete and Integrated*, M. S. Ghausi, Holt, Rinehart & Winston, New York, p. 808, 1985. (I)

- *Metal-Semiconductor Contacts*, E. H. Rhoderick and R. H. Williams, Oxford University Press, Oxford, 1988. (A)

- *Semiconductor Devices, Physics and Technology*, S. M. Sze, John Wiley & Sons, New York, 1985. (I)

(5) Sensors

- *Automotive Sensors*, M. H. Westbrook and J. Turner, IOP Publishing, Bristol, p. 272, 1994.

- *Biomedical Transducers and Instruments*, T. Togawa, CRC Press, Boca Raton, 1997.

- *Biosensors: A Practical Approach*, A. E. G. Cass, ed., Oxford University Press, Oxford, 1990. (A)

- *Biosensors: Fundamentals & Applications*, A. Turner, I. Karube, and G. Wilson, ed., Oxford University Press, Oxford, 1987. (A)

- *Biosensors: Theory and Applications*, D. G. Buerk, Technomic Publishing Co., Lancaster, 1993.

- *Chemical and Biochemical Sensing with Optical Fibers and Waveguides*, G. Boside and A. Harmer, Artech House, Boston, 1996.

- *Chemical Sensing with Solid State Devices*, M. J. Madou and S. R. Morrison, Academic Press, New York, 1989.

- *Fiber-Optic Sensors*, E. Udd, ed., John Wiley & Sons, New York, 1991.

- *Flow-Through (Bio)Chemical Sensors*, M. Valcarcel and M. D. Luque de Castro, Elsevier, Amsterdam, 1994.

- *Hall Effect Devices: Magnetic Sensors and Characterisation of Semiconductors*, R. S. Popovic, IOP Publishing, Bristol, p. 320, 1991.

- *Handbook of Chemical and Biological Sensors*, R. Taylor, A. D. Little, J. S. Schultz, IOP Publishing, Bristol, p. 604, 1996.

- *Handbook of Modern Sensors*, J. Fraden, Springer, New York, p. 556, 1997.

- *Handbook of Transducers*, H. N. Norton, Prentice Hall, New York, 1989. (I)

- *Physical Sensors for Biomedical Applications*, R. H. Neuman *et al.*, CRC Press, Boca Raton, 1980.

- *Position Sensors*, H. Walcher, VDI Verlag, 1985.

- *Pressure Sensors*, D. Tandeske, Marcel Dekker, New York, 1991.

- *Sensors, a Comprehensive Review*, W. Gopel, J. Hesse and J. N. Zemel, in 8 volumes, 1989-1998, Wiley-VCH, Weinheim.

- *Sensor Materials*, P. T. Moseley and A. J. Crocker, IOP Publishing, Bristol, p. 227, 1996.

- *Sensors: Principles & Applications*, P. Hauptmann, Prentice Hall, New York, 1991. (I)

- *Solid State Gas Sensors*, P. Moseley and B. Tofield, eds., IOP Publishing, Bristol, 1987.

- *Stannic Oxide Gas Sensors*, K. Ikohura and J. Watson, CRC Press, Boca Raton, 1994.

- *Surface-launched Acoustic Wave Sensors: Chemical Sensing and Characterisation*, M. Thompson, John Wiley & Sons, New York, 1997.

- *Thermal Sensors*, G. C. M. Miejer and A. W. van Herwaarden, IOP Publishing, Bristol, p. 304, 1994.

- *Thin Film Resistive Sensors*, P. Ciureanu and S. Middelhoek, eds., IOP Publishing, Bristol, p. 495, 1992.

(6) Instrumentation, interfacing and signal conditioning

- *Control Engineering*, C. C. Bissell, Chapman & Hall, New York, 1994. (I)

- *From Instrumentation to Nanotechnology*, J. W. Gardner and H. T. Hingle, Gordon & Breach Science Publishers, Philadelphia, p. 336, 1991.

- *Instrumentation Reference Book*, B. E. Noltingk, ed., Butterworth, London, 1988.

- *Interfacing Sensors to the IBM PC*, W. J. Tompkins and J. G. Webster, eds., Prentice Hall, New York, 1988.

- *Principles of Electronic Instrumentation*, A. de Sa, Edward Arnold, London, 1990.

- *Sensors & Signal Conditioning*, R. Pallas-Areny and J. G. Webster, John Wiley & Sons, Chichester, 1991.

- *Transducers and Interfacing*, B. R. Bannister and D. G. Whitehead, Van Nostrand, New York, 1986.

(7) Microsensors and silicon sensors

- *Biosensors: Microelectrochemical Sensors*, M. Lambrechts and W. Sansen, IOP Publishing, Bristol, p. 320, 1992.

- *Integrated Optics, Microstructures and Sensors*, M. Tabib-Azar, Kluwer Academic Publishers, Dordrecht, p. 399, 1995. (P)

- *Microsensors*, R. S. Muller, R. T. Howe, S. T. Senturia, R. L. Smith and R. M. White, eds., IEEE Press, New York, 1990. (P)

- *Microsensors: Principles & Applications*, J. W. Gardner, John Wiley & Sons Ltd, Chichester, p. 331, 1994. (I)

- *Sensor Technology & Devices*, L. Ristic, ed., Artech House, 1994.

- *Sensor Technology in The Netherlands: State of the Art*, A. Van den Berg and P. Bergveld, eds., Kluwer Academic Publishers, Dordrecht, p. 325, 1998. (P)

- *Silicon Sensors*, S. Middelhoek and S. A. Audet, Academic Press, New York, 1989.

- *Silicon Sensors and Circuits: On-Chip Compatibility*, Kluwer Academic Publishers, Dordrecht, p. 336, 1995. (A)

(8) Signal and data processing

- *Introduction to Multivariate Analysis*, C. Chatfield and A. J. Collins, Chapman & Hall, New York, 1980.

- *Using Multivariate Statistics*, B. G. Tabachnick and L. S. Fidell, Harper, 1989. (A)

- *Computer Systems that Learn*, S. H. Weiss and C. A. Kulikowski, Morgan Kanfman Publishers, Inc. San Francisco, 1989. (I)

- *Fundamentals of Neural Networks*, L. Fausett, Prentice Hall, New York, 1994. (I)

(9) Smart/intelligent sensors and sensor systems

- *Electronic Noses: Principles and Application*, J. W. Gardner and P. N. Bartlett, Oxford University Press, New York, p. 250, 1999. (I)

- *Handbook of Intelligent Sensors for Industrial Automation*, N. Zuech, ed., Addison-Wesley, p. 521, 1991.

- *Integrated Smart Sensors: Design and Calibration*, G. van der Horn and J. H. Huijsnz, Kluwer Academic Publishers, Dordrecht, 1998.

- *Intelligent Sensor Technology*, R. Ohba, ed., John Wiley & Sons, London, 1992.

- *Intelligent Sensor Systems*, J. Brignell and N. White, Iop Publishing, Bristol, 1994.

- *Robot Sensors Vol. 1 Vision*, A. Pugh, ed., IFS Ltd, UK, 1986.

- *Sensors & Sensory Systems for an Electronic Nose*, J. W. Gardner and P. N. Bartlett, eds., Kluwer Academic Publishers, Dordrecht, p. 327, 1991. (P)

- *Smart Sensors*, P. W. Chapman, Research Triangle Park, N.C., p. 162, 1996.

- *Understanding Smart Sensors*, R. Frank, Artech House, Boston, p. 269, 1996. (I)

(10) Actuators and Microactuators

- *Actuators for Control*, H. Funakubo, ed., Gordon & Breach Science Publishers, New York, p. 429, 1991.

- *Advances in Actuators*, A. P. Dorey and J. H. Moore, IOP Publishing, Bristol, p. 228, 1995.

- *Microactuators*, M. Tabib-Azar, Kluwer Academic Publishers, Dordrecht, p. 287, 1988.

(11) MEMS, MST and smart electronics

- *Advanced Microsystems for Automotive Applications 98*, D. Ricken and W. Gessner, eds., Springer, New York, p. 350, 1998. (P)

- *Micromechanical Sensors, Actuators and Systems*, D. Cho *et al.*, eds., DSC-Vol. 32, ASME, New York, p. 362, 1991. (P)

- *Microsystems Technology and Microrobotics*, S. Fatikow and Rembold, Springer, New York, 408, 1997. (I)

- *Microtransducer CAD*, A. Nathan and H. Baltes, Springer, New York, p. 300, 1998.

- *Smart Electronics and MEMS*, A. Hariz, V. K. Varadan and O. Reinhold, eds., Vol. 3242, Proc. SPIE, p. 398, 1997. (P)

(12) Smart materials and structures

- *Active Materials and Adaptive Structures*, G. Knowles, ed., IOP Publishing, Bristol, p. 940, 1992. (P)

- *Adaptive Structures: Dynamics and Control*, John Wiley & Sons, New York, p. 467, 1996.

- *Fiber Optic Smart Structures*, E. Udd, ed., John Wiley & Sons, New York, p. 671, 1995.

- *Smart Material Structures: Modelling and Estimation and Control*, H. T. Bauks, John Wiley & Sons, New York, p. 467, 1996.

- *Smart Materials and Structures*, M. V. Gandhi and B. S. Thompson, Chapman and Hall, London, p. 309, 1992. (I)

- *Smart Materials and Structures*, G. R. Tomlinson and W. A. Bullough, eds., IOP Publishing, Bristol, p. 852, 1998. (P)

- *Smart Structures and Materials*, B. Culshaw, Artech House, Boston, p. 207, 1996.

Appendix L

Webography

Readers may find the following list of websites available of interest to them.

Table L.1 List of some important journals and periodicals in sensors and MEMS

Title	Host web site
Analytical Chemistry	*http://pubs.acs.org/journals/*
Biosensors and Biomechanics	*http://www.elsevier.nl*
IEEE Sensors Journal	*http:/ieee.org*
Journal of the Electrochemical Society	*http:/ecs.electrochem.org/*
Journal of Micro-Electro-Mechanical Systems	*http:/ieee.org*
Journal of Micromechanics and Microengineering	*http:/iop.org/journals/*
Measurement, Science and Technology	*http:/iop.org/journals/*
Nanobiology	*http://www.carfax.co.uk*
Nanotechnology	*http:/iop.org/journals/*
Proc. IEE: Science, Measurement and Technology	*http:/iee.org*
Sensors and Actuators	*http://www.elsevier.nl*
Sensors and Materials	Scientific publishing division of MY K.K., Tokyo, Japan.
Sensor Technology	*http:/www.wiley.com*
Sensors Update	*http:/www.vch-wiley.com*
Smart Materials and Structures	*http:/iop.org/journals/*

Table L.2 List of some important conference series in Sensors and MEMS. The list excludes many National conferences

Title	Frequency	Duration
European Conference on Solid-State Sensors and Actuators (*aka* Eurosensors)	Annual	1987–now
Hilton Head Solid-State Sensor and Actuator Workshop	Biennial	1984–now
International Meeting on Chemical Sensors	Biennial	1983–now
International Conference on Solid-State Sensors and Actuators (*aka*Transducers)	Annual	1981–now
MEMS	Annual	1989–now

Table L.3   List of some 20 universities active in the fields of microsensors and MEMS. University website is given because the activities tend to be spread across many departments or within research centres

University	Country	Website
Barcelona	Spain, Europe	*http://www.upc.es*
Berkeley	California, US	*http://berkeley.edu*
Berlin	Germany, Europe	*http://www.tu-berlin.de*
Birmingham	UK, Europe	*http://www.bham.ac.uk*
Cornell	US	*http://cornell.edu*
Delft	Netherlands, Europe	*http://www.tudelft.nl*
Glasgow	UK, Europe	*http://gla.ac.uk*
Imperial College	UK, Europe	*http://ic.ac.uk*
Linkoping	Sweden, Europe	*http://www.liu.se*
Mainz	Germany, Europe	*http://www.uni-mainz.de*
Munich	Germany, Europe	*http://www.tu-muenchen.de*
Nanyang Technological	Singapore Asia	*http://ntu.ac.sg*
Neuchatel	Switzerland, Europe	*http://www.neuchatel.ch*
North Eastern	US	*http://neu.edu*
PennState	Pennsylvania, US	*http://psu.edu*
Singapore	Singapore Asia	*http://nus.ac.sg*
Southampton	UK, Europe	*http://soton.ac.uk*
Stanford	California, US	*http://stanford.edu*
Texas	Texas, US	*http://utexas.edu*
Tokyo	Japan	*http://www.tokyo.ac.jp*
Tokyo Institute of Technology	Japan	*http://www.titech.ac.jp*
Twente	Netherlands, Europe	*http://www.utwente.nl*
Warwick	UK, Europe	*http://warwick.ac.uk*

Table L.4   List of manufacturers, design house and competence centres in Europe

Name	Type	Country	Web address
MCI	Manufacturer	Germany	*www.europractice.com/MCI*
MEMSOI	Manufacturer	France	*www.tronics-mst.com*
NORMIC	Manufacturer	Norway	*www.normic.com*
MC4	Manufacturer	Switzerland	*www.csem.ch*
MAGFAB	Manufacturer	Germany	–
ACREO	Design house	Sweden	*www.acreo.se*
CEA-LETI	Design house	France	*www-dta.cea.fr/home_leti-uk.htm*
CNM	Design house	Spain	*www.cnm.es*
AML	Design house	UK	*www.aml.co.uk*
IMT	Design house	Switzerland	*www-imt.unine.ch*
ISIT	Design house	Germany	*www.isit.fthg.de*
LAAS	Design house	France	*www.laas.fr*
MIC	Design house	Denmark	*www.mic.dtu.uk*
MST-DESIGN	Design house	UK	*www.mst-design.co.uk*
NMRC	Design house	Ireland	*www.nmrc.ucc.ie*
SINTEF	Design house	Norway	*www.sintef.no/ecy*

Table L.4   (*Continued*)

Name	Type	Country	Web address
TMP	Design house	Netherlands	*www.microproducts.nl*
Physical Measurement Systems	Competence centre	Germany	*www.isit.fthg.de*
MOEMS	Competence centre	France	*www-dta.cea.fr*
MEDICS Biomedical devices	Competence centre	Germany	*www.medics-network.com*
Microactuator centre	Competence centre	UK	*www.cmf.rl.ac.uk/CCMicro/*
Microfluidic centre	Competence centre	Netherlands	*www.MicroFluidicCenter.com*
Liquid handling competence centre	Competence centre	Germany	*www.microfluidics.com*
Medical microinstrument competence centres	Competence centre	Italy	*www.pont-tech.it/MMICC.htm*

Appendix M

List of Worked Examples

The book contains a number' of different worked examples that describe the *fabrication* details of microstructures, microsensors, microactuators and MEMS devices. These worked examples are spread across a number of chapters and here we provide a list of them all as a quick reference guide. Of course, many other types of devices are described but not together with their full process flow.

Table M.1   Worked examples of process flow given in the book in chronological order

Reference no.	Device	Device type
4.1	Bipolar transistor	Microelectronic
4.2	Basic MOSFET	Microelectronic
4.3	Lateral DMOS n-channel MOSFET	Microelectronic
5.1	Mechanical velcro	Micromechanical structure
5.2	Undoped silicon cantilever beam	Micromechanical structure
5.3	Formation of an array of thin membranes	Micromechanical structure
5.4	Fabrication of (thin) cantilever beams	Micromechanical structure
6.1	Freestanding polysilicon beam	Micromechanical structure
6.2	Linear motion microactuator	Microactuator
6.3	Rotor on a centre-pin-bearing	Microactuator
6.4	Rotor on a flange bearing	Microactuator
6.5	Silicon condenser microphone	Microsensor
6.6	Centre-pin-bearing side-drive micromotor	Microactuator
6.7	Gap comb-drive resonant actuator	Microactuator
6.8	Integration of air gap capacitive pressure sensor and digital readout	Microsensor (smart)
6.9	Micronozzles	Micromechanical structure
6.10	Overhanging microgripper	Microactuator
8.1	Magnetostrictive strain gauge	Microsensor
8.2	Silicon resistive gas sensor based on microhotplate	Microsensor
14.1	MEMS-IDT accelerometer	MEMS

Index